EAI/Springer Innovations in Communication and Computing

Series Editor

Imrich Chlamtac, *European Alliance for Innovation, Ghent, Belgium*

The impact of information technologies is creating a new world yet not fully understood. The extent and speed of economic, life style and social changes already perceived in everyday life is hard to estimate without understanding the technological driving forces behind it. This series presents contributed volumes featuring the latest research and development in the various information engineering technologies that play a key role in this process. The range of topics, focusing primarily on communications and computing engineering include, but are not limited to, wireless networks; mobile communication; design and learning; gaming; interaction; e-health and pervasive healthcare; energy management; smart grids; internet of things; cognitive radio networks; computation; cloud computing; ubiquitous connectivity, and in mode general smart living, smart cities, Internet of Things and more. The series publishes a combination of expanded papers selected from hosted and sponsored European Alliance for Innovation (EAI) conferences that present cutting edge, global research as well as provide new perspectives on traditional related engineering fields. This content, complemented with open calls for contribution of book titles and individual chapters, together maintain Springer's and EAI's high standards of academic excellence. The audience for the books consists of researchers, industry professionals, advanced level students as well as practitioners in related fields of activity include information and communication specialists, security experts, economists, urban planners, doctors, and in general representatives in all those walks of life affected ad contributing to the information revolution.

Indexing: This series is indexed in Scopus, Ei Compendex, and zbMATH.

About EAI - EAI is a grassroots member organization initiated through cooperation between businesses, public, private and government organizations to address the global challenges of Europe's future competitiveness and link the European Research community with its counterparts around the globe. EAI reaches out to hundreds of thousands of individual subscribers on all continents and collaborates with an institutional member base including Fortune 500 companies, government organizations, and educational institutions, provide a free research and innovation platform. Through its open free membership model EAI promotes a new research and innovation culture based on collaboration, connectivity and recognition of excellence by community.

Jerzey Feliks · Radim Lenort · Martin Straka
Editors

The 12th Carpathian Logistics Congress

CLC 2024

Springer

Editors
Jerzey Feliks
Faculty of Management
AGH University of Krakow
Kraków, Poland

Radim Lenort
Department of Production, Logistics
and Quality Management
Skoda Auto University
Mladá Boleslav, Czech Republic

Martin Straka
Institute of Logistics and Transport
Technical University of Košice
Košice, Slovakia

ISSN 2522-8595 ISSN 2522-8609 (electronic)
EAI/Springer Innovations in Communication and Computing
ISBN 978-3-032-28930-8 ISBN 978-3-032-28931-5 (eBook)
https://doi.org/10.1007/978-3-032-28931-5

This Springer imprint is published by the registered company Springer Nature Switzerland AG
The registered company address is: Gewerbestrasse 11, 6330 Cham, Switzerland

Contents

Supply Chain Management (SCM) Strategy and Planning

Supply Chain Management and Logistics Megatrends: A Literature Review 3
Radim Lenort and Pavel Wicher

Selected Strategies for Integrating Hydrogen Fuel into Transportation Systems 11
Katarzyna Gdowska, Karolina Osielczak, and Roger Książek

Effective Evacuation Strategies in Crisis Management Logistics 19
Oľga Glova Végsöová

Experimental Modelling and Simulation of Production and Logistics Systems in Testbed 4.0 28
Miriam Pekarcikova, Peter Trebuna, Marek Kliment, and Jana Kronova

Purposeful Mindset Enhanced by On-Line IT Solution for Long Term Competitive Advantage in the Automotive Supply Chain 38
David Holman, Dalibor Fibiš, and Venuše Dolejšová

Comparative Analysis of Fashion and Military Clothing Supply Chains 47
Justyna Świerczek and Agnieszka Tubis

The Role of Digital Maturity in Improving Safety and Reliability in Wood-Based Supply Chains—Case Study 53
Sylwia Werbinska-Wojciechowska, Robert Giel, and Natalia Gnacy

Sustainable Logistics Processes

Environmental Strategy and Key Indicators Analysis in Sustainability ESG Reporting in Automotive Industry 63
Dorota Burchart, Radim Lenort, Marcin Staniek, Josef Horák, and Iga Przytuła

Environmental and Economic Impacts of Introducing Hydrogen Buses—A Case Study of Krakow's Public Transport 70
Katarzyna Gdowska and Karolina Osielczak

Management's Green Motivation in Supply Chain Management—A Case Study of Green Supply Chain 78
Agnieszka Jagoda, Anna Baraniecka, and Joanna Lyszkiewicz–Potaczek

Pro-Environmental Behaviour Factors in Choosing the Last Mile Delivery Method. A Gender Perspective 86
Tomasz Kołakowski, Agnieszka Jagoda, Katarzyna Cheba, and Magdalena Mucowska

The Management of the Technological Process According to the Product Lifecycle and Circular Economy Concept 93
Olena Stryhunivska, Rafał Rumin, Jerzy Feliks, and Olena Paliienko

Advanced Logistics Technologies

Lean Logistics Methods and Sustainable Growth: Case Study of the Assembly Workplace 105
Helena Fidlerova, Jarmila Blahová, and Helena Makyšová

Risk Management in Logistics: Digital Twins and Their Role in Operational Threat Assessment 115
Sylwia Werbinska-Wojciechowska, Robert Giel, and Klaudia Winiarska

Enhancing Manufacturing Logistics Efficiency through 3D Simulation in Visual Components 124
Simona Spirkova, Martin Straka, and Jakub Kovalcik

Cause-Effect Analyses in the Operation of AGV Systems 132
Honorata Poturaj

Specific Methodology for Modeling a Logistic System Using an Educational Robot 138
Jakub Kovalčík, Martin Staka, and Peter Kačmáry

Modeling Human Interaction with a Logistic System 147
Marek Ondov, Andrea Rosova, Marian Sofranko, and Lukas Horizral

Excellence in Logistics and ERP Project Management—A Contemporary Framework for Digital Transformation and Operational Excellence 155
Marian Krupa and Marek Karkula

AI-based Logistics Applications

Internal Auditing in Logistics: AI-driven and Advanced Analytics Solutions ... 169
Hasan Güney, Ahmet Ustaoğlu, Ahmet Yesevi Türker, Tuğçe Elçi, and Deniz Kantar

Evaluation of Artificial Intelligence Methods for Lead Time Prediction in Non-Cycled Areas of Automotive Production ... 176
Cornelius Hake, Jonas Weigele, Frederik Reichert, and Christian Friedrich

Artificial Intelligence in Logistics: New Opportunities for Optimization and Innovation with Chatgpt ... 189
Kristína Kleinová, Martin Straka, and Peter Kačmáry

Transport Optimization and Planning

Development of Public Transport Use in Selected Cities within Slovakia ... 199
Peter Kačmáry, Simona Špirková, Kristína Kleinová, and Maroš Mesároš

Preliminary Analysis of the Hydrogen Refuelling Station Location Problem in Rail Transport ... 211
Katarzyna Gdowska, Roger Książek, Radosław Kapłan, Magdalena Dudek, and Andrzej Raźniak

Implementation of Unit Load AGV Technology in Streamlining of the In-House Logistics of an Automotive Assembly Hall ... 219
Tomáš Malčic and Jiří Záruba

Operational Transport Planning Using a Heuristic Approach ... 228
Patrícia Muchová and Marek Ondov

Transport Packaging in the Cargo Bike Delivery Sector ... 237
Mateusz Rydlewski and Szymon Czapkowski

Maturity Model for Transport System Assessment ... 244
Emilia T. Skupień

The Optimization of Fuel Delivery Routes to Gas Stations—A Sensitivity Analysis of The Solution to a Change in The Level of Acceptable Risk ... 253
Grzegorz Tarczyński, Jacek Kaleta, and Marek Karkula

Socioeconomic Aspects of Hyperloop Development Based on The Central Communication Port (CPK) 270
Rafał Rumin, Olena Stryhunivska, Kinga Bieniek, Dawid Pękała, Szymon Gach, and Mateusz Milczanowski

Author Index 281

Supply Chain Management (SCM) Strategy and Planning

Supply Chain Management and Logistics Megatrends: A Literature Review

Radim Lenort(✉) and Pavel Wicher

Skoda Auto University, Mladá Boleslav 293 01, Czech Republic
{Radim.lenort,pavel.wicher}@savs.cz

Abstract. In the rapidly evolving global market, supply chain management and logistics play a critical role in ensuring the efficient flow of goods, information, and resources. This paper presents a systematic literature review focused on the megatrends shaping the future of supply chain management and logistics. In an era marked by unprecedented technological advancements and global disruptions, understanding these megatrends is crucial for both academic researchers and industry practitioners. The review synthesizes findings from a wide array of scholarly articles, identifying and classifying key supply chain management and logistics megatrends. The findings underscore the transformative potential of these megatrends and the need for proactive adaptation to maintain robustness and sustainability in supply chain and logistics operations.

Keywords: Supply chain management · Logistics · Megatrend · Literature review

1 Introduction

In an increasingly interconnected and volatile global market, addressing the megatrends influencing supply chain management (SCM) and logistics is more crucial than ever. These long-term transformative forces dictate not only operational efficiencies but also strategic resilience in the face of disruptions such as geopolitical shifts, technological advancements, and climate challenges. Understanding and proactively responding to these megatrends is vital for businesses aiming to sustain competitive advantages and foster innovation. Furthermore, this knowledge equips academic researchers and industry practitioners with the tools to develop adaptive, forward-thinking strategies that enhance the robustness, sustainability, and agility of supply chain networks.

Naisbitt [1] was the first to introduce the concept of a megatrend, outlining 10 key megatrends intended to forecast significant social, economic, and political transformations that would influence the future. The current definition of megatrends encompasses long-lasting, transformative forces that shape global society across various domains, including social, economic, and technological spheres [2]. Key characteristics of megatrends are [3, 4]:

- Global Impact: Megatrends are not confined to specific regions; they influence global dynamics, affecting everything from economies to social structures.

J. Feliks et al. (Eds.): CLC2024 2024, 2026.
https://doi.org/10.1007/978-3-032-28931-5_1

- Drivers of Change: They arise from various influences, such as technological advancements, demographic shifts, and geopolitical changes.
- Long-Term Orientation: Megatrends are characterized by their enduring nature, often spanning decades and requiring forward-thinking strategies for adaptation.

SCM and logistics megatrends can be defined as significant, long-term shifts that reshape how supply chains and logistics operate globally. These trends reflect substantial changes in technology, economic structures, consumer expectations, environmental considerations, and geopolitical dynamics.

The main aim of this paper is to conduct a systematic literature review aimed at identifying and analysing megatrends affecting contemporary SCM and logistics. The paper focuses on the discussion of these megatrends and their impact on SCM and logistics development and strategy in the context of global change and technological advances. The sub-objectives can be formulated as follows:

- Identify and classify the key megatrends affecting SCM and logistics.
- Compare the approaches of different authors to the definition and analysis of megatrends, including differences in their conception of exogenous and endogenous factors.
- Discuss how these megatrends affect the strategic direction of companies and their ability to adapt to dynamic environmental changes.

2 Literature Review

A systematic literature review was used to analyse SCM and logistics megatrends in detail. This was carried out in the Web of Science, Scopus and Google Scholar databases. The keywords used were "supply chain", "logistics", "megatrends" and their combinations. Only research articles that defined specific SCM and logistics megatrends were selected for that analysis. Thus, a total of 11 studies were analysed (see Table 1).

Bowersox et al. [5] identify ten megatrends expected to transform the landscape of SCM. These megatrends highlight the movement toward more integrated, technology-driven, and customer-centric supply chain practices. Bowersox et al. emphasize the impact of digitalization, including advancements in information technology and communication systems, which enhance real-time data sharing and operational efficiency. Additionally, the megatrends point to a shift toward more strategic partnerships, global logistics networks, and agile supply chain models that respond dynamically to market changes. The study underscores that these developments will not only streamline logistics operations but also create more responsive, resilient, and adaptable supply chain systems capable of meeting the evolving demands of the global economy. This study was followed by Sweeney [6], who studied the new status of megatrends defined by Bowersox in an Irish context.

Straube et al. [7] explore the evolving landscape of global logistics and outlines key strategies businesses can adopt to stay competitive. They identify megatrends such as the globalization of supply chains, the increasing importance of emerging markets, and the shift toward more sustainable practices. The study emphasizes the integration of advanced technologies like automation and real-time data systems to improve efficiency and decision-making. Additionally, the study highlights the significance of strategic flexibility and collaboration among supply chain partners to respond effectively to complex, interconnected global challenges. Straube et al. provides insight into how these megatrends shape logistics strategies and drive the need for innovative solutions to enhance resilience and competitiveness in a rapidly changing environment.

Gernandt [3] discusses the significant global megatrends that are reshaping the logistics industry. He outlines megatrends such as technological innovation, demographic shifts, urbanization, and the growing emphasis on sustainability. These megatrends are influencing logistics by driving the adoption of automated solutions, enhancing supply chain transparency, and fostering more resilient and adaptive networks. The paper emphasizes that logistics providers must rethink their strategies to incorporate sustainable practices, leverage data-driven decision-making, and adapt to rapidly changing consumer expectations. Gernandt concludes that understanding and responding proactively to these megatrends is essential for logistics companies aiming to maintain their competitive edge and ensure long-term success.

Kersten et al. [4] explore how global megatrends contribute to increasing complexity within logistics systems. The study identifies key megatrends such as technological advancements, globalization, and sustainability pressures, and examines their multifaceted impact on supply chain operations. The authors emphasize that these megatrends drive higher levels of interconnectivity and create challenges related to managing data, ensuring transparency, and maintaining flexibility in logistics networks. Additionally, the study highlights the need for advanced strategies and innovative solutions to handle this growing complexity effectively. Kersten et al. argue that understanding and addressing these megatrends is crucial for logistics providers to develop more resilient, adaptive, and efficient systems capable of thriving in an evolving global landscape.

Stank et al. [8] revised the megatrends defined by Bowersox et al. (2000) and extend them with five new megatrends. They explore how these megatrends—ranging from technological advancements and the digitization of supply chains to the heightened focus on sustainability and risk management—are evolving in the context of modern challenges. The article emphasizes the interconnected nature of these megatrends and their collective impact on creating more agile, resilient, and customer-focused supply chains. Stank et al. argue that for businesses to remain competitive, they must rethink traditional logistics strategies and adopt innovative, technology-driven approaches that prioritize transparency, collaboration, and rapid adaptability to disruptions. The insights provided aim to guide supply chain leaders in preparing for the significant shifts that will define the next era of logistics management.

Galińska [9] explores the dominant megatrends impacting the structure and function of modern supply chains. She identifies megatrends such as technological innovation, globalization, urbanization, and the rising importance of sustainability and environmental consciousness. The paper emphasizes that these megatrends push supply chains toward adopting advanced digital tools and sustainable practices while facing increasing pressure for speed, efficiency, and adaptability. Galińska points out that responding effectively to these shifts requires strategic changes that enhance flexibility, leverage

Table 1 . Identified SCM and logistics megatrends and their characteristics

Author, year	Identified megatrends	Type of megatrends	Focus of the study
Bowersox et al., 2000 [5], Sweeney, 2013 [6]	Customer service to relationship management; Adversarial to collaborative; Forecast to endcast; Experience to transition strategy; Absolute to relative value; Functional to process integration; Vertical to virtual integration; Information hoarding to sharing; Training to knowledge-based learning; Managerial accounting to value-based management	Endogenous	SCM
Straube et al. 2010 [7]	Internationalization; Security; Ecological sustainability	Exogenous	Logistics
Gernandt, 2012 [3]	Production and value creation: globalization—localization—glocalization, virtual goods; The development of society: the demographic change, global population and urbanization; Technology: communication and connectivity, increasing speed of innovation; Sustainability, resources and recycling: climate change and natural resources, energy, water and food	Exogenous	Logistics
Kersten et al. 2014 [4]	Globalization; Customer expectations/individualization; Cost pressure; Container traffic; Demographic change; Talent shortfall; Sustainability/green logistics; Reverse logistics; New technologies; Electric and alternative-energy vehicles; RFID; IT-integration; Transparency in supply chains; E-Business/online shopping	Combination	SCM
Stank et al. 2015 [8]	Inertia to innovation to transformation; "Dumb" technology to "Smart" technology to autonetic technology; Local optimization to global optimization to glocal optimization; Risk agnostic to risk management to risk prognosis; Untenability to sustainability to prostainability	Endogenous	SCM

(*continued*)

new technologies, and foster more resilient networks. The author concludes that understanding and integrating these megatrends is vital for businesses to secure a competitive advantage and align with the rapidly evolving demands of the global market.

Table 1 . (*continued*)

Author, year	Identified megatrends	Type of megatrends	Focus of the study
Galińska, 2018 [9]	Changes in business centres locations; Growing influence of modern, innovative and intelligent technologies; Application of ICT technology; Globalization and regionalization	Combination	SCM
Kalaitzi et al. 2021 [10]	Protectionism; Political stability; Supranationalism; Global trade shift; Digital economy; Financial Innovation; Population growth; Demographic change; Urbanization; Change in consumption pattern; Individualism; Digital natives; Knowledge based economy; Digital transformation; Technology development and automatization; Electrification of transport; Renewable energy sources; Consumer protection laws; Social and environmental regulations; Climate change; Resource scarcity; Catastrophic events/hazards	Exogenous	SCM
Zondervan et al. 2022 [11]	Resiliency; Digitalization; Sustainability	Endogenous	Logistics
Pessot et al. 2023 [12]	Aging population; Growing urbanization; Shifts in consumer demands; Geopolitical shifts; Depletion of natural resources; Climate change	Exogenous	SCM
Frias et al. 2023 [13]	Globalization; Sustainability; Technology evolution	Combination	Logistics

Kalaitzi et al. [10] explore megatrends – referred to as transformative global forces – that impact supply chains. The authors identify 23 major megatrends analysing them through a PESTLE (Political, Economic, Social, Technological, Legal, and Environmental) framework. This systematic literature review, complemented by expert workshops, reveals how these forces create both challenges and opportunities for future supply chain innovation. The study emphasizes the need for businesses to adapt to rapidly evolving environments, leveraging these insights to anticipate future disruptions and opportunities in SCM.

Zondervan et al. [11] investigate the impacts of the COVID-19 pandemic on logistics megatrends, focusing on the acceleration of certain innovations within the sector. The study employs text mining to analyse over 5,000 abstracts of logistics-related articles published between 2016 and 2021. Key findings include a notable increase in the importance of resiliency and digitalization in logistics, driven by advancements in technologies like blockchain, IoT, and automation (e.g., drones, robots). However, the research found no significant acceleration in sustainability efforts, despite shifts in focus towards low-carbon, hydrogen, and electric vehicles during the pandemic. The paper suggests that logistics firms, particularly SMEs, should assess their readiness for digital transformation to stay competitive and resilient, and it calls for policymakers to support these efforts through information and funding.

Pessot et al. [12] identify and analyse several megatrends that significantly impact global supply chains. These include demographic shifts (e.g., aging populations and urbanization), environmental pressures (like climate change and resource depletion), and technological advancements, particularly the rise of digitalization and automation. The authors highlight how these megatrends create both challenges and opportunities for supply chains, emphasizing the role of Industry 4.0 technologies – such as artificial intelligence, IoT, and automation – In enabling supply chains to adapt and thrive in this complex, evolving environment. These technologies help improve resilience, flexibility, and sustainability, which are essential for navigating the disruptions caused by these global forces.

Frias et al. [13] identify three major megatrends shaping the future of logistics: technological evolution, globalization, and sustainability. First, the rapid advancement of technologies such as AI, autonomous vehicles, drones, and the Internet of Things (IoT) is revolutionizing supply chains, offering significant efficiency gains but also raising concerns about automation's potential to displace jobs. Second, globalization continues to expand markets and lengthen supply chains, which, while fostering economic growth, also increases vulnerability to disruptions from geopolitical events, natural disasters, and other global crises. Finally, the increasing focus on sustainability calls for greener logistics practices, including the adoption of sustainable fuels and circular economy principles. While these innovations promise environmental benefits, they may also introduce challenges in terms of higher costs and potential inefficiencies. Together, these megatrends highlight the complex, evolving landscape in logistics, necessitating careful consideration of both their opportunities and risks.

The authors of analysed studies have approached the definition of megatrends in two primary ways, or a combination of both (see Table 1):

1. They have identified exogenous megatrends that represent the changes in the broader business environment.
2. They have developed endogenous megatrends that reflect the evolving capabilities and strategies within SCM and logistics.

The perception of megatrends in terms of their number and degree of generality varies considerably between studies. A minority of authors focus on a small number of broadly defined megatrends (three to six) while others attempt to define an exhaustive list of specific megatrends (10–23). On the contrary, in terms of the focus on SCM and logistics, there is a balanced interest of researchers in both areas.

3 Discussion

The analysis of the research on SCM and logistics megatrends reveals significant insights into how these transformative forces shape the modern landscape of SCM and logistics. The studies reviewed have approached the concept of megatrends from varied perspectives, focusing on both exogenous factors that influence the broader business environment and endogenous factors that reflect internal developments within supply chains. This dual perspective allows for a comprehensive understanding of the dynamic changes in the sector.

The synthesis of findings underscores that megatrends such as digitalization, sustainability, globalization, demographic shifts, and technological innovation are not only reshaping logistics operations but are redefining strategic priorities for businesses. For instance, advancements in IoT, blockchain, and automation have enhanced the transparency, efficiency, and responsiveness of supply chains, yet they bring challenges such as higher costs and the need for skilled labour. These innovations promote resilience and adaptability, critical for navigating disruptions and sustaining competitive advantage.

Furthermore, sustainability has emerged as a pivotal megatrend, pushing companies to adopt greener practices and integrate circular economy principles into their operations. The pressure to balance economic performance with environmental responsibility has redefined success criteria in logistics and SCM. The shift toward sustainability aligns with global efforts to combat climate change and resource depletion, highlighting the need for long-term strategies that prioritize environmental stewardship.

The reviewed literature also points out the complexities introduced by geopolitical shifts, protectionist policies, and changing consumer expectations. These exogenous trends compel businesses to remain agile and invest in risk management and scenario planning. The research emphasizes that while globalization has extended market reach and growth opportunities, it has also exposed supply chains to increased vulnerabilities from global disruptions.

Studies differ in the scope and number of megatrends they identify, with some focusing on a select few broad categories and others presenting exhaustive lists. This variation illustrates that megatrends can be perceived differently depending on the research focus and industry context. Regardless, the consensus remains that understanding these megatrends and preparing for their impacts are critical for strategic decision-making.

4 Conclusions

The research affirms that adapting to megatrends is essential for the future viability of supply chains and logistics. Companies must leverage technology, foster sustainability, and build resilience to thrive in an environment characterized by rapid change and uncertainty. Strategic shifts should include investing in digital tools, enhancing data-driven decision-making, and promoting collaborative approaches that integrate various stakeholders.

In conclusion, the path forward for SCM and logistics lies in embracing these megatrends with a proactive and innovative mindset. Organizations that anticipate and adapt to these global forces will be better equipped to maintain competitiveness and contribute positively to the sustainable development of the global economy.

References

1. Naisbitt, J.: Megatrend: Ten New Directions Transforming our Lives. Warner Books, New York (1982)
2. Naughtin, C.K., Schleiger, E., Bratanova, A., Terhorst, A., Hajkowicz, S.: Forty years in the making: a systematic review of the megatrends literature. Futures. **157**(103329) (2024)
3. Gernandt, K.: Megatrends and their impact on logistics. In: The Logistics Institute—Asia Pacific Whitepaper Series, vol. 12, (2012)
4. Kersten, W., von See, B., Skirde, H.: Identification of megatrends affecting complexity in logistics systems. In: Next Generation Supply Chains: Trends and Opportunities. Proceedings of the Hamburg International Conference of Logistics (HICL), pp. 3–27. Epubli GmbH, Berlin (2014)
5. Bowersox, D., Closs, D.J., Stank, T.P.: Ten mega-trends that will revolutionize supply chain logistics. J. Bus. Logist. **21**(2), 1–16 (2000)
6. Sweeney, E.: Supply chain "mega-trends": current status and future trends. J. Chart.Ed Inst. Logist. Transp., 31–34 (2013)
7. Straube, F., Nagel, A., Rief, D.: Trends and strategies in global logistics. In: Waters, D. (ed.) Global Logistics, pp. 31–48. Kogan Page, London (2010)
8. Stank, T., Autry, C., Daugherty, P., Closs, D.: Reimagining the 10 megatrends that will revolutionize supply chain logistics. Transp. J. **54**(1), 7–32 (2015)
9. Galińska, B.: Logistics megatrends and their influence on supply chains. In: Business Logistics in Modern Management: Proceedings of the 18th International Scientific Conference. Josip Juraj Strossmayer University of Osijek, Osiek (2018)
10. Kalaitzi, D. et al.: Megatrends and trends shaping supply chain innovation. In: Fornasiero, R., et al. (eds.) Next Generation Supply Chains Lecture Notes in Management and Industrial Engineering, pp. 3–34. Sringer (2021)
11. Zondervan, N.A., Tolentino-Zondervan, F., Moeke, D.: Logistics trends and innovations in response to COVID-19 pandemic: an analysis using text mining. PRO. **10**(12), 2667 (2022)
12. Pessot, E., Zangiacomi, A., Marchiori, I., Fornasiero, R.: Empowering supply chains with industry 4.0 technologies to face megatrends. J. Bus. Logist. **44**(4), 609–640 (2023)
13. Frias, A., Simöes-Marques, M., Água, P.B., Correia, A.: Logistics future trends and their transformative impact. Human factors and systems interaction. In: AHFE International Conference, pp. 217–226. AHFE International, USA (2023)

Selected Strategies for Integrating Hydrogen Fuel into Transportation Systems

Katarzyna Gdowska(✉), Karolina Osielczak, and Roger Książek

AGH University of Krakow, Krakow, Poland
{kgdowska,roger}@agh.ed.pl, osielczak@student.agh.edu.pl

Abstract. This paper examines the transformative potential of hydrogen fuel in transportation, covering its diverse types—green, grey, blue, yellow, pink, turquoise, and black/brown—and its production methods, including steam methane reforming, electrolysis, gasification, and pyrolysis. It also explores storage solutions, such as compressed, liquid, and materials-based systems, that are critical for widespread adoption. Strategic initiatives, such as Poland's hydrogen strategy and EU policies, including the Green Deal and Fit for 55, are highlighted for advancing hydrogen infrastructure and reducing emissions. By addressing production, storage, and policy, the paper underscores hydrogen's role in enabling a sustainable, low-carbon transportation future.

Keywords: Hydrogen fuel · Transport · Fit for 55 · Transport management · Strategic management

1 Introduction

Hydrogen is a promising clean energy carrier, producing only water as a by-product in fuel cells. Its versatility, derived from diverse sources such as natural gas, nuclear power, biomass, and renewables, makes it suitable for transportation, electricity generation, and other applications. Hydrogen plays a crucial role in energy storage, transportation, and delivery, thereby enhancing its significance in the global energy landscape [1]. Hydrogen's classification into various "colors" (green, blue, gray, etc.) reflects its diverse production methods and environmental impacts. Renewable hydrogen, particularly green hydrogen, produced through water electrolysis using renewable energy, offers a pathway to climate neutrality. However, high production costs, public awareness, and infrastructure gaps pose significant challenges, particularly in regions with limited adoption of modern energy technologies [2, 3]. Globally, hydrogen adoption is on the rise, driven by the expansion of refueling infrastructure and the increasing use of hydrogen across various sectors, including road, air, and maritime transport.

Challenges in transport, storage, and combustibility limit hydrogen's widespread adoption. Although hydrogen refueling is time-efficient, its production requires 35–70 kWh to produce 1 kg of hydrogen, enough to power a car for 100 km, compared to the 30 kWh needed for electric vehicles to travel the same distance. Overcoming energy efficiency and production cost barriers is crucial to hydrogen's broader use.

J. Feliks et al. (Eds.): CLC2024 2024, 2026.
https://doi.org/10.1007/978-3-032-28931-5_2

China leads globally in hydrogen refueling stations, followed by Japan, South Korea, Germany, and the U.S. Europe has seen rapid growth, with 178 stations as of May 2023, led by Germany. The European hydrogen market, valued at $36.5 billion in 2021, is projected to grow at an annual rate of 3.5% through 2032 [4, 5]. Hydrogen's potential as a lightweight fuel for shipping and aviation is expanding, as demonstrated by DHL's hydrogen-powered vans with a 500 km range [6]. Hydrogen's versatility as a clean energy carrier is crucial for decarbonizing hard-to-abate sectors. Still, its widespread adoption depends on addressing challenges such as production costs, efficiency, and public perception [7].

This paper examines the production, applications, infrastructure, and challenges to the adoption of hydrogen, highlighting its potential in achieving global energy sustainability.

2 Hydrogen – the Fuel of the Future?

In 2019, the transportation sector accounted for 25% of the EU's CO_2 emissions, with road transport contributing 71.7%, primarily from passenger cars (61%). Transport emissions increased by 33.5% from 1990 to 2019. Strategies such as car sharing, public transportation, cycling, and walking can significantly reduce these emissions [8]. To address climate change and promote sustainability, a shift to renewable energy is essential, including renewable hydrogen produced from wind, solar, or hydro-power. By 2050, hydrogen could meet up to 20–50% of transportation and 5–20% of industrial energy needs in the EU [9]. Hydrogen is a key player in the clean energy transition as an efficient and climate-friendly energy carrier and storage medium. When combusted, it generates more heat than fossil fuels like oil or natural gas and produces no harmful by-products. In 2020, conventional methods dominated hydrogen production in the EU, EFTA countries, and the UK, accounting for 95.7% of total capacity. Alternative approaches, including electrolysis by-products (3.7%), reforming with carbon capture (0.5%), and power-to-hydrogen (0.1%), played a minor role. Conventional methods, including steam reforming, partial oxidation, gasification, and by-product hydrogen production, yielded 11.4 million tons annually across 387 facilities [10]. Similarly, in Poland, 95% of hydrogen is derived from natural gas, with only 5% classified as environmentally sustainable [11].

A classification system – presented in Fig. 1 – using a „color palette" distinguishes hydrogen production methods by energy source and climate impact, as follows:

- Green hydrogen is produced via water electrolysis using renewable energy like wind, solar, or hydropower, making it carbon-neutral. However, high production costs remain a barrier to large-scale adoption [12, 13].
- Gray hydrogen is produced through fossil fuel-based steam reforming, emitting about 10 tons of CO_2 per ton of hydrogen produced. It accounts for 40% of hydrogen production as a by-product of industrial processes. Despite environmental drawbacks, gray hydrogen remains a cost-effective option for industries such as petrochemicals and ammonia production [14].
- Blue hydrogen relies on steam methane reforming and incorporates Carbon Capture and Storage (CCS) to sequester CO_2, offering a low-carbon alternative. However,

concerns about CCS's long-term viability, including risks of CO_2 leakage, persist [10, 14].

- Yellow hydrogen is produced through electrolysis powered by grid electricity, including solar energy. Its environmental impact depends on the grid's energy mix, being lower when renewable energy dominates and higher when fossil fuels are used [15].
- Pink hydrogen is produced from electrolysis powered by nuclear energy, with variants such as purple hydrogen (atomic power combined with heat) and red hydrogen (nuclear thermal energy used for high-temperature catalytic water splitting). Countries with atomic grids, such as France, are well-suited for producing pink hydrogen [14].
- Turquoise hydrogen is produced via methane pyrolysis, splitting natural gas into hydrogen and solid carbon. This method can be CO2-neutral if powered by renewables and if the solid carbon is securely stored. However, scaling the process and ensuring environmental safety remain challenges [16].
- Black and brown hydrogen are produced via coal gasification, with brown hydrogen using lignite coal and black hydrogen using bituminous coal. These methods emit about 20 kg of CO_2 per kg of hydrogen, making them the most carbon-intensive and environmentally harmful options [14, 15].

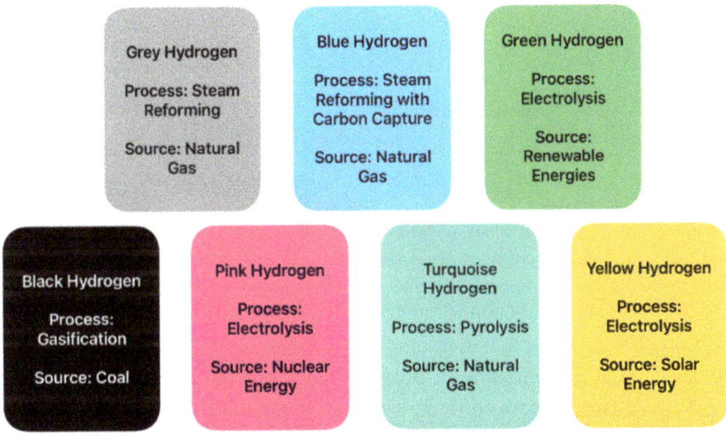

Fig. 1. Classification of hydrogen. Source: own elaboration based on [10, 12–16].

Despite its potential as a clean fuel, hydrogen faces hurdles in production efficiency, storage, and cost. Producing 1 kg of hydrogen requires 35–70 kWh of energy, enough to power a car for approximately 100 km, which is comparable to the efficiency of electric vehicles. Hydrogen's energy density (33.3 kWh/kg) surpasses that of traditional fuels, such as gasoline and diesel, making it a promising option for heavy-duty transport and aviation [17]. Photovoltaic electrolysis is among the most environmentally friendly methods, but costs about USD 9 per kilogram. Solar thermolysis and water electrolysis, including nuclear hydrolysis, cost approximately USD 8 and USD 6 per kilogram, respectively. Nuclear thermolysis offers a more cost-effective solution at USD 2.4 per kilogram. Steam methane reforming (SMR) with carbon capture and storage (CCS) costs about USD 2.6 per kilogram, slightly more than SMR without CCS, which costs USD 0.20 less. Coal gasification, while the most economical method, has the highest environmental impact. Currently, hydrogen production predominantly relies on hydrocarbons, such as natural gas, resulting in "gray hydrogen" with significant carbon emissions. By 2030, green and blue hydrogen are expected to be cost-competitive with gray hydrogen, and by 2050, they may become cheaper alternatives. Advancements in CCS technology and declining costs for electrolyzers and renewable energy will drive this shift. Key factors enabling a competitive hydrogen supply chain include access to hydrocarbons, affordable renewable energy sources, strong domestic demand, a strategic location, and experience in the energy market. Investments in carbon capture and related technologies are critical to maintaining competitiveness as the industry evolves (Fig. 2).

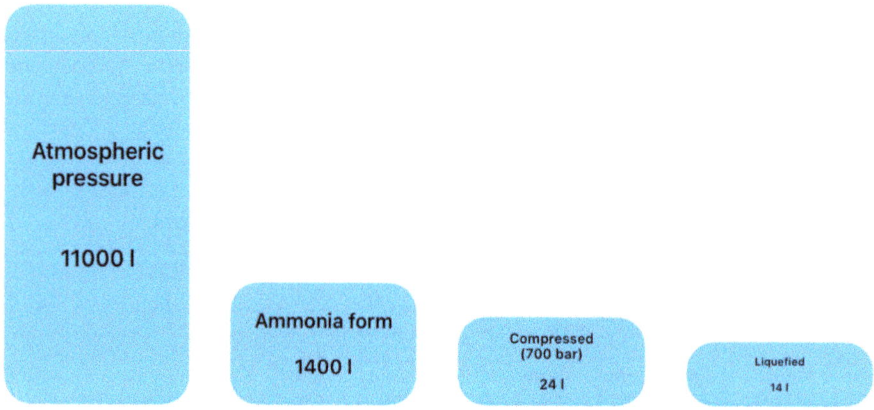

Fig. 2. Storage capacity of 1 kg of hydrogen across different methods. Source: own elaboration.

Hydrogen storage presents challenges due to its low volumetric energy density, which requires significantly more space than other gases, such as natural gas. For example, storing 1 kg of hydrogen at atmospheric pressure requires approximately 11 m^3 of space [18]. Cryogenic storage is essential, as hydrogen's boiling point is near absolute zero. Although hydrogen is non-corrosive, it can cause metal cracking under certain conditions [19]. Despite its high energy per mass, hydrogen necessitates advanced storage methods to achieve higher energy density and practicality across applications. Storage methods, including atmospheric pressure storage (requiring 11,000 l for 1 kg of hydrogen), ammonia (1,400 l), compression at 700 bars (24 l), and liquefaction (14 l per kg), significantly affect the volume requirements [18]. Ongoing advancements aim to enhance the energy density, efficiency, and usability of hydrogen storage, thereby supporting its integration into various technologies and applications.

3 Hydrogen Strategies Affecting Transportation

The energy landscape is shifting towards hydrogen as a key element in sustainable power. This chapter examines Polish and European strategies for hydrogen integration, focusing on Poland's alignment with European directives and its role in the broader hydrogen transition. The European Green Deal aims for carbon neutrality by 2050, promoting hydrogen as a renewable energy source, particularly in transportation to replace fossil fuels.

Poland's „Hydrogen Strategy until 2030 with an Outlook until 2040″ outlines goals to advance the hydrogen economy, focusing on renewable hydrogen production, infrastructure expansion, and the development of a robust value chain to support climate neutrality and economic competitiveness. As the third-largest hydrogen producer in Europe, Poland faces challenges, with only a small portion of hydrogen produced from renewable sources [20]. The strategy includes actions such as leveraging research potential and supporting hydrogen-related industries, as well as key initiatives to create a Hydrogen Valley Ecosystem and a Hydrogen Technology Center, and to promote education and awareness. It also plans to develop hydrogen infrastructure and fuel cell production for heating and transport [21]. A significant emphasis is on hydrogen as an alternative fuel for transportation, with plans to replace fossil fuels in urban, road, and railway transport, as well as maritime vessels. Starting in 2025, cities with populations exceeding 100,000 will be required to purchase zero-emission buses, aiming to deploy 800–1,000 eco-friendly buses by 2030. The strategy also supports the development of hydrogen refueling networks to facilitate this transition.

The European Green Deal aims to achieve EU-wide climate neutrality by 2050, promoting cross-sector collaboration in climate, energy, transport, industry, and agriculture [22]. The Fit for 55 package, a preparatory step towards the Green Deal, aims for a 55% reduction in greenhouse gas emissions by 2030, focusing on areas such as renewable fuel infrastructure, reductions in vehicle emissions, and cuts in methane emissions in the energy sector. Hydrogen-powered buses, with better fuel efficiency and longer ranges than electric buses, are key to achieving these goals in transport [23].

The European Climate Law legally mandates a 55% reduction in net emissions by 2030, with transportation contributing nearly 70% of EU pollution. Hydrogen-powered buses can enhance sustainability and improve air quality in urban areas, thereby helping to meet EU climate goals. The adoption of hydrogen buses is expected to grow internationally as cities seek sustainable transport solutions [24].

The EU Industrial Emissions Directive targets pollution from large industrial plants. Hydrogen plays a significant role in industrial sectors such as ammonia production, methanol synthesis, and steelmaking, offering a low-emission alternative to conventional hydrogen production. Widespread adoption of hydrogen is expected to reduce costs, making hydrogen-powered buses more accessible [24].

The Hydrogen and Decarbonised Gas Market Package sets rules to reduce the gas market's carbon footprint, encouraging the shift from natural gas to renewable and low-carbon gases. This package includes objectives such as planning hydrogen infrastructure, developing a regulatory framework, and establishing the Clean Hydrogen Partnership to support the production and distribution of renewable hydrogen. These actions are crucial for hydrogen-powered public transport fleets and a sustainable EU transportation future [24].

The Alternative Fuels Infrastructure Regulation (AFIR), part of the Fit for 55 project, ensures access to infrastructure for recharging and refueling vehicles with alternative fuels, including hydrogen. Targets for 2025 and 2030 include refueling stations every 60 km and hydrogen stations in urban areas. This infrastructure will support the broader adoption of hydrogen buses in cities, making them more practical and efficient for urban transport systems [24].

4 Conclusion

Hydrogen has significant potential to decarbonize sectors like transportation, industry, and energy storage. Its versatility, as a fuel derived from renewable sources, nuclear power, biomass, and natural gas, supports the global shift to sustainable energy. Hydrogen's clean combustion and role in energy storage underscore its importance, although challenges in production efficiency, cost, infrastructure, and storage must be addressed for widespread adoption. In the EU, hydrogen-powered vehicles, including buses and trucks, can help reduce transportation emissions, especially in sectors that are difficult to electrify, such as heavy transport and aviation. However, cost and infrastructure barriers must be overcome. Poland's „Hydrogen Strategy until 2030" aligns with EU climate goals, focusing on renewable hydrogen production and innovation, but faces challenges due to its reliance on fossil-fuel-based hydrogen. The EU's Green Deal and supporting initiatives, such as the Clean Hydrogen Partnership and the Hydrogen and Decarbonised Gas Market Package, provide regulatory frameworks to accelerate hydrogen deployment. As green hydrogen becomes more cost-competitive and storage technologies advance, the adoption of hydrogen in transportation and industry is expected to grow, contributing to a low-carbon economy.

Acknowledgements. The research was supported by Europe's Rail Flagship Project 4 – Sustainable and green rail systems, which has received co-funding from the Horizon Europe Framework Program (HORIZON) under grant agreement no. 101101917 and the Polish Ministry of Education and Science under grant agreement no. MEiN 5337/HE/2023/2 under the program titled International Co-Financed Projects. The project is implemented at the AGH University of Science and Technology in Krakow from 2022 to 2026.

References

1. Addcomposites Oy: How Hydrogen Fuel Cell Technology Is Revolutionizing Transport. https://www.addcomposites.com/post/how-hydrogen-fuel-cell-technology-is-revolutionizing-transport (2023) Last accessed 05 Nov 2024
2. Robert Bosch GmbH Corporate Department Communications & Governmental Affairs: Bosch Tech Compass 2023. https://www.bosch.com/stories/technology-report-tech-compass-2023/ (2023) Last accessed 05 Nov 2024
3. Ingaldi, M., Klimecka-Tatar, D.: People's attitude to energy from hydrogen—from the point of view of modern energy technologies and social responsibility. Energies. **13**, 6495 (2020)

4. European Alternative Fuels Observatory: Fuel Price Comparison. https://alternative-fuels-observatory.ec.europa.eu/consumer-portal/fuel-price-comparison (2023). Last accessed 05 Nov 2024
5. Global Market Insights: Global Hydrogen Market Size. https://www.gminsights.com/industry-analysis/europe-hydrogen-market (2023) Last accessed 05 Nov 2024
6. National Grid: The Hydrogen Colour Spectrum. https://www.nationalgrid.com/stories/energy-explained/hydrogen-colour-spectrum (2023) Last accessed 05 Nov 2024
7. OSIELCZAK, K.: Hydrogen Fuel Production and Supply Chain Management for Sustainable Transportation: a Production Engineering Perspective. Bachelor's Thesis, AGH University of Krakow, Krakow (2024)
8. European Parliament: CO_2 Emissions from Cars: Facts and Figures. https://www.europarl.europa.eu/news/en/headlines/society/20190313STO31218/co2-emissions-from-cars-facts-and-figures-infographics (2019) Last accessed 05 Nov 2024
9. European Parliament: How the EU Is Boosting Renewable Energy. https://www.europarl.europa.eu/news/en/headlines/society/20221128STO58001/how-the-eu-is-boosting-renewable-energy (2022) Last accessed 05 Nov 2024
10. Hydrogen Europe: Hydrogen Europe in a Nutshell. https://hydrogeneurope.eu/in-a-nutshell/ (2023) Last accessed 05 Nov 2024
11. Agencja Rozwoju Przemysłu S.A: Oddział Katowice. H2POLAND.Forum H2POLAND: Wodór tu i teraz Wodór paliwem XXI w. [online]. [viewed: 2024-11-05]. Available from: https://h2poland.eu/documents/15/3XW_1_kwartał_2023.pdf Last accessed 05 Nov 2024 (2023)
12. ACCIONA: What Are The Colours Of Hydrogen And What Do They Mean? https://www.acciona.com.au/updates/stories/what-are-the-colours-of-hydrogen-and-what-do-they-mean/?_adin=02021864894 (2023) last accessed 05 Nov 2024
13. Dash, S.K., Chakraborty, S., Roccotelli, M., Sahu, U.K.: Hydrogen fuel for future mobility: challenges and future aspects. Sustainability. **14**, 8285 (2022)
14. Arcos, J.M.M., Santos, D.M.F.: The hydrogen color Spectrum: techno-economic analysis of the available Technologies for Hydrogen Production. Gases. **3**, 25–46 (2023)
15. National Grid: What Is Hydrogen?. https://www.nationalgrid.com/stories/energy-explained/what-is-hydrogen (2023) Last accessed 05 Nov 2024
16. EWE Aktiengesellschaft: The Colours of Hydrogen. https://www.ewe.com/en/shaping-the-future/hydrogen/the-colours-of-hydrogen (2023) Last accessed 05 Nov 2024
17. Wiącek, D.W., jako paliwo przyszłości.: Autobusy: technika, eksploatacja, systemy transportowe. **12**(10), 446–452 (2011)
18. TEREGA: Hydrogen storage: a challenge for the development of the industry. https://www.terega.fr/en/our-activities/hydrogen/hydrogen-storage-a-challenge-for-the-development-of-the-industry (2023) Last accessed 05 Nov 2024
19. Willige, A.: 4 ways of storing hydrogen from renewable energy. Spectra. https://spectra.mhi.com/4-ways-of-storing-hydrogen-from-renewable-energy (2022) Last accessed 05 Nov 2024
20. Środowiska, M., Klimatu, I.: Polska Strategia Wodorowa do roku 2030. Warszawa. https://www.gov.pl/web/klimat/polska-strategia-wodorowa-do-roku-2030 (2021) Last accessed 05 Nov 2024
21. Platforma Przemysłu Przyszłości: Polskie plany wodorowe, https://przemyslprzyszlosci.gov.pl/polskie-plany-wodorowe/ Last accessed 05 Nov 2024

22. European Commission: The European Green Deal. Brussels. https://www.consilium.europa.eu/en/policies/green-deal/ (2019) Last accessed 05 Nov 2024
23. KPMG International: The European Green Deal & fit for 55. https://assets.kpmg.com/content/dam/kpmg/xx/pdf/2021/11/green-deal-and-fit-for-55-slip-sheet_v5_web.pdf (2021) Last accessed 05 Nov 2024
24. European Commission: European Climate Law. Brussels. https://climate.ec.europa.eu/eu-action/european-climate-law_en (2021) Last accessed 05 Nov 2024

Effective Evacuation Strategies in Crisis Management Logistics

Oľga Glova Végsöová(✉)

Institute of Earth Resources, BERG Faculty, Technical University of Kosice, Kosice, Slovakia
olga.glova.vegsoova@tuke.sk

Abstract. This paper focuses on the key aspects of logistics in crisis management, emphasizing the planning of evacuation strategies and ensuring public safety during incidents involving hazardous substances. The study examines risk zone modelling and identifies procedures that enhance effective decision-making and coordination among response units. Particular attention is given to the challenges posed by rapidly changing conditions and the need for integrated approaches that bolster the adaptability of crisis management. The findings demonstrate that comprehensive logistical plans, based on precise modelling and well-developed evacuation strategies, are essential for minimizing risks and protecting the public in urban areas. The study underscores the importance of continuous improvement in planning processes and the integration of modern technologies to support safety and enhance the effectiveness of crisis response.

Keywords: Logistics · Risk Modelling · Evacuation Strategies · Hazardous Substance · Safety · Public Protection

1 Introduction

1.1 A Subsection Sample

The introduction should provide a clear statement of the study, the relevant literature on the study subject and the proposed approach or solution. Logistics in crisis management is an indispensable component of public protection during industrial accidents, particularly in incidents involving hazardous substance releases. These events are characterized by a high degree of unpredictability, placing immense demands on the ability to coordinate resources, implement effective evacuation strategies, and ensure public safety in real-time [1, 2]. Hazardous substances such as ammonia pose acute threats to human health and the environment. Timely and efficient logistics can mitigate the consequences of such events through carefully planned and adaptable procedures [3, 4].

A significant practical application of crisis logistics can be observed in the context of an ice rink in Slovakia, a venue of considerable importance as both a sports and community centre. Such facilities are particularly at risk due to the use of ammonia in their refrigeration systems—a highly toxic substance that, in the event of malfunction or damage, can pose serious hazards to visitors, staff, and the surrounding community.

J. Feliks et al. (Eds.): CLC2024 2024, 2026.
https://doi.org/10.1007/978-3-032-28931-5_3

The simulation of scenarios at the ice rink included two models: one with stable atmospheric conditions and low wind speed, allowing for slower dispersion of the substance, and another involving a high-concentration release coupled with adverse weather conditions, which significantly expanded the risk zone. This study highlighted that ice rinks, due to their infrastructure and high population density during sports and cultural events, require precise evacuation planning, continuous monitoring of safety systems, and rapid response capabilities.

Modern tools such as the ALOHA (Areal Locations of Hazardous Atmospheres) simulation software enable the prediction of hazardous substance dispersion in the air and the delineation of risk zones. These predictive models are invaluable for planning evacuations and managing the response of rescue units [5, 6]. However, research in this field is constantly evolving, as dynamic conditions such as changing weather and diverse terrain demand more sophisticated tools and integrated approaches [7, 8]. Combining ALOHA with advanced technologies such as Geographic Information Systems (GIS), artificial intelligence (AI), and drones could significantly enhance predictive capabilities and decision-making processes in crisis scenarios [9, 10].

This study aims to analyse existing approaches to managing hazardous substance releases and propose opportunities for the future development of logistics processes and technologies in crisis management. It places particular emphasis on the synergistic integration of risk modelling, the implementation of evacuation strategies, and the incorporation of innovative technologies, which together enable swift and accurate responses to emergencies [11, 12].

2 Materials and Methods

This study focuses on crisis management logistics in the context of hazardous substance leaks, specifically ammonia, and on the effective planning and implementation of measures to minimize their impact on people and the environment. The methodology combines modern simulation tools, systematic risk analysis, and detailed planning of logistics processes to identify optimal procedures and minimize both time and material demands during crisis situations.

The primary analytical tool used in this study was ALOHA (Areal Locations of Hazardous Atmospheres) software. This tool allows the modelling of hazardous substance dispersion in the air based on meteorological conditions and the physical properties of the substance. Simulations were conducted based on the parameters summarized in Table 1 ALOHA Simulation Parameters for Three Scenarios, which considered three distinct scenarios representing varying severity levels.

To facilitate the analysis, risk zones were defined according to the internationally recognized Emergency Response Planning Guidelines (ERPG). These categories allowed a detailed assessment of the impact of leaks on human health and the environment, as well as the design of preventive and rescue measures. The impacts and measures are summarized in Table 2 ERPG Categories—Impacts and Measures.

Another crucial aspect of the analysis was the evaluation of crisis management logistics, which included evacuation planning, required material and technical resources, and the time needed for response operations. These aspects are detailed in Table 3 ERPG Categories—Crisis Management and Logistics.

Table 1 ALOHA Simulation Parameters and Scenario Conditions.

Parameter	Scenario 1.1 value	Scenario 1.2 value	Scenario 2 value	Note
Substance	Ammonia (NH_3)	Ammonia (NH_3)	Ammonia (NH_3)	Identical chemical properties across all scenarios
Molecular weight	17.03 $g.mol^{-1}$	17.03 $g.mol^{-1}$	17.03 $g.mol^{-1}$	0.12–0.42
Boiling point	−33 °C	−33 °C	−33 °C	33.6–33.5
Wind speed	1 $m.s^{-1}$	2 $m.s^{-1}$	5 $m.s^{-1}$	Determines substance volatility
Air temperature	20 °C	25 °C	25 °C	Stable conditions; temperature variations between 1.1 and 1.2 have minimal effect on volatility
Substance release volume	300 kg	500 kg	1500 kg	Gradual increases in simulated volumes reflect varying severities
Tank pressure	1 MPa	1 MPa	1 MPa	Tank pressure remains constant, influencing release speed
Release duration	15 min	30 min	10 min	Scenario 2 represents a sudden and intense release

(*continued*)

These simulation results and analyses provide valuable insights into assessing the effectiveness of crisis management strategies. They enable a thorough examination of the

Table 1 (*continued*)

Parameter	Scenario 1.1 value	Scenario 1.2 value	Scenario 2 value	Note
ERPG zone categories	**ERPG-1** $\leq$ 25 ppm **ERPG-2** 25–150 ppm **ERPG-3** $\geq$ 150 ppm	**ERPG-1** $\leq$ 25 ppm **ERPG-2** 25–150 ppm **ERPG-3** $\geq$ 150 ppm	**ERPG-1** $\leq$ 25 ppm **ERPG-2** 25–150 ppm **ERPG-3** $\geq$ 150 ppm	Similar categories, but zone extents vary due to differing conditions
Risk zone extent (m)	**ERPG-1**: 500 **ERPG-2**: 200 **ERPG-3**: 100	**ERPG-1**: 895 **ERPG-2**: 362 **ERPG-3**: 110	**ERPG-1**: 1350 **ERPG-2**: 740 **ERPG-3**: 450	Zone extents are influenced by wind speed, release volume, and atmospheric stability

Adapted from [13] with modifications

Table 2 ERPG categories and their effects on humans, the environment, and proposed measures

Category and concentration	Impact on humans	Impact on the environment	Preventive and rescue measures
ERPG-1 $\leq$ 25 ppm	Mild irritation of eyes, nose, and throat	No significant environmental impact	Regular inspection of cooling systems, installation of gas detectors, relocation of people
ERPG-2 25–200 ppm	Significant respiratory irritation, coughing, shortness of breath	Damage to nearby vegetation	Staff training for hazardous material handling, oxygen therapy
ERPG-3 $>$ 200 ppm	Acute respiratory issues, risk of suffocation, vision impairment	Contamination of soil and water	Implementation of safety protocols, artificial respiration, immediate hospitalization

Adapted from [13] with modifications

extent of risk zones, the time required for interventions, and the logistical demands, contributing to improved understanding and optimization of response strategies in hazardous substance release scenarios.

Table 3 Logistics and crisis response framework for ERPG levels.

Category and concentration	Proposed measures and logistics	Required resources and equipment	Response time
ERPG-1 $\leq$ 25 ppm	Public notification via local radio, electronic boards, and mobile apps	Informational leaflets, basic protective gear	1–2 h
ERPG-2 25–200 ppm	Activation of local warning systems (sirens, SMS alerts), organization of evacuation corridors	Evacuation vehicles, oxygen supplies	3–6 h
ERPG-3 > 200 ppm	Immediate activation of national alert systems, detailed evacuation instructions	Decontamination units, rescue teams	6 + hours

Adapted from [13] with modifications

3 Results

The ammonia release simulations conducted using the ALOHA software provided essential insights into crisis management and logistics, highlighting key factors affecting the efficiency of interventions and population protection. The three analysed scenarios enabled a detailed examination of hazardous zones and the identification of the most effective solutions for managing emergency situations.

In Scenario 1.1, which simulated stable atmospheric conditions with a wind speed of 1 m s^{-1} and the release of 300 kg of ammonia, the critical hazard zone (ERPG-3) was confined to a radius of 100 m. This scenario underscored the importance of preparedness for localized risks, allowing effective management aligned with the available capacities of emergency units. Logistical measures included the swift mobilization of local emergency teams equipped with chemical protective gear, the establishment of clearly defined evacuation points, and clear communication with the population through municipal broadcasts, sirens, and mobile applications.

Scenario 1.2, involving increased wind speed (2 m s^{-1}) and a release volume of 500 kg, demonstrated the significant expansion of hazard zones. The ERPG-1 zone extended to a radius of nearly 895 m, necessitating more sophisticated coordination between components of the integrated rescue system. The primary logistical challenge was organizing a supply chain to evacuate a larger population, provide protective equipment, and establish evacuation routes tailored to dynamic conditions. Drones, integrated with GIS, played a crucial role in monitoring ammonia dispersion, enabling command centres to adapt evacuation plans in real time. The simulation model for these evacuation scenarios is illustrated in Fig. 1, which visualizes the evacuation zones and hazard areas identified during the analysis. This visual representation aids in understanding the spatial distribution of hazardous zones and supports the planning of evacuation strategies.

The most critical scenario was Scenario 2, simulating a massive release of 1500 kg of ammonia with a wind speed of 5 m s^{-1}. The critical hazard zone (ERPG-3) reached

Fig. 1. Evacuation model for ammonia release simulation in scenario 1 [13].

a radius of 450 m, while the ERPG-1 zone spanned up to 1350 m, encompassing dense residential areas and public institutions. Logistical efforts involved the extensive deployment of personnel and technologies for decontamination, evacuation, and the protection of critical infrastructure. Figure 2 illustrates the model showing the planned route of rescue services for this scenario, highlighting the critical paths and logistical considerations required to manage the emergency effectively.

The results of this study emphasize that effective crisis management and logistics must not only address local needs but also align with broader global challenges. With the growing impact of climate change, the frequency and intensity of industrial accidents are expected to increase due to extreme weather events and infrastructure disruptions. For instance, higher temperatures may amplify the volatility of hazardous chemicals, while strong winds or storms can complicate the spread and containment of dangerous substances.

Additionally, this study demonstrates that modelling hazardous releases and integrating them with technologies such as drones and GIS aligns with the rising trend of leveraging artificial intelligence and automation in crisis management. These technologies enable faster responses and more accurate predictions, minimizing harm to public health and environmental damage.

International collaboration is another crucial aspect. Crises like chemical releases often transcend national borders, requiring coordination among countries, the exchange of best practices, and the standardization of technologies and procedures. Integrating methodologies such as this study's approach with international frameworks (e.g., the EU Civil Protection Mechanism or UN standards) could significantly enhance nations' abilities to respond to large-scale emergencies.

Fig. 2. Representation of rescue service routes for scenario 2 [13].

This integrated approach to crisis management reflects not only the current demands for protecting populations but also long-term trends toward sustainable and technologically advanced emergency management. Its implementation can serve as a model for effectively managing crises in a dynamically changing global environment.

4 Conclusion

The ALOHA software stands out as a significant tool in the realm of crisis management and logistics, particularly in addressing emergencies involving the release of hazardous substances such as ammonia. Its primary strength lies in its ability to deliver accurate short-term predictions, enabling precise delineation of hazard zones and supporting evacuation planning. Its user-friendly design and flexibility make it effective in various phases of crisis management, from preliminary planning to real-time emergency response. Moreover, its capability to generate clear and actionable data rapidly is critical for decision-making during high-pressure situations.

Despite its undeniable advantages, ALOHA also has certain limitations. Its reliance on static data, including pre-configured meteorological and environmental inputs, can reduce its effectiveness in situations where conditions change rapidly. Another shortcoming is its focus on short-term impacts, with limited attention to the long-term consequences of chemical releases, which may include substantial environmental and health implications. Furthermore, the absence of integration with modern technologies such as drones, GIS, or artificial intelligence algorithms restricts its adaptability to dynamically evolving scenarios.

To enhance ALOHA's role in crisis management and logistics, several innovations are recommended. Integrating real-time monitoring through drones and environmental sensors would significantly improve its ability to adapt to changing conditions. Incorporating artificial intelligence algorithms could enable the processing of large datasets in real time, increasing prediction accuracy and optimizing evacuation strategies. Adding GIS functionality would provide a spatial dimension to hazard visualization, greatly facilitating logistical planning. Expanding its chemical database would broaden its applicability to various industrial scenarios, and combining ALOHA with complementary tools like PHAST (Process Hazard Analysis Software Tool) could establish a comprehensive system for assessing both short-term and long-term risks.

The findings underscore that modernizing ALOHA would not only enhance its effectiveness in managing ammonia-related incidents but also make it a more versatile tool for addressing diverse industrial emergencies. Integration with advanced technologies would transform it into a robust and flexible platform capable of meeting the demands of today's dynamic crisis management environment. This approach aligns not only with the current needs for public health and environmental protection but also with the long-term trend toward technologically advanced and sustainable emergency management.

This study highlights the imperative of continuous improvement and innovation in crisis management tools. Only through such advancements can we effectively address emerging challenges posed by climate change, globalization, and the growing complexity of industrial systems. By implementing the proposed innovations, ALOHA could serve as a model for technologically sophisticated and efficient crisis management, combining accurate predictions, flexible logistics, and rapid response to critical events.

Acknowledgements. This research was supported by the project KEGA, grant number 010TUKE-4/2023, titled Application of Educational Robots in the Process of Teaching the Study Program Industrial Logistics.

References

1. Bae, H., Simmons, D.R., Polmear, M.: Promoting the quarry workers' hazard identification through formal and informal safety training. Occup. Health Saf. **12**, 317–323 (2021)
2. Samerei, S.A., Aghabayk, K., Mohammadi, A., Shiwakoti, N.: Data mining approach to model bus crash severity in Australia. J. Saf. Res. **76**, 73–82 (2021)
3. Anjana, N.S., Amarnath, A., Harindrananathan Nair, M.V.: Toxic hazards of ammonia release and population vulnerability assessment using geographical information system. J. Environ. Manag. **210**, 201–209 (2018)
4. Shirali, G.A., Asl, Z.M., Jahani, F., Ahangar, A.S., Etemad, S.A.: Modelling the effect of ammonia leakage from ammonia reservoirs using ALOHA software and developing an emergency response program in one of process industries. J. Occup. Hyg. Eng. **5**, 12–19 (2018)
5. Soltanzadeh, A., Mahdinia, M., Nikbakht, N., Hosseinzadeh, K., Sadeghi-Yarandi, M.: Evaluation of human vulnerability and toxic effects of chronic and acute occupational exposure to ammonia: a case study in an ice factory. Work **78**, 1021–1033 (2024)
6. Kim, M., Gi-Hyoug, C.: Influence of evacuation policy on clearance time under large-scale chemical accident: an agent-based modelling. Int. J. Environ. Res. Public Health **17**, 9442 (2018)

7. Atabi, F., Ghorbani, R., Jabbari, M.: Assessment of safe distance for five toxic materials commonly in the accidents of chemical road transportation using ALOHA and PHAST software and CEI index: case of Tehran-Qazvin highway. Iran Occup. Health J. **14**, 36–50 (2017)
8. Hellas, M.S., Rachid, C., Zerrouki, H.: Contribution to the analysis of the toxic ammonia cloud dispersion associated with pipeline catastrophic rupture: case study of Arzew. Oran. Alger. J. Res. Technol. **6**, 29–35 (2022)
9. Son, V.L., Phong, V.N., Kien, T.T.: Application of the ALOHA simulation model to assess ammonia leakage from the coal-fired power station of Duyen Hai 2 thermal power plant. IOP Conf. Ser. Earth Environ. Sci. **1226**, 012014 (2024)
10. Polorecka, M., Kubas, J., Danihelka, P., Petrlova, K., Repkova Stofkova, K., Buganova, K.: Use of software on modelling hazardous substance release as a support tool for crisis management. Sustainability **13**, 438 (2021)
11. Cetinyokus, S., Dinc, D., Ata, S.: Consequences analysis of a natural gas pipeline: the case of the trans-Anatolian natural gas pipeline. Process. Saf. Prog. **43**, 321–332 (2024)
12. Reis, J., Ribeiro, C., Rodrigues, D., Rodrigues, A.: Accident scenario modelling with hazardous materials in transportation infrastructures. Proc. Inst. Civ. Eng.–Forensic Eng. **172**, 133–149 (2020)
13. Glova Végsöová, O., Grabara, J.K.: Assessing crisis management tools for sustainability of industrial safety. Appl. Sci. **14**, 10037 (2024)

Experimental Modelling and Simulation of Production and Logistics Systems in Testbed 4.0

Miriam Pekarcikova(✉), Peter Trebuna, Marek Kliment, and Jana Kronova

TUKE—Technical University of Kosice, Kosice, Slovak Republic
{miriam.pekarcikova,peter.trebuna,marek.kliment,
jana.kronova}@tuke.sk

Abstract. Modelling and simulation of production and logistics concepts and processes makes sense primarily in terms of the possibilities of variability in terms of parameter settings and environment, which allows you to fine-tune and optimize prototypes in the virtual world before their physical implementation and will be reflected in increasing the quality of the system. Although there are many positive aspects to adopting digitalization in the sense of Industry 4.0 and 5.0, it is necessary to consider several technological, software and other challenges that companies have to face during implementation. The article deals with the issue of using modern software tools that help companies optimize production and logistics flows in the Testbed 4.0 experimental workplace environment. Testbed 4.0 was designed as a trial test line, or rather. As a complex set of equipment, equipped with the most modern equipment and technology. Thanks to these conditions, it is possible to design, verify, evaluate products, processes and technologies of digitalization.

Keywords: Modelling · simulation · logistic · Testbed 4.0

1 Introduction

Designing and operating complex logistics and production systems is accompanied by numerous risks and challenges. The sheer number of alternatives and the difficulty of evaluating them often prevent managers and designers from identifying truly optimal solutions using traditional analytical tools. This phenomenon, commonly referred to as the local optimisation effect, arises not only during system operation but also at the design stage. The issue becomes even more complicated when optimisation is considered only at the level of a single production line or shop floor, rather than across the enterprise as a whole. Narrow perspectives and limited criteria frequently dominate system design, and when projects prove too expensive, compromises are introduced simply to make implementation possible [1–4].

J. Feliks et al. (Eds.): CLC2024 2024, 2026.
https://doi.org/10.1007/978-3-032-28931-5_4

Additional constraints—such as uncertainty in future production requirements, strict time schedules, financial limitations, and the lack of advanced design tools—make comprehensive optimisation of system parameters extremely difficult. As a result, deficiencies are often embedded in the system architecture from the outset, restricting its full potential. Once the system is in operation, further modifications are typically required, which in turn increase costs and reduce efficiency.

Computer simulation offers a practical way to address these shortcomings. Each simulation study can be organised as an independent project, applying the principles of project management. This involves defining specific tasks, arranging them in logical and temporal order, allocating human and computational resources, estimating funding requirements, and calculating the payback period for invested resources [5–7].

Two fundamental rules guide the application of simulation:

- The benefits obtained through simulation must exceed the costs of implementation and system improvement.
- The decisive factor for adopting simulation in practice is the tangible value of the benefits achieved.

In industrial practice, simulation projects are justified only when the measurable benefits clearly exceed the costs involved. These benefits can be both quantitative and qualitative, ranging from efficiency gains to improved reliability. Importantly, the purpose of simulation is not always tied to direct financial outcomes. For example, suppliers of production or transport systems often rely on simulation to demonstrate the functionality or dependability of their solutions, thereby strengthening their market position.

The timing of simulation is critical. When applied during the early stages of system design, simulation offers the greatest potential for improvement. At later phases—implementation or operation—the scope for modifications is limited, and adjustments tend to be costly, sometimes outweighing the expected benefits [8–12].

The growing importance of simulation in contemporary industry is driven by several interconnected factors. Classical analytical methods such as queuing theory, service networks, or linear programming are often inadequate for complex, non-linear, and dynamic systems. Enterprises face rising solution complexity, higher capital intensity, and shorter decision cycles, which increase the risk of suboptimal decisions. Rapid technological change further intensifies the need for predictive models that can support managers in making timely and reliable choices.

Another decisive factor is the exponential growth of available data and knowledge. With countless possible project configurations, advanced modelling tools are required to process and interpret multidimensional datasets effectively. Organisational changes also contribute: decentralised and modular structures, together with collaborative approaches such as concurrent engineering, demand flexible simulation environments capable of supporting dynamic reconfiguration.

Finally, advances in computing power and affordable software platforms have democratized access to simulation. What was once limited to specialized applications has now become a practical and economically viable tool for a wide range of industrial contexts. As a result, simulation is increasingly used not only for modelling and analysis but also as a strategic instrument for optimisation, decision support, training, and conceptual validation.

2 Experimental Modelling and Simulation Testbed 4.0

Testbeds can be viewed as dedicated experimental workspaces designed either as specialised test benches or as complex facilities equipped with advanced technological tools. They provide controlled environments that enable the design, validation, and assessment of products, processes, and industrial digitalization technologies.

Testbeds are essential for safe experimentation and for testing new approaches without risking negative impacts on real production operations. This is particularly critical for verifying the functionality, reliability, and benefits of digital twin solutions before they are deployed in live industrial systems. By allowing systems to be tested in a controlled environment, testbeds also facilitate scalability—once a digital twin or automated process performs successfully in a testbed, its implementation in real-world operations becomes faster, easier, and more cost-efficient. In addition, they serve as open research infrastructures where academic institutions and industry partners can share experimental results, benchmarks, datasets, and methodological approaches. Shared testbed platforms further promote interoperability and standardisation, acting as reference environments for verifying system compatibility, communication protocols, and data interfaces.

Testbeds play a crucial role in supporting Industry 4.0 concepts by providing companies, particularly small and medium-sized enterprises with limited resources, a platform to experiment with new technologies. They enable businesses to explore optimal designs and workflows without requiring full in-house expertise, competitive pressure, or large financial investments. This "test before invest" approach allows companies to submit requirements and proposals and receive comprehensive strategic and technological insights into the processes being evaluated.

The scope of modern testbeds is extensive, covering areas such as cybersecurity, robotics, automation, data processing, serial and mass production solutions, artificial intelligence, and virtual or augmented reality. Furthermore, many testbeds are designed for interconnectivity, allowing cross-industry collaboration—for example, engineering testbeds can communicate and collaborate with medical, civic, or military sectors—fostering multi-industry innovation. In summary, testbeds are key infrastructures for research, development, and validation of innovations in digital twin technologies and Industry 4.0. They provide a safe, flexible, and scalable environment for experimentation, support knowledge sharing across academia and industry, and ensure interoperability and standardisation, making them indispensable tools for modern industrial transformation.

Objectives of the Testbed:

- To verify and validate the behaviour of a PLC-based control system against a virtual model of the production line or machine.
- To enable virtual commissioning before physical deployment, reducing risk, time, and costs.
- To capture the data flow between the physical PLC, sensors, and the digital model (digital twin) for analysis, optimisation, or prediction.
- To provide a modular infrastructure where components (PLC programs, sensors, simulation, models) can be changed for experimentation and research.

The overview of testbeds in Slovakia, the Czech Republic, and Poland is presented in the Table 1, focusing on Industry 4.0, manufacturing digitisation, PLC integration, and digital twins.

Testbed 4.0 at the Technical University of Košice, Faculty of Mechanical Engineering, was established as a response to the increasing demands of enterprises undergoing digital transformation. Its mission is to provide companies with tailored services that support the design, testing, and evaluation of efficient solutions aligned with Industry 4.0 principles. By offering this environment, businesses can explore alternative scenarios, gain a deeper understanding of potential investment outcomes, and identify the most suitable strategies for implementation.

The scope of Testbed 4.0 extends across the entire product lifecycle. From early product development and methodological design through logistics management and production operations, the facility encourages systematic data collection and robust information flows. Expert guidance combined with advanced technological infrastructure ensures that companies can detect overlooked inefficiencies, analyze them thoroughly, and transform raw data into actionable insights. A central element of the testbed's concept is the use of digital twin technology.

Digital twins enable enterprises to design, optimize, and validate processes within a virtual ecosystem, reducing risks and improving decision-making. To maintain long-term adaptability, hardware components that might quickly become obsolete are replaced with virtualized devices, ensuring that the infrastructure remains relevant over time. When designing and implementing testbeds in general, several critical aspects must be considered. These include clear definition of objectives, balancing realism with controllability, modularity and expandability, synchronization in Hardware-in-the-Loop (HIL) and real-system interactions, stability and reliability, continuous maintenance, isolation and security of experiments, comprehensive data monitoring, and reproducibility of results. Realistic environments provide valuable insights but often increase complexity, requiring carefully chosen compromises.

Modularity and scalability allow new modules to be integrated and outdated components replaced, while synchronization and stability safeguard the accuracy of experiments. Security and isolation protect sensitive industrial data, and reproducibility ensures scientific validity by allowing experiments to be replicated and verified. In summary, Testbed 4.0 is more than a safe environment for experimentation. It is a comprehensive workplace for applied industrial testing, research, and innovation. By combining modular design, reproducible experiments, expert knowledge, and digital twin frameworks, it supports enterprises in successfully adopting Industry 4.0 while minimizing risks and maximizing both strategic and operational benefits.

Table 1 TestBeds–Slovakia, Czech republic and Poland

Country	Testbed name	Location/institution	Focus and technologies	Partners/projects	Source links
Slovakia	TestBed 4.0	TU Košice + SOVA digital	Automation, PLC control, digital twin, 5G campus, AR/VR, smart manufacturing	Industry 4.0 Slovakia, MIRRI SR, SOVA Digital	Testbed.sk/hlavnespravy.sk/industry4um.sk
Czech Republic	Testbed for industry 4.0 (CIIRC ČVUT)	ČVUT Prague	Robotics, cyber-physical systems, digital twins, flexible manufacturing cells, 5G network	RICAIP, Siemens, rockwell, Škoda auto	Ciirc.cvut.cz /ricaip.eu
Czech Republic	CEITEC BUT testbed for industry 4.0	VUT Brno (CEITEC)	Smart factory, digital twin, collaborative robotics, automation for SMEs	RICAIP, Siemens, ABB robotics	Ceitec.eu/ricaip.eu
Poland	Industry 4.0 technology testing centre	Gliwice–APA group	IIoT, automation, process digitization, SCADA, MES, NAZCA 4.0 platform	APA Group, Silesian University of Technology	Test.apagroup.pl/polsl.pl

3 Case Study: Creation of a Simulation Model in Tecnomatix Plant Simulation with a Connection to Digital Elements in TestBed 4.0

The case study explored the possibilities of connecting the designed model in TX Plant Simulation software with digital elements such as a light gate, a machine and a button, based on which it is possible to control the simulation. An important step was to correctly set up the data and write the code for all processes and elements of the simulation model. It is important to work with the KEPServerEX program, which is used to connect digital elements via Siemens PLC with the created model in TX Plant Simulation. TX Plant Simulation software allows connecting digital elements with physical ones through the OPC UA user interface, which was used to achieve the goal. This phase required knowledge and understanding of the functioning of technologies, as well as the ability to effectively configure and integrate systems to achieve our goals. To create a credible simulation model (Fig. 1), it was necessary to create 2D/3D models that were arranged in a "U"-shaped line, which is most often used in production and approximately copies one of the workplaces in the Testbed 4.0 laboratory.

3D model of a light gate was also added, which will be an active element within the model and passing through the gate will control the launch of the simulation and the physical model, which is part of the Testbed 4.0 equipment. KEPServerEX was used to integrate the light gate, the machine, and the button that will start the production process on the line. This program sends signals to a Siemens PLC controller to which our digital elements and computer are connected. The PLC essentially acts as a "black box" that controls all the devices on the production line. It changes variables like 1's and 0's to reflect the current state of the devices and processes. Thanks to this integration, we can observe changes in the operation and make sure that everything is working correctly (Fig. 2). KEPServerEX is run to start the integration process. The result is a final simulation that runs in parallel with the digital elements to the maximum extent, which stops when passing through a light gate and which resumes when the reset button is pressed.

The 3D simulation also includes a light diode that changes its colour depending on what is currently happening on the machine. If the red light is on, the machine is blocked, if the yellow light is on, the machine is waiting for the arrival of a part, and if the green light is on, the work is in full swing. Creating a simulation model of production in the TX Plant Simulation program and its subsequent connection with digital elements in a specialized Testbed 4.0 laboratory is a key part of the case study (Fig. 3). This model allows us to visually evaluate the efficiency of the production process and all its components without having to disrupt real production. This approach seriously saves our time and resources. Thanks to the capabilities of the TX Plant Simulation software, it is possible to accurately simulate the flow of materials in production. Support for 2D/3D conversion and the ability to insert custom 3D objects allows us to create more realistic models of production processes. This allows us to more accurately analyse and optimize processes, prevent potential errors and improve results.

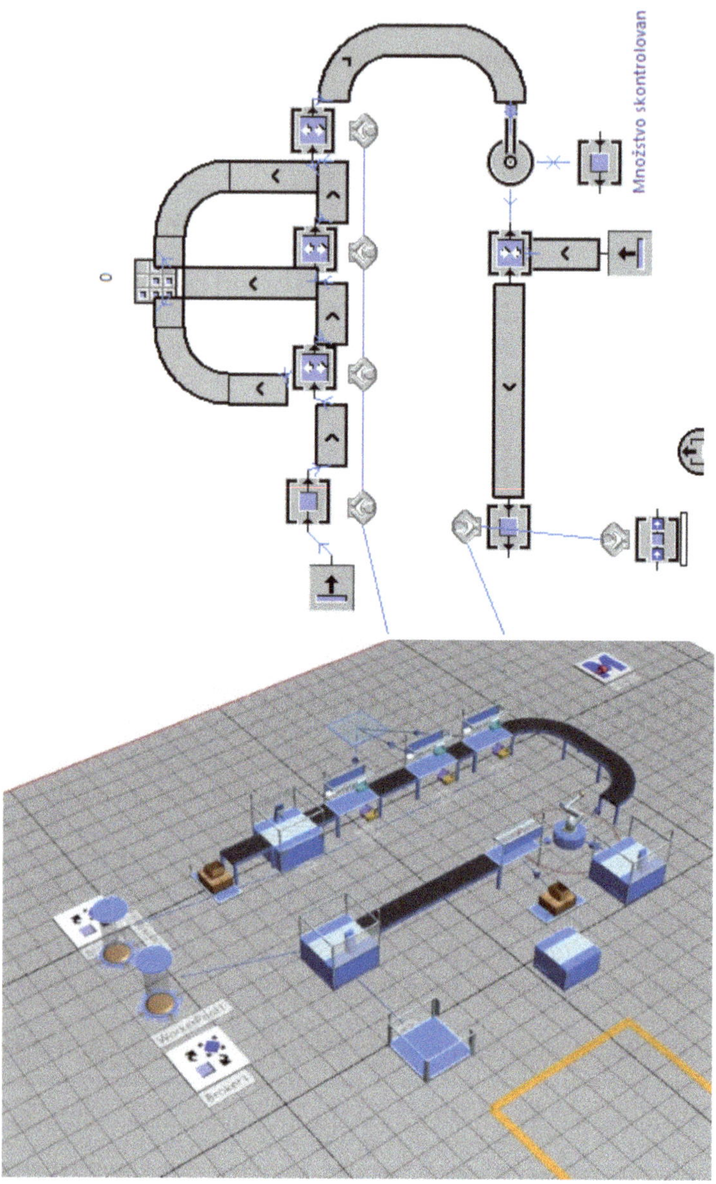

Fig. 1. 2D and 3D model created in the shape of a "U" and a light gate, software TX plant simulation

As part of this work, it is also looked at the possibilities of implementing digitalization in various production and logistic processes and discussed their significance. This allowed to understand the significance of digitalization for modern production and identify potential benefits that can be obtained through this technological approach.

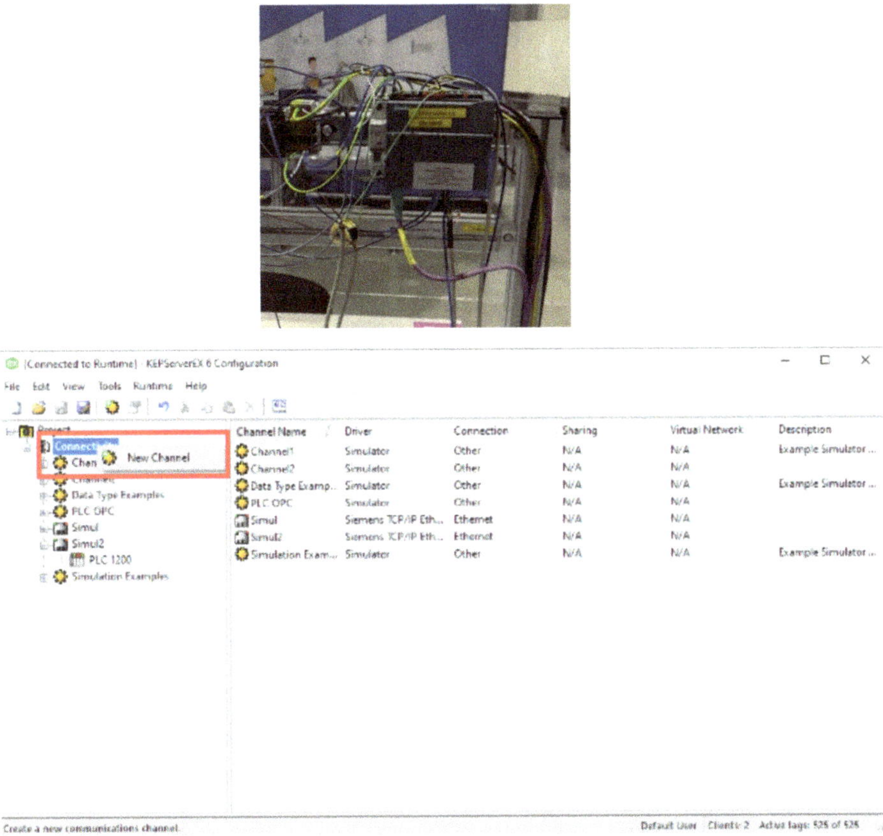

Fig. 2. Digital automated workplace in KEPServerEX home area

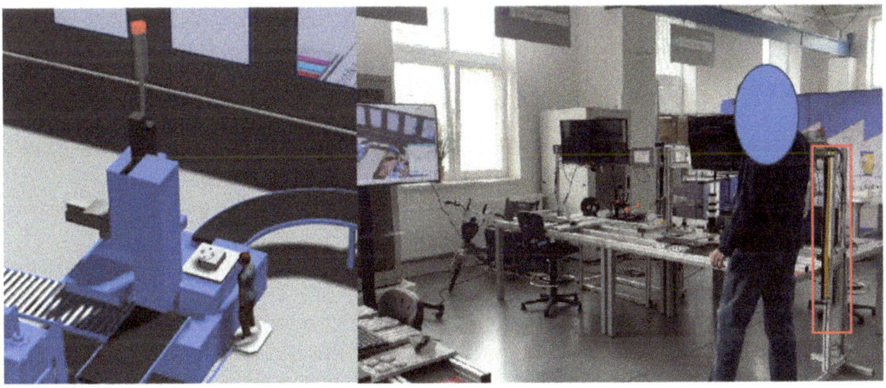

Fig. 3. Connecting digital and physical elements via TX plant simulation software in testbed

4 Conclusion

Digitalization in logistics and production is fundamentally reshaping how systems operate, introducing cyber-physical components that interact in real time and generate vast streams of data. The creation of secure and structured data spaces becomes essential, as these environments allow storage, exchange, and effective utilization of information across interconnected processes. For enterprises, the transformation is not only about adopting new technologies but also about rethinking organizational structures. Flexible concepts and adaptive frameworks must replace rigid models, enabling companies to respond quickly to changing demands. High levels of connectivity, transparency, and predictive capabilities are therefore critical for maintaining competitiveness. At the same time, the reliance on algorithms and artificial intelligence raises important questions about how future Internet of Things networks will be coordinated and how autonomous decision-making will evolve. These developments inevitably bring risks—uncertainty, complexity, and volatility—that must be anticipated and managed proactively. The implications of digitalization extend beyond technical or economic dimensions [13–15]. They signal a profound societal shift, where traditional human labor is increasingly complemented or replaced by intelligent machines. For businesses, the challenge lies in embracing this transition responsibly: ensuring that technological progress enhances efficiency and resilience while also supporting sustainable growth and social balance.

Acknowledgements. This article was created by the implementation of the grant project APVV-19-0418 "Intelligent solutions to enhance business innovation capability in the process of transforming them into smart businesses", VEGA 1/0508/22 "Innovative and digital technologies in manufacturing and logistics processes and system", VEGA 1/0383/25 "Optimizing the activities of manufacturing enterprises and their digitization using advanced virtual means and tools", KEGA 020TUKE-4/2023 "Systematic development of the competence profile of students of industrial and digital engineering in the process of higher education", KEGA 003TUKE-4/2024 "Innovation of the profile of industrial engineering graduates in the context of required knowledge and specific capabilities for research and implementation of intelligent systems of the future".

References

1. Grznar, P., Gregor, M., Gola, A., Nielsen, I., Mozol, S., Seliga, V.: Quick workplace analysis using simulation. Int. J. Simul. Modell. **21**(3), 465–476 (2022)
2. Bangsow, S.: Tecnomatix Plant Simulation—Modeling and Programming by Means of Examples, 2nd edn. Springer, Cham, Switzerland (2020)
3. Straka, M., Sofranko, M., Glova, J., Vegsoova, O., Kovalcik, J.: Simulation of homo-geneous production processes. Int. J. Simul. Modell. **21**(2), 214–225 (2022)
4. Malkus, T., Kozina, A.: The features of negotiations within reverse logistics cooperation. Acta Logistica **10**(1), 111–119 (2013). https://doi.org/10.22306/al.v10i1.364
5. Knapcikova, L., Behunova, A., Behun, M.: Using a discrete event simulation as an effective method applied in the production of recycled material. Adv. Prod. Eng. Manage. **15**(4), 431–440 (2020). https://doi.org/10.14743/apem2020.4.376
6. Marasova, D., Saderova, J., Ambrisko, L.: Simulation of the use of the material handling equipment in the operation process. Open Eng. **10**(1), 216–22 (2020) https://doi.org/10.1515/eng-2020-0015

7. Saniuk, S., Grabowska, S., Straka, M.: Identification of social and economic expectations: contextual reasons for the transformation process of industry 4.0 into the industry 5.0 concept. Sustainability **14**(3), 1–20 (2022)
8. Straka, M.,Saderová, J., Bindzár, P., Malkus, T., Lis, M.: Computer simulation as a means of efficiency of transport processes of raw materials in relation to a cargo rail terminal: a case study. Acta Montanist. Slovaca **24**(4), 307–317
9. Edl, M., Lerher, T., Rosi, B.: Energy efficiency model for the mini-load automated storage and retrieval systems. Int. J. Adv. Manuf. Technol. **70**(1–4), 1–19 (2014). https://doi.org/10.1007/s00170-013-5253-x
10. Siemens.: Advanced Planning and Scheduling. White papers. [online]. (2022). [viewed: 2024–06–24]. Available from: https://www.plm.automation.siemens.com/global/en/products/manufacturing-operations/advanced-planning-scheduling.htm
11. Siemens.: Digital Industries Software. Tecnomatix Plant Simulation software featured in SMART Manufacturing magazíne. [online]. (2017). [viewed: 2024–06–24]. Available from: https://blogs.sw.siemens.com/tecnomatix/tecnomatix-plant-simulation-software-featured-in-smart-manufacturing-magazine/
12. Siemens.: Manufacturing Operations Transformation. White papers (2022). [viewed: 2024–06–24]. Available from: https://www.plm.automation.siemens.com/global/en/products/manufacturing-operations/manufacturing-execution-system.htm
13. Hnilica, R., Jankovsky, M., Dado, M., Messingerova, V.: Experimental evaluation of combined effects of risk factors in work environment. 12th Interantional Scientific Konference Engineering for Rural Development, MAY 23–24, 2013, Jelgava, LATVIA, pp. 577–583
14. Wozniak, W., Sasiadek, M., Stryjski, R., Mielniczuk, J., Wojnarowski, T.: An Algorithmic Concept for Optimising the Number of Handling Operations in an Intermodal Terminal Node, VISION 2020: Innovation Management, Development, Sustainability and Competitive Economic Growth 2016, VOLS I–VII, Soliman, KS, pp. 1490–1500, 28th International Business-Information-Management-Association Conference, Seville, Spain
15. Végsoová, O., Khouri, S., Straka, M., Rosová, A., Kacmáry, P., Betus, M.: Using Technical Means and Logistics Principle Applications to Solve Ecological Water Course Accidents. Polish J. Environ. Stud. **28**(5), 3875. ISSN 1230-1485. https://doi.org/10.15244/pjoes/95183

Purposeful Mindset Enhanced by On-Line IT Solution for Long Term Competitive Advantage in the Automotive Supply Chain

David Holman[1](✉), Dalibor Fibiš[1], and Venuše Dolejšová[2]

[1] Skoda Auto University, Mladá Boleslav, Czech Republic
David.Holman@savs.cz
[2] Manufactus GmbH, 6a, 82319 Starnberg, Germany

Abstract. The foundation of competitive advantage in the 20th century was largely predicated on know-how, allowing organizations to sustain an edge through specialized expertise and proprietary competencies. However, the 21st-century global context, defined by unprecedented interconnectedness and rapid technological dissemination, has rendered know-how widely accessible and insufficient for maintaining a competitive position. While traditional reactive approaches supported by Maximization mindset and robust IT infrastructures, such as Enterprise Resource Planning (ERP) systems, have evolved to incorporate know-why capabilities for enhanced analysis and strategic forecasting, these advancements remain inadequate for securing sustained competitive advantage, particularly when accounting for the environmental and social imperatives of sustainable development.

Achieving and sustaining competitiveness in the 21st century necessitates a shift from a reactive framework to a proactive paradigm underpinned by Understanding of Wholeness. This advanced intellectual capacity surpasses the earlier construct of holistic understanding by enabling organizations to effectively manage complex and evolving challenges. A proactive, Purposeful mindset based on Wholeness Systems Thinking (WST) cultivates adaptability, fostering processes that are not only efficient for current operational needs but are effective for strategic continuous adaptation and evolution. These processes can be augmented through advanced IT solution such as Integrated Kanban System (IKS), which embody this proactive philosophy by facilitating the seamless and value-centric flow of operations across globally interconnected networks. A specific application of Understanding of Wholeness supported by IKS in the automotive supply chain (consisting of 1st tier supplier and automotive producer) will be presented and described.

Keywords: Automotive Supply Chain · Wholeness Systems Thinking · Purposeful Mindset · Maximization Mindset · Adaptability · IT solutions · ERP

J. Feliks et al. (Eds.): CLC2024 2024, 2026.
https://doi.org/10.1007/978-3-032-28931-5_5

1 Introduction

In the dynamic and competitive landscape of the automotive industry, long-term success depends on resilient and adaptive supply chain management (SCM) practices. Traditional IT systems, such as ERP solutions, have been instrumental in providing foundational support for SCM by integrating various functions and streamlining processes. However, as the industry confronts challenges in volatile, uncertain, complex, and ambiguous (VUCA) environments, there is a growing need to adopt more flexible, online solutions capable of real-time response and adaptation. Innovative online solutions, like the Integrated Knowledge System (IKS), offer new potential for SCM by supporting data-driven decision-making and enhancing transparency across supply chain networks, thereby promoting a holistic approach to managing supply chains in the automotive sector [1, 2].

This paper aims to examine the role of online web solutions, such as IKS, in supporting SCM in the automotive industry, contrasting their benefits and limitations with traditional ERP solutions. In doing so, it explores the interplay between IT capabilities, reactive and proactive management practices, and the underlying mindset applied within contemporary automotive supply chains. Currently, many organizations continue to operate with a mindset focused on maximizing individual supply chain components and interactions, often overlooking the broader external environment, and failing to integrate these parts cohesively. This Maximization mindset can limit overall system effectiveness and adaptability in a fast-evolving market [3, 4]. This article introduces the concept of a Purposeful Mindset, inspired by the emerging WST. This mindset shifts focus from isolated optimizations to a broader view that recognizes the external environment, defining a clear Purpose as essential for effective performance. By aligning supply chain components and interactions with this Purpose, the Purposeful Mindset promotes not only efficient performance of individual elements but also enhances the overall effectiveness, resilience, and adaptability of the supply chain to respond to VUCA challenges. Through this exploration, the paper advocates for the application of Purposeful Mindset principles and IKS online solution to transform SCM practices for sustainable competitive advantage. Concrete application of real time data sharing rather than planning according to the prediction will be presented based on the concrete data describing the interaction between central European automotive producer and 1st tier supplier.

In today's VUCA business environment, organizations require agile and responsive systems to manage resources effectively. Traditional ERP systems and online web-based solutions like IKS from companies such as Manufactus offer distinct approaches to addressing these challenges [5, 6]. ERP and MRP systems integrate various business processes, including finance, supply chain, human resources, and production, into a unified platform. This integration provides a comprehensive view of organizational operations, facilitating informed decision-making [5, 6]. However, traditional ERP systems often operate on batch processing and may lack real-time responsiveness, potentially limiting their agility in rapidly changing environments [5, 7]. Customization and scalability can also be challenging, leading to longer implementation times and reduced flexibility [7]. Web-based solutions like IKS are designed to enhance lean manufacturing and just-in-time (JIT) production methodologies. They focus on real-time inventory management and production control, enabling immediate responses to change in demand and supply.

For instance, the IKS e-Kanban system by Manufactus offers functionalities such as dynamic electronic Kanban boards, real-time visualization of Kanban movements, and integration with external suppliers through web portals [8]. These features provide high transparency and flexibility, allowing organizations to adapt swiftly to VUCA conditions.

2 Methodology

2.1 Wholeness Systems Thinking (WST)

WST is a wholistic methodological approach that emphasizes understanding and managing systems as integrated wholes rather than as isolated parts. Rooted in systems theory and extended by recent frameworks like 21st EWA Edu model, the DIKUW (Data, Information, Knowledge, Understanding of Wholeness, Wisdom) hierarchy, this approach encourages viewing systems within the context of their external environment and defined Purpose. In practice, WST comprises two main phases: Wholeness Synthesis (WS) and Wholeness Analysis (WA). This method is particularly suited for research where system interactions with external factors play a critical role, as in sustainability and SCM contexts [9].

2.2 Maximization and Purposeful Mindset

This study presents Maximization Mindset and Purposeful Mindset as contrasting methodological approaches to explore decision-making frameworks and their impact on efficiency and effectiveness within SCM. These approaches are analyzed within the context of the DIKHW (Data, Information, Knowledge, Holistic Understanding, Wisdom) and DIKUW (Data, Information, Knowledge, Understanding of Wholeness, Wisdom) hierarchies, which serve as theoretical frameworks [9].

The **Maximization Mindset** is grounded in a traditional, knowledge-based educational approach, emphasizing the accumulation and application of data and information to optimize specific measurable outcomes. This mindset operates within the DIKHW hierarchy, focusing on achieving the highest possible efficiency in isolated systems or processes without fully considering their broader context. Key characteristics follows: Strives for maximum output, efficiency, and measurable results. Often overlooks external environmental influences and long-term sustainability. Aligns closely with Reductionist Systems Thinking (RST), optimizing system components without considering the wholistic performance of the entire system. Relies on the analysis and synthesis of data and knowledge to improve specific parts of the system. Measures success through isolated KPIs, often leading to fragmented and siloed decision-making. Promotes short-term gains, frequently at the expense of long-term adaptability and sustainability [9].

The **Purposeful Mindset**, inspired by WST, extends beyond efficiency to prioritize effectiveness within a broader external context. It aligns with the DIKHUW hierarchy, integrating "Understanding of Wholeness" and "Wisdom" as higher intellectual capacities. This mindset emphasizes aligning all system components with a clearly defined Purpose derived from the external environment. The key characteristic of the Purposeful Mindset follows. Focuses on long-term sustainability and aligning actions with a clearly defined Purpose. Considers external environmental factors and interactions between system components. Encourages systemic thinking, integrating efficiency and effectiveness to achieve wholistic goals. Employs WS to identify and understand the external environment and derive a system's Purpose. Uses WA to ensure alignment of internal components with the defined Purpose [9].

This dual-methodological framework allows for a comparative analysis of the two mindsets, highlighting their distinct roles in shaping organizational strategies and performance within SCM. By transitioning from a Maximization Mindset to a Purposeful Mindset, organizations can achieve greater alignment with long-term effectiveness connected with strategic goals including sustainability. Promotes adaptability in volatile, uncertain, complex, and ambiguous (VUCA) environments by balancing efficiency with strategic effectiveness. Facilitates integrated decision-making that considers ecological, social, and economic dimensions of sustainability.

2.3 IKS On-Line Software Solution

The IKS system is a web-based application that provides transparency and management of internal and external processes, including in-house production, warehouse movements, product deliveries, customer shipments, intermediate production at suppliers, material consolidation, and more. IKS enables the implementation of a pull-based Kanban system across the entire supply chain, from suppliers to customers. Compared to traditional Kanban systems, IKS operates on a digital platform where all data is uploaded via barcode scanning or high-frequency identification chips. The system provides a real-time overview of all scanned items, tracking their need for production, movement, and current location—all centralized in one place [8].

3 Experimental Part

3.1 Supply Chain Process Introduction

This section introduces the supply chain members. No concrete company names will be presented. Data applied to the software tool described the real situation from the August of 2024. Supplier Beta operated two production lines dedicated to these trays. The production capacity of the two manufacturing lines in Automotive producer's production hall is 1,520 finished batteries per day. Each truck can transport 14 pallets, with each pallet containing 7 battery trays, regardless of the variant (short, medium, or long). This results in a total capacity of 98 trays per truck. Given the one-to-one relationship between battery trays and finished batteries, the daily supply requirement from the supplier to Automotive producer amounts to approximately 15.5 fully loaded trucks. For practical purposes, this is rounded up to 16 trucks per day. Under the current direct delivery transport concept, the supplier is contractually obligated to deliver the ordered materials within a 24-h window without a fixed delivery time. This flexibility allows the supplier to ship all required materials for the day in a single dispatch, potentially leading to multiple trucks arriving in quick succession. To ensure uninterrupted production, Automotive producer maintains a substantial inventory buffer equivalent to 2.7 days of production.

This inventory includes a fixed safety stock covering 1.7 days of production and an additional 1 day for operational processes such as consolidation, empty container return, loading/unloading times, and internal logistics operations. This buffer mitigates the risks associated with unpredictable delivery schedules and ensures consistent production flow.

3.2 Actual Status

The data originates from internal Automotive producer sources and was collected during August, 2024. The documents were then filtered to include only parts relevant to the processing of this work. The Fig. 1 illustrates the average inventory levels of items in the product mix (blue bars), the planned consumption of battery trays, and thus the production of finished batteries (black line), the maximum inventory threshold (red line). From the Fig. 1, it can be observed that the inventory levels are consistently higher than the planned production of batteries.

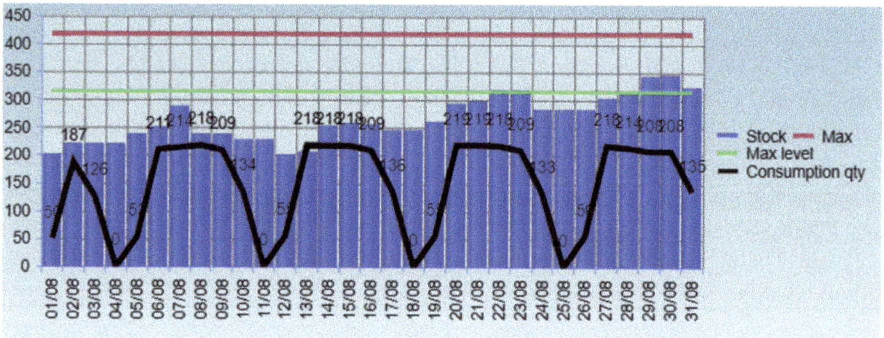

Fig. 1. Actual warehouse level of boxes of studied component, one box contains 7 components, (IKS).

The observed warehouse inventory levels represent the buffer stock required under the current interaction management practices between suppliers and manufacturers. These practices are characterized by a relatively low frequency of information sharing, influenced by traditional processes and IT solutions. Specifically, call-off frequencies with suppliers are scheduled at intervals of six months, monthly, and weekly, with weekly quantities further distributed as daily averages. Consequently, higher safety stock volumes are necessary to compensate for the infrequent data updates and to ensure continuity in operations.

3.3 Proposed Status

The proposed optimization for the delivery of battery trays from supplier Beta to the automotive manufacturer involves the implementation of an online electronic kanban system within the IKS framework [8]. This system maintains the same overall time period but adjusts deliveries to reflect the actual consumption of battery trays. Given the requirement of 16 truck deliveries per day, the delivery schedule is divided into four intervals, each consisting of four trucks. This approach will necessitate the allocation of specific, regular time slots for deliveries (Fig. 2).

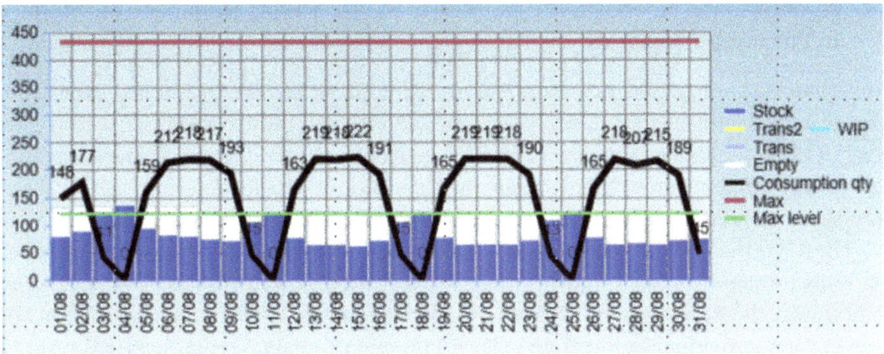

Fig. 2. Possible warehouse level based on application of on-line solution (IKS)

The warehouse level reflects the real situation on the assembly line, which is further influenced by the demand for EVs, which is currently very volatile. Nevertheless, (See, Fig. 1) presents the improved results within the same period, 08/2024. Average inventory level was reduced by 65% from 1,7 day to 0,6 day. The consumption quantities remain obviously the same for previous status and for proposed status.

4 Results

The results section will present a comparison of traditional interaction management between 1st tier supplier and automotive manufacturers, relying on IT solutions such as ERP systems, with an approach utilizing online, real-time data sharing. The results will be presented across two dimensions: first, evaluating specific outcomes related to warehouse inventory levels, and second, examining the corresponding mindsets associated with each IT solution, highlighting their influence on SCM practices.

4.1 The Role of Chosen IT Solution

Instead of relying on traditional, robust IT solutions, the IKS real-time online system was tested, yielding significant results, and identifying substantial improvements. Traditional processes and IT systems primarily focus on mitigating effects, such as adjusting warehouse levels and logistics unit costs, using the latest sophisticated solutions. While these approaches achieve savings in unit costs related to handling and storage, they often lack a direct correlation with significant reductions in overall warehouse levels. This outcome reflects the influence of the Maximization Mindset, as discussed later. Conversely, the real-time data-sharing approach demonstrated considerable enhancements in organizational processes, leading to substantial reductions in warehouse levels, the number of costly specialized containers, required warehouse space, capital costs, and other related expenses. The potential reduction in warehouse levels by up to 65% highlights the opportunity for significant cost savings across these areas.

4.2 Application of Intellectual Pyramids in SCM: A Shift from Maximization to Purposeful Mindsets

The application of intellectual pyramids in SCM illustrates the distinction between a Maximization Mindset, represented by the DIKHW intellectual pyramid, and a Purposeful Mindset, aligned with the DIKUW intellectual pyramid. These frameworks fundamentally influence how supply chain systems address challenges, particularly when comparing traditional ERP systems to real-time, online data-sharing solutions.

The Maximization Mindset, guided by the DIKHW intellectual pyramid, emphasizes the transformation of Data into Information, Knowledge, and Holistic Understanding to optimize the entire system. Holistic Understanding focuses on the interactions of system parts, providing an integrated view of the supply chain. Traditional ERP systems exemplify this approach, utilizing batch-processed data and predefined algorithms to align internal operations. However, the reliance on historical patterns and retrospective analysis limits the system's ability to respond to dynamic external changes. For example, while ERP systems can efficiently schedule production and manage inventory, they fail to account for disruptions like unexpected demand surges or geopolitical risks, as these require predictive capabilities beyond past trends. The external environment is considered insufficiently, reducing the system's ability to solve root causes of supply chain inefficiencies, such as lack of visibility and poor alignment with market dynamics. As a result, these systems excel at solving effects like particular logistics operations unit costs, but struggle to address the systemic causes of supply chain disruptions influencing the quality of logistics processes further resulting in long-term warehouse level.

The Purposeful Mindset operates with a fundamentally different logic than the Maximization Mindset, emphasizing reverse thinking to enable systemic changes that solve causes rather than merely addressing effects. Instead of beginning with Data, Information, Knowledge, Holistic Understanding, and Wisdom, the Purposeful Mindset starts by studying the external environment, deriving the system's Purpose, and applying this Purpose to guide decisions at the Knowledge, Information and Data levels. This approach aligns with the DIKUW intellectual pyramid, which replaces Holistic Understanding with Understanding of Wholeness. This shift enables systems to fully account for dynamic external environments. Understanding of Wholeness integrates the system's Purpose, derived from its interaction with external systems, ensuring meaningful limits are set for its parts and interactions to drive alignment.

Online real-time data-sharing solutions exemplify the Purposeful Mindset by capturing live Data from sensors and connected devices, transforming it into actionable Information via dynamic dashboards. These systems leverage Knowledge to anticipate disruptions and develop adaptive strategies, guided by the overarching Purpose. By incorporating Understanding of Wholeness, such solutions align supply chain operations with shifting external factors, such as market demands, environmental changes, and geopolitical trends. This alignment allows supply chains to address root causes of inefficiencies, including supplier variability and coordination challenges, while ensuring resilience and sustainability. Unlike traditional systems, real-time solutions dynamically adjust to external changes, solving both effects and causes within a purposeful, adaptable framework (Table 1).

Table 1. Comparison of traditional ERP systems and Real-time online solutions in SCM.

Aspect	Traditional ERP maximization mindset (DIKHW)	Online, Real-time data sharing purposeful Mindset (DIKUW)
Focus	Internal optimization with limited consideration of external factors	System-wide integration aligned with external realities
External environment	Insufficiently considered; relies on historical trends	Sufficiently integrated; adapts to dynamic changes
Understanding	Holistic Understanding: focuses on parts and their interactions	Understanding of Wholeness: aligns system Purpose with external drivers
Reaction time	Batch-based, delayed responses to changes	Real-time, dynamic adaptation
Problem-solving	Resolves effects (e.g., inventory shortages, production delays)	Addresses causes (e.g., misaligned supply-demand strategies)
Sustainability	Prioritizes short-term efficiency	Incorporates long-term sustainability and resilience

This analysis underscores the necessity of shifting from a Maximization Mindset to a Purposeful Mindset in SCM. By adopting the Purposeful Mindset based on DIKUW intellectual pyramid, organizations can align their operations with external realities, enabling adaptive, sustainable, and Purpose-driven supply chains. This approach transforms SCM from isolated optimizations to Wholistic solutions that address systemic challenges in a rapidly evolving global environment in comparison with Holistic solutions covering the entire systems without sufficient consideration of external environment. The shift from a Holistic Approach based on Holistic Understanding to a Wholistic Approach grounded in Understanding of Wholeness represents a crucial evolution in systems thinking. The Holistic Approach integrates system parts and interactions, relying on historical practices to optimize internal dynamics but treating the external environment as static, limiting adaptability. In contrast, the Wholistic Approach embeds the system's Purpose and aligns it with a dynamic external environment, enabling meaningful limits and proactive responses. This ensures resilience and addresses root causes, making it a more effective framework for managing complex, Purpose-driven systems.

5 Conclusion

This study underscores the transformative impact of adopting a Purposeful Mindset in SCM compared to the traditional Maximization Mindset. Traditional ERP systems, aligned with the DIKHW intellectual pyramid, focus on internal optimization through

Holistic Understanding. While effective for mitigating short-term effects, these systems often fail to sufficiently address dynamic external environments, limiting their ability to resolve root causes such as supply chain disruptions and misalignment with market demands. In contrast, real-time online data-sharing solutions, guided by the DIKUW intellectual pyramid, integrate Understanding of Wholeness to align systems with their Purpose and dynamic external factors. By considering real-time data, these solutions address systemic causes, resulting in significant improvements, including up to 65% reductions in warehouse levels, decreased container reliance, and minimized capital costs. This shift from a Holistic to a Wholistic Approach represents a critical evolution in SCM, ensuring resilience, adaptability, and sustainable alignment with long-term goals in a VUCA global environment.

References

1. Ivanov, D., Dolgui, A.: A digital supply chain twin for managing the disruption risks and resilience in the era of industry 4.0. Prod. Plan. Control. **31**(6), 444–457 (2020)
2. Kumar, A., Singh, R.K., Modgil, S.: Exploring the relationship between ICT, SCM practices and organizational performance in Indian manufacturing organizations. J. Adv. Manag. Research. **17**(2), 161–185 (2020)
3. Davenport, T.H.: Putting the enterprise into the enterprise system. Harv. Bus. Rev. **76**(4), 121–131 (1998)
4. Slack, N., Brandon-Jones, A., Johnston, R.: Operations Management, 8th edn. Pearson (2016)
5. SAP. What is MRP? [online]. Available from: https://www.sap.com/products/erp/what-is-mrp.html (2024). Viewed 14 Nov 2024
6. Oracle. What is ERP? [online]. Available from: https://www.oracle.com/erp/what-is-erp/ (2024). Viewed 14 Nov 2024
7. TechTarget. Material requirements planning (MRP) [online]. Available from: https://www.techtarget.com/searcherp/definition/material-requirements-planning-MRP (2024). Viewed 14 Nov 2024
8. Manufactus. Integrated Kanban System (IKS). [online]. Available from: https://www.manufactus.com/iks-integrated-kanban-system/?lang=cs (2024). Viewed 14 Nov 2024
9. Holman, D., Švejdarova, E.: Unifying effectiveness and efficiency in sustainable supply chain Management in Automotive: the 21st EWA Edu model enhanced by wholeness systems thinking for lifelong learning. Rocznik Ochrona Środowiska. **26**, 539–557 (2024)

Comparative Analysis of Fashion and Military Clothing Supply Chains

Justyna Świerczek(✉) and Agnieszka Tubis

Department of Technical Systems Maintenance and Operation, Faculty of Mechanical Engineering, Wrocław University of Science and Technology, 27 Wyspiańskiego Str, 50–370 Wrocław, Poland

{justyna.swierczek,agnieszka.tubis}@pwr.edu.pl

Abstract. Publications on the management of apparel supply chains focus primarily on organising a fast, flexible, and reliable supply of materials and finished goods at subsequent stages of processing and distribution. A critical issue is implementing the concept of 'rapid response', which, based on the information integration of partners, allows a rapid response to changing market demand for finished goods. This is mainly due to the specifics of the clothing market and customers' consumption behaviour, which often leads to impulsive reactions to fashion novelties. The supply chains responsible for supplying military clothing are quite different. The specifics of customer behaviour and the relationships established by links in military clothing supply chains do not necessitate the implementation of solutions using Quick Response concepts. However, these chains report a demand for improvements to current material flow control systems, which should consider their safety, comfort, protection and performance requirements. This article will provide a comparative analysis of traditional (fashion) and military apparel supply chains. The analysis aims to identify the differences in the material flow management systems in the chain. The research used a narrative literature review method and a case study of a selected supply chain serving the military apparel market. The results enable the identification of potential opportunities for enhancing cooperation mechanisms in military clothing supply chains and provide further directions for ongoing research.

Keywords: Clothing market · Supply chain management

1 Introduction

Today's garment supply chains are a complex network of global connections that integrate raw material and material suppliers, manufacturers, and distributors. The way they function, the management of material and information flows along the links, and the optimisation techniques implemented directly impact production costs, product quality, lead times and environmental impact. As early as the 1980s, there was a segmentation of the clothing industry into Low Fashion (fashion with a two-season rhythm, the introduction of new solutions aimed at reducing production costs) and Fast Fashion (the shortest possible time to transfer new trends to shops, frequent replacement of collections, an

J. Feliks et al. (Eds.): CLC2024 2024, 2026.
https://doi.org/10.1007/978-3-032-28931-5_6

attempt to keep up with changing fashions) [1]. Supply chains serving both segments are characterised by different priorities and ways of organising product flows.

The research subject presented in this article is two types of apparel supply chains: traditional (fashion) and military apparel. Although both supply chains have the same goal—to deliver the finished product to the final customer—they differ in product specifics, customer requirements and external factors affecting the realised logistics service process. The military clothing market is closer to a low-fashion supply chain than to a fast-fashion supply chain. However, it is impossible to attribute 100% of the characteristics of this sector to it. Therefore, it is reasonable to conduct analyses to better understand the characteristics of its functioning, allowing dedicated solutions to be developed for the specific needs of the supply chains serving this market.

This article aims to provide a comparative analysis of the supply chains serving the fashion and military apparel markets. The article is structured in four sections. Section 1 presents the research area and indicates the purpose of the prepared article. Section 2 formulates the research questions and presents the stages of the research carried out, as well as the methods used at the subsequent stages of the analysis. The results obtained through a comparative analysis are presented in Sect. 3. Section 4, on the other hand, summarises the obtained results and formulates the most important conclusions.

2 Methodology

The first step of the work was to formulate the research questions to be answered by the analysis: Q1: What are the main differences between fashion and military clothing supply chains? Q2: What is the dependence of both chains on prevailing fashion trends, and how does this affect the distribution planning process? The questions posed made it possible to order the sequence of research activities undertaken, as shown in Fig. 1. For each activity, the research method used is indicated.

Regarding fashion supply chains, a wealth of industry and academic literature describes the characteristics of material flows, production requirements, and customer needs. However, the literature review on military apparel supply chains shows that, in this case, the range of available research results and industry publications is minimal. Therefore, the characterisation of the military apparel supply chain was prepared based on the industrial research of a selected case study.

3 Results

The fashion apparel industry has, since the 1980s, been heavily focused on improving the logistical handling of material flows, aimed primarily at reducing material flow times by optimising storage and downstream processing [2]. Therefore, modern apparel companies use several methods to improve internal and external processes, such as:

- Lean Management, or 'lean' management, aims to reduce inventory, reduce unnecessary transport costs, reduce defects and over-processing [3]
- Quick Response has revolutionised supply chain management at Europe's largest fashion suppliers [4]. The concept is to enhance a company's operational efficiency by accelerating the supply chain's responsiveness to changing market conditions.

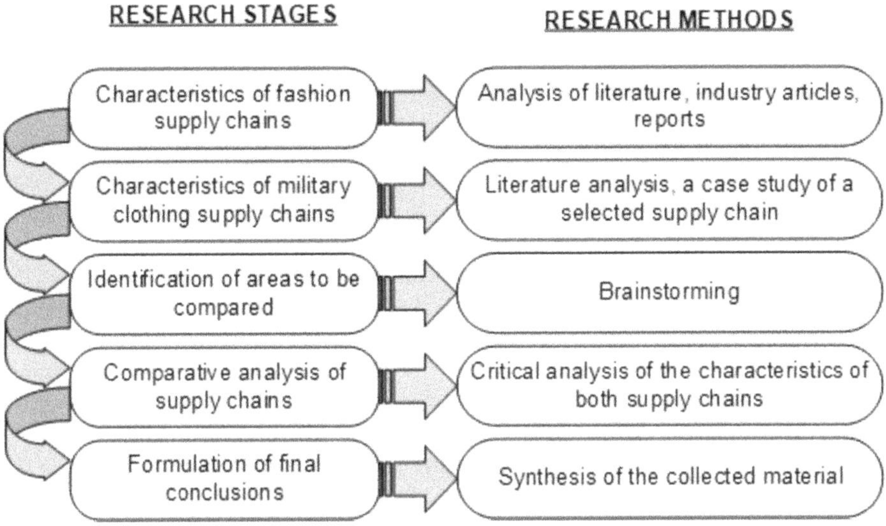

Fig. 1. Research stage and methodology

On the other hand, military clothing must meet very high standards in terms of material strength, functionality and safety. This translates into entirely different requirements for materials, production technology, and the necessary testing and certification. Aspects relating to responsiveness play a secondary role, as the clothing market is not so influenced and is rapidly changing in this case.

The analysis focused on six criteria to identify the main characteristics that differentiate the two chains: location of production, dependence on trend, design, construction, and prototyping, materials, distribution network, and demand planning, as well as the flow diagram.

Location of production

For many years, fashion supply chains have relied on Asian factories for production. The reason for this was the much lower costs, e.g., person-hours [5]. However, analyses have shown that fashion companies benefit from relocating part of their production to Europe. Although the cost of production is higher than that of production in Asia, the shorter lead time ensures greater product availability in distribution centres and the ability to react quickly to changing demand. This allows for a competitive advantage [6].

The situation is different in the military clothing industry. European producers of military clothing aimed at professional customers (military, special units) want to rely exclusively on indigenous/European production and have total control over each design and production stage. On the other hand, military clothing aimed at a non-professional audience (e.g. shooting enthusiasts) must also meet similar high-quality requirements but at a price that corresponds to the financial possibilities of the average consumer. However, selecting suppliers based only on the lowest price offer is out of the question. Only some suppliers of semi-finished products or components and only some clothing factories can meet the industry's requirements. Long-term relationships with suppliers are therefore necessary.

Dependence on trends

Military clothing is not as subject to prevailing trends as regular clothing and is characterised by a high degree of timelessness. Many products remain unchanged for years or are only subject to minor improvements. The colour scheme has remained within a rigid framework for years. Above all, earthy colours (beiges, browns, greens, navy blues, etc.) and garments covered in military camouflage dominate. This stems from the belief that one should be as inconspicuous as possible to survive for as long as possible under challenging conditions, e.g., on the battlefield. Military camouflages usually have a genesis related to specific circumstances. The patterns of official camouflage remain unchanged for decades. Military clothing enthusiasts are well-educated in this regard and have specific expectations of the product. In addition to the official military camouflages of individual countries and regions, alternative versions are also available on the commercial market.

We can encounter, among others, Polish camouflage, such as Wz. 93, or panther. This camouflage was introduced in 1993 and adopted by the Polish Army. It has been present on regulation uniforms for more than 30 years. Another example is Flecktarn—a German camouflage from the 1970s, or US Woodland, created in the 1980s for the US armed forces. All of these and many others are still present in military clothing. Despite the wide choice of camouflage and colours derived from nature, the possibilities are limited when compared with the design and colours of classic clothing. The fashion market outdoes itself in the wealth of patterns, colours and forms. Its possibilities are endless. In military clothing, the aim is to be invisible; in fashion clothing, the aim is to stand out from the crowd.

Design, construction, and prototyping

Despite the limited influence of fashion on military clothing designs, their production is by no means straightforward. Quite the contrary, this is because the quality of the execution and the quality of the materials used are crucial. Suppliers of military clothing have to meet exacting requirements and adhere to a high level of quality control. Military designs are characterised by structural complexity, which should result in maximum wearer comfort. A decisive aspect of evaluating military clothing is its functionality, which refers to the design of the equipment and the additional reinforcement of selected parts of the garment. Additional functionalities are hidden in all garment parts (e.g., gloves, headgear, combat blouse, bulletproof vest, combat belt, backpack), as they may prove crucial in a life-threatening situation.

Although the methodology for working with such garments is the same as that of the fashion supply chain (design, construction, sewing, testing, production), the specifics are quite different. Therefore, the transition from design inception to realisation is time-consuming. At the prototyping stage, in-depth testing is conducted, which can take several months. Once the first sewing has been prepared, an expert inspects the garment. Sometimes, the garment is checked straight away by individuals specifically and is even worn for a trial period. If the expert reports an abnormality, the sewing is repeated until a positive opinion is obtained.

The situation is different in the fashion supply chain. Everyone comes into contact with classic garments on a daily basis. The quality of the joints, pocket or leg finishing presented by well-known fashion chains is imperfect. However, consumers tend to pay less attention to this due to the lower price and the understanding that these garments are intended to last only one season and will likely go out of fashion afterwards. Users of military clothing have a different approach. There is no room for mistakes here.

Materials

Materials also play an important role. The materials used are windproof, waterproof, abrasion-resistant, quick-drying, fireproof, thermally active, and IRR-coated (invisible in night vision). At the same time, they offer adequate flexibility, lightness, and durability. The technology departments of leading military clothing companies spend considerable time researching and developing ever-new and more functionally advanced materials. It is necessary to conduct research in specialised laboratories, such as laboratories with security and defence accreditation. Such companies must also guard their output by registering their trademarks, designs and patents.

Distribution network and demand planning

Distribution networks for fashion and military clothing also differ. While the distribution network of, e.g., Zara relies almost entirely on its subsidiaries, the distribution networks of military clothing are not as extensive and thus consist of independent entities. Therefore, implementing demand control tools in the fashion supply chain is much easier. Using tools such as Vendor Managed Inventory (VMI) or a shared inventory management system in vintage clothing enables precise planning and scheduling of production and delivery processes [7]. Despite the availability of numerous planning and forecasting tools, there is still no clearly defined method to determine demand in the clothing industry with high reliability.

Military clothing shows fluctuating demand. To prepare effectively for the market's needs, new tools for planning and optimising the military supply chain need to be developed. It is necessary not only to regularly control stock levels but also to monitor global events. Fluctuations in demand can result, for example, from the prevailing political, social, or economic situation in the countries where any of the supply chain elements are located. Fluctuations can be caused by armed conflicts or even the mere threat of them, pandemics, natural disasters or terrorist attacks.

At the time of the two most significant armed conflicts since the Second World War, the war in Ukraine and Israel's war with Hamas, which occurred on the verge of the end of the COVID-19 pandemic, these fluctuations are obvious. They have had an impact not only on the demand for military clothing and accessories but also on international logistics, such as the issue of navigating the Suez Canal. An additional factor that makes planning difficult is the lead time for projects. In the case of military clothing, the time required is relatively long, often amounting to several months. In contrast, in the fashion supply chain, the goal is to realise the full logistics service (going from design, production, and organisation of delivery to the arrival of the goods in shops) in no more than a few weeks.

Flow diagram

The analysis of flows in the military clothing supply chain must take into account its various stages, including the procurement of materials and components (such as zippers, Velcro, buttons, webbing, and other accessories), production (sewing and processing), quality control, transportation, storage, and delivery to distribution centres. Reverse logistics, otherwise known as environmentalism, is also becoming increasingly important. It can be interpreted as the flow of goods occurring in the distribution channel, having a reverse flow direction. This means that the goods move from the customer to the producer. Reverse logistics saves money, realises the product's full potential and has a real impact on reducing textile waste in the environment. Military clothing, even if it has clear signs of use, will be suitable for purchase again once it has been refurbished because of the high quality of the materials.

4 Conclusion

The topic of military apparel supply chain management is an interesting but poorly recognised area of research. Meanwhile, a comparative analysis of fashion and military supply chains shows many differences. Comparative analysis criteria indicate the need for different flow control models in both types of chains. Military clothing belongs to specialised clothing. There is no shortage of interesting and practical solutions in fashion supply chains, but these cannot be implemented directly for handling military clothing, which is a specialised category. Therefore, the comparative analysis is a starting point for further research into the possibilities of streamlining material flows in military clothing supply chains. Based on the knowledge of managing apparel supply chains, it is possible to identify potential methods and principles that, when appropriately modified, will yield the expected results in the logistical handling of military apparel flows.

References

1. Ciszweska-Kulwińska, E.: Logistyka branży odzieżowej – wyzwania i szanse rozwoju. WILiŚ Politechnika Gdańska. (2016) Available from: https://wilis.pg.edu.pl/. Accessed 30 Sept 2024
2. Palewicz, J., Baran, J.: Organizacja łańcucha dostaw w branży odzieżowej. Logistyka. **6** (2012)
3. Nayak, R.: Lean Supply Chain Management in Fashion and Textile Industry. Springer, Singapore (2022)
4. Woźniak, H.: (Nie)zwykła logistyka ZARY. Logistyka. **6**, 59–66 (2010)
5. Nenni, M.E., Giustiniano, L., Pirolo, L.: Demand forecasting in the fashion industry: a review. Int. J. Eng. Bus. Manag. **5** (2013)
6. Nagurney, A., Yu, M.: Fashion supply chain management through cost and time minimization from a network perspective. In: Industrial Engineering: Concepts, Methodologies, Tools, and Applications, pp. 1382–1401. IGI Global (2013)
7. Shen, B., Chan, H., Chow, P., Thoney-Barletta, K.: Inventory management research for the fashion industry. Int. J. Inven. Res. **3**(4), 297–317 (2017)

The Role of Digital Maturity in Improving Safety and Reliability in Wood-Based Supply Chains—Case Study

Sylwia Werbinska-Wojciechowska(✉), Robert Giel, and Natalia Gnacy

Wroclaw University of Science and Technology, Wroclaw, Poland
sylwia.werbinska@pwr.edu.pl

Abstract. This paper explores the concept of digital maturity in relation to the performance of the supply chain of the wood-based industry, emphasizing the importance of ensuring safety and maintaining system reliability. The ongoing digital transformation is reshaping all components of supply networks, progressively replacing traditional approaches with digitally enabled processes. Such a shift creates new possibilities for improving efficiency and operational effectiveness while mitigating the risks inherent to supply chain operations. Accordingly, the paper examines the potential for advancing digital maturity in a selected enterprise that forms part of the wood products supply network. The research approach begins with a focused review of the relevant literature, followed by an analysis of major operational challenges characteristic of wood-based supply chains. Based on these insights, a preliminary assessment of digital maturity is carried out for the chosen company. The paper concludes with a summary of the main outcomes and recommendations for further investigations.

Keywords: Wood-Based Supply Chain Reliability And Safety · Digital Maturity Assessment · Digital Supply Chain

1 Introduction

Supply chains (SCs) can be understood as complex systems of interrelated processes involving the planning, coordination, and management of material and service flows between suppliers and end customers [1]. Traditionally, their structure has depended on a network of spatially dispersed facilities that support transportation and distribution activities. Yet, in the modern global economy, many supply chains have lost their self-sufficiency, struggling with instability caused by globalization, economic fluctuations, material shortages, and the growing complexity of market interactions. To remain competitive, companies are compelled to adapt swiftly to such turbulence—for instance, by reinforcing sustainability practices or sourcing locally to offset labor and resource constraints. Executing these adaptive measures effectively calls for the deployment of advanced technologies, which enhance precision and efficiency while meeting stakeholder and customer demands. Accordingly, supply chains are evolving toward greater intelligence, driven by the principles of digitalization, transparency, and collaborative networking [2].

J. Feliks et al. (Eds.): CLC2024 2024, 2026.
https://doi.org/10.1007/978-3-032-28931-5_7

The adjective 'digital' denotes the use of electronic datasets, in contrast to analog ones [3]. Digital transformation is defined as the systematic process utilized to restructure economies, institutions, organizations, and society at a systemic level [4, 5]. Indeed, it can be perceived as changes in the system that result in implementing technological improvements in order to increase supply chain efficiency [1]. In this context, digital technologies refer to the adoption of such approaches as Industry 4.0 [6], as well as Supply Chain 4.0 [7], or Logistics 4.0 [8], which collectively aim to integrate cyber-physical systems, automation, and data-driven decision-making across supply chain operations. The notion of a digital supply chain is strongly associated with the concept of digital maturity and the extent to which digital tools are adopted and effectively applied [9]. Maturity, in general terms, can be interpreted as the degree to which a process or organization has evolved toward an advanced, optimized, or fully developed state. Within a business context, maturity models serve as structured instruments that enable the evaluation of an organization's advancement in a specific domain according to predefined criteria [4]. Because of the diversity of applications, supply chain digital maturity models do not follow a single standardized format. According to [10], such models usually consist of five fundamental elements:

- Assessment domains— reflecting the technological or managerial areas under evaluation,
- Assessment indicators— specifying measurable aspects used to determine progress,
- Maturity stages— representing successive levels of digital advancement, typically ranging from three to six,
- Framework structure— integrating domains, indicators, and maturity stages into a coherent analytical matrix,
- Evaluation procedure—describing how the assessment process should be performed in practice.

Extensive comparative analyses of existing supply chain maturity models can be found in [4, 11], where 21 and 28 frameworks are systematically reviewed and contrasted. Further discussions on Industry 4.0 maturity assessment approaches are available in [12] and expanded in [13], offering valuable insights and recommendations for continued research in this area.

The growing push toward digital transformation arises from the escalating demands of stakeholders, who increasingly expect reliability, availability, and sustainability in product and raw material supply chains. This tendency is particularly visible in the forestry and wood-based sectors [14]. Wood-based products rely on both primary materials, such as roundwood, and secondary materials derived from production residues (e.g., sawdust, shavings, or chips). Managing these complex material flows requires addressing challenges linked to the physical nature of wood (e.g., variable density, irregular dimensions) as well as the integration of digital solutions into operational practices. Therefore, this paper highlights the major barriers faced by wood-based supply chains in their pursuit of digital maturity, emphasizing implementation constraints and the need for employee upskilling and digital competence development.

2 Digitalization Issues of a Wood-Based Supply Chain

The wood-based supply chain (WBSC) contains the networks and flow systems for products where 'recycled wood' or timber waste is utilized. This includes residuals from sawn timber production (e.g., sawdust, shavings, woodchips) and round timber that undergoes processing. Key examples of the final wood-based product types are laminated boards, worktops, as well as chipboards. The functionality of such a WBSC is best exemplified by examining a production company (along with its supply and distribution connections) operating in the global arena. The material journey within this SC begins with the process of timber extraction from forests (Fig. 1). In this supply chain, the forestry enterprise serves as the starting link, responsible for obtaining and supplying raw timber to the sawmill. The wood extraction process demands careful organization, taking into account factors such as the available harvesting volume and the maturity of the forest stands. Once felled, the logs are temporarily stored before being forwarded to the processing facility for further treatment. The residual material from wood processing serves as the base raw material for the wood-based products industry. Furthermore, products derived from initial wood processing, such as sawn timber, are also subject to subsequent transformation. During these later stages of manufacturing, additional waste is also produced, which then becomes a feedstock for the wood-based products sector. An additional source of raw materials for the subject company is the municipal waste recycling sector. Recycling companies collect municipal waste, including discarded furniture, and process it into intermediate wood-based materials. The case company's finished products, such as wooden boards, are mainly supplied to the furniture and construction industries. From there, they are either delivered to secondary wood-processing enterprises or distributed through wholesale channels. In the final stage, retailers make these products available to end consumers.

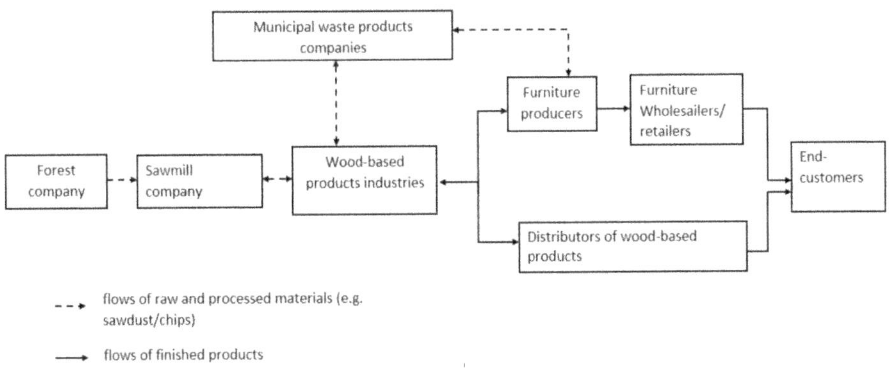

Fig. 1. A scheme for the case company's wood-based supply chain

At the same time, the examined organization faces several key challenges related to the management of both material and information flows. The main difficulties in handling physical materials are primarily associated with procurement, internal logistics, and distribution activities. Regarding information management, the organization struggles with:

- Inadequate communication between departments, including production, laboratory, customer service, and logistics,
- Limited understanding of the importance of specific information for other areas of the company. Departments often do not recognize which data are critical or fail to give them proper priority. For example, information about material density, which directly impacts warehousing, customer service, and logistics, is not consistently shared across the internal supply chain.

Within procurement, a significant challenge lies in securing raw materials for production, influenced by factors such as geopolitical instability and rising coal prices, which in turn drive increased demand for pellets.

From the distribution perspective, the high costs associated with transport operations present a major obstacle. Product characteristics, such as size, weight, and value, restrict the efficiency of transport planning. The broad diversity of products further complicates efforts to adhere to the 5Rs principle. Additionally, the escalating expenses of transportation services lead to a situation where the ratio of product prices to delivery costs becomes disproportionately high, negatively affecting overall operational efficiency.

Within the area of in-house logistics, operational hurdles arise during the execution of handling, transport, and storage operations, largely owing to the characteristics of the material and finished product (e.g., non-standardized dimensions of raw materials and products). On one hand, it is imperative to consider operator safety and environmental protection during handling activities. On the other hand, the expenses and duration of storage processes and internal movements are considerable. These involve a large volume of handling tasks, which consequently extends the transport time and elevates material flow costs (making the implementation of the FIFO principle practically unfeasible). It is also important to consider that losses resulting from material damage or product defects during storage and internal handling can account for approximately 3% of the company's total inventory value. This challenge is further intensified by the company's extensive product range. The diversity of products and finishes, including variations in size, texture, and surface treatment, significantly complicates efficient warehouse management and storage optimization.

Operational performance is additionally hindered by inadequate communication between departments and a limited understanding of which data are critical for other units. For instance, variations in product density affect actual weight, which can alter the number of items per pallet and necessitate adjustments in transport and truckload planning.

Overall, enhancing operational efficiency in the company requires the digitalization of logistics processes, with a particular focus on information flows between suppliers and customers. Implementing such solutions would improve the accuracy of demand forecasting, production planning, weight calculation procedures, and load optimization.

3 Digital Transformation in the Case Company Vs. its Safety and Reliability

Several studies in the literature investigate the adoption of Industry 4.0 technologies within wood supply chains. For example, [15] proposes a comprehensive framework that provides an integrated perspective on Forestry 4.0, focusing on the forest supply chain. This framework is structured around four main components: the digital technologies relevant to each business activity in the supply chain, the necessary network infrastructure, advanced system intelligence, and a collaborative digital ecosystem for forestry supply chains. Similarly, [16] reviews the current state-of-the-art technologies applied across forest supply chains, highlighting the progress and gaps in implementation. The adoption of Industry 4.0 in the wood sector is further examined in [14], while [17] offers practical guidelines for designing effective Industry 4.0 implementation strategies specifically for the furniture industry, based on survey data collected from furniture manufacturers.

Considering the wood-based sector as a whole, it is evident that digital models will inevitably become integral to supply chain operations; however, full-scale deployment across all stages remains limited. The analysis of the selected company, along with its interactions with suppliers and customers, indicates that the necessary digital infrastructure to achieve a high level of maturity is still insufficient. In addition, the workforce requires more extensive training. Participants within the analyzed wood-based supply chain exhibit varying levels of digital maturity and adopt advanced technologies independently, primarily to optimize their internal operations rather than the performance of the overall supply chain.

For example, in the investigated case company, the operational performance is supported by two main computerized systems – SAP and Hydra. In business affairs, there is a digital solution dedicated to the biggest company's customer, who is able to place an order in the special online application, which transfers data to the SAP system. There is no need to contact workers or write an email. It optimizes order time and avoids making a mistake. Additionally, some of the containers used in the company have a GPS localization (Global Positioning System) to provide the possibility of more efficient operation and to prevent stealing. In the future, there is a plan to refit all the containers. In addition, there are other programs used for the production areas (like AutoCAD for construction designing processes support) and transportation areas (like Transporeon for expedition processes support). Basic logistic processes are generally automated and digitalized, such as supply process planning, cost conversion, material requirements planning, or selected customer relationship management.

The company's developing digital maturity has significant potential to enhance safety and reliability across its operations. The current implementation of ERP systems like SAP and production management tools like Hydra provides a strong foundation for improving efficiency and reducing errors, directly influencing reliability. Automating business processes streamlines production and logistics, reducing human error and operational delays, thereby ensuring timely deliveries.

Enhancing digital maturity supports advancements in predictive maintenance, supply chain resilience, process automation, and data-informed decision-making. In the wood-based industry, predictive maintenance solutions that utilize sensor-generated data allow organizations to anticipate equipment failures and plan maintenance activities proactively, minimizing downtime and maintaining uninterrupted production. Additionally, condition-based monitoring systems contribute to greater reliability by continuously assessing the operational status of machinery, facilitating timely interventions, prolonging equipment lifespan, and preventing expensive repairs.

Greater digital maturity also improves supply chain visibility, crucial in the wood-based industry where delays in raw material delivery can disrupt production. Real-time tracking and supply chain management systems optimize inventory, streamline procurement, and ensure timely replenishment, minimizing stockouts and production delays. Predictive analytics enable demand planning, aligning production with market needs and reducing inefficiencies.

Automated systems like SAP and Hydra reduce manual errors in data entry, order processing, and scheduling, ensuring consistent, reliable operations. Automation enhances precision in production, reducing defects and ensuring high-quality outputs. As digital maturity grows, the company becomes better equipped to scale operations while maintaining reliability, with flexible systems that adapt to changing needs.

Digital tools also enhance safety by automating compliance with occupational health and safety regulations, tracking training, and enforcing safety protocols. GPS tracking on containers mitigates theft risks and improves logistical safety, while expanded tracking systems promise further security benefits. Automated safety checks integrated into ERP and production tools ensure compliance before tasks are performed, reducing risks.

Challenges such as insufficient infrastructure, skill gaps, and uneven digital maturity across the supply chain must be addressed to fully realize these benefits. Advanced maintenance systems, greater integration of digital tools, and expanded training will further enhance safety and reliability, enabling the company to maintain consistent performance and meet customer demands.

The company's increasing digital maturity positively impacts safety and reliability by automating and optimizing various operational areas. Nevertheless, achieving the full potential of these benefits depends on addressing infrastructure and workforce limitations, as well as promoting more comprehensive adoption of digital technologies across the entire supply chain.

4 Conclusions

This article provides a brief analysis of digitalization within the wood-based sector, focusing on the supply chain of the selected production company. The literature review indicates that digital maturity in wood-based SCs remains underdeveloped, particularly in terms of technological capabilities. In the case company, key operational processes are currently supported by systems such as SAP, Hydra, Transporeon, and AutoCAD. Nevertheless, the overall supply chain infrastructure is not yet equipped to achieve a high or even medium level of digital maturity. Looking ahead, the company plans to upgrade its remaining containers with GPS tracking, and in 2025, the implementation of SAP HANA is scheduled.

In addition, the authors' future research works will focus on developing a digital maturity model for wood-based supply chains that includes such assessment aspects as business process, customer support, sustainability goals, and resilience. The main areas for digital analysis focus on technology and interoperability & cooperation. The proposed model will allow the development of a new proactive approach for resilience assessment in the area of supply chain performance, with a particular focus on wood-based sector problems.

References

1. Buntak, K., Kovačić, M., Mutavdžija, M.: Measuring digital transformation maturity of supply chain. Tehnički glasnik. **15**(2), 199–204 (2021)
2. García-Reyes, H., Avilés-González, J., Avilés-Sacoto, S.V.: A model to become a supply chain 4.0 based on a digital maturity perspective. Procedia Comput. Sci. **200**(2019), 1058–1067 (2022)
3. Rachinger, M., Rauter, R., Müller, C., Vorraber, W., Schirgi, E.: Digitalization and its influence on business model innovation. J. Manuf. Technol. Manag. **30**(8), 1143–1160 (2019)
4. Hellweg, F., Janhofer, D., Hellingrath, B.: Towards a maturity model for digital supply chains. Logist. Res. **16**(1) (2023)
5. Unruh, G., Kiron, D.: Digital transformation on purpose. MIT Sloan Manag. Rev (2017)
6. Gharaibeh, L., Eriksson, K.M., Lantz, B., Matarneh, S., Elghaish, F.: Toward digital construction supply chain-based industry 4.0 solutions: scientometric-thematic analysis. Smart and sustainable. Built Environ (2022)
7. Lee, C.K.M., Chung, S.Y.: From industry 4.0 to supply chain 4.0: a systematic review. Smart Manuf. **1**(1) (2022)
8. Paksoy, T., Koçhan, C., Ali, S.S.: ogistics 4.0: Digital Transformation of Supply Chain Management. CRC Press (2021)
9. Yin, W.: Identifying the pathways through digital transformation to achieve supply chain resilience: an fsQCA approach. Environ. Sci. Pollut. Res. **30**(4), 10867–10879 (2023)
10. Wernicke, B., Stehn, L., Sezer, A.A., Thunberg, M.: Introduction of a digital maturity assessment framework for construction site operations. Int. J. Constr. Manag. **23**(5), 898–908 (2021)
11. Hellweg, F., Lechtenberg, S., Hellingrath, B., Thomé, A.M.T.: Literature review on maturity models for digital supply chains. Braz. J. Oper. Prod. Manag. **18**(3), 1–12 (2021)
12. Felch, V., Asdecker, B., Sucky, E.: Maturity models in the age of Industry 4.0—Do the available models correspond to the needs of business practice? In: Proceedings of the Annual Hawaii International Conference on System Sciences, pp. 5165–5174 (2019)
13. Walter, O.M.F.C., Paldini, E.P., Hennig, E., Konrath, A.C.: Industry 4.0 maturity models: review and classification as a support for Industry 4.0 implementation (2020)
14. Molinaro, M., Orzes, G.: From forest to finished products: the contribution of industry 4.0 technologies to the wood sector. Comput. Ind. **138**, 103637 (2022)
15. Feng, Y., Audy, J.F.: Forestry 4.0: a framework for the forest supply chain toward industry 4.0. Gestão Produção. **27**(4), 1–21 (2021)
16. He, Z., Turner, P.: A systematic review on technologies and industry 4.0 in the Forest supply chain: a framework identifying challenges and opportunities. Logistics. **5**(4), 88 (2021)
17. Červený, L., Sloup, R., Červená, T., Riedl, M., Palátová, P.: Industry 4.0 as an opportunity and challenge for the furniture industry—a case study. Sustainability (Switzerland). **14**(20) (2022)

Sustainable Logistics Processes

Environmental Strategy and Key Indicators Analysis in Sustainability ESG Reporting in Automotive Industry

Dorota Burchart[1(✉)], Radim Lenort[2], Marcin Staniek[1], Josef Horák[2], and Iga Przytuła[1]

[1] Faculty of Transport and Aviation Engineering, Silesian University of Technology, Krasińskiego 8, 40–019 Katowice, Poland
dorota.burchart@polsl.pl

[2] Škoda Auto University, Na Karmeli 1457, 29301 Mladá Boleslav, Czech Republic

Abstract. The growing importance of environmental issues for organizations has led to a growing number of guidelines regarding sustainability indicators in the transportation sector. Many organizations are now required to prepare sustainability reports (ESG—Environmental, Social, and Governance—reporting) in accordance with the Corporate Sustainability Reporting Directive (CSRD). At the same time, stakeholders such as customers, investors, and regulators are placing increasing pressure on companies to demonstrate not only compliance but also proactive engagement in sustainable development. Currently, many organizations in the automotive industry are developing their own ESG reporting framework, creating dedicated methodologies and defining their own indicators for specific aspects of sustainability. This reflects the dynamic nature of the sector, where technological innovation, environmental performance, and social responsibility are closely interconnected. This paper presents key indicators for the three main pillars, including environmental, economic, and social aspects, and outlines the principles for their effective implementation, using the example of Škoda Auto as a case study.

Keywords: ESG reporting · Key Performance Indicators (KPIs) · carbon footprint · automotive industry

1 Introduction

One of the main goals of the European Commission (EC) is to achieve climate neutrality by 2050 (i.e., achieving net-zero greenhouse gas emissions), which is intended to be one of the fundamental development directions for European Union countries. To achieve this goal, the EC is introducing various guidelines and regulations, including the European Green Deal [1], which is primarily intended to reduce greenhouse gas (GHG) emissions, introduce technological innovations, and develop strategies in many economic sectors, including the transport sector. Therefore, it is necessary to first develop strategic plans for each sector and develop Key Performance Indicators (KPIs) for individual organizations.

J. Feliks et al. (Eds.): CLC2024 2024, 2026.
https://doi.org/10.1007/978-3-032-28931-5_8

The main document encompassing new guidelines and obligations for many organizations in the European Union regarding reporting on key aspects of sustainability is the Corporate Sustainability Reporting Directive (CSRD) [2]. The CSRD directive clearly outlines the obligations for companies regarding the three main aspects of sustainability and the reporting of indicators for each of the three dimensions: environmental (E), economic (G), and social (S). All performance indicators, according to the CSRD guidelines, should be summarized in an ESG report. ESG reporting is a new non-financial reporting requirement for many organizations, which poses significant challenges, particularly with environmental indicators such as carbon footprint. For organizations in the logistics and transport sector, implementing ESG reporting also requires understanding the importance of reporting and familiarizing themselves with numerous new rules and procedures of ESG [3, 4]. Transport industry significantly contributes to greenhouse gas emissions, adopting ESG reporting standards can lead to improved energy efficiency, the adoption of cleaner technologies, and the optimization of logistics processes and supply chains [5]. Additionally, ESG reporting also considers the Global Reporting Initiative (GRI) Standards and the Sustainability Accounting Standards Board (SASB) guidelines [6, 7]. The article characterizes the main KPIs (Key Performance Indicators) for ESG reporting and presents the key indicators in ESG reporting in the transport sector using the example of Skoda Auto.

2 ESG Key Performance Indicators

ESG Key Performance Indicators (KPIs) have become crucial in assessing an organization's efforts towards sustainable development. Investors demand greater transparency in reporting, and a company's ability to present effective strategies and measurable progress in ESG areas is essential to attract interest and maintain stakeholder trust. ESG KPIs, which cover various aspects of sustainability, provide companies and investors with reliable metrics for evaluating progress and performance. Based on a literature review and ESG guidelines, the main Key Performance Indicators for ESG reporting in organizations have been identified (Fig. 1.) [8–10].

ESG KPIs are used to quantify aspects of sustainability and ethical practices within an organization. KPIs assist organizations in identifying sustainability goals and indicate the organization's commitment to reducing environmental impact, supporting good social contribution and promoting best management practices. Key environmental performance indicators (Environmental KPIs) demonstrate a company's actions towards combating climate change, conserving resources, and protecting biodiversity. Key social performance indicators (Social KPIs) focus on the social aspect, promotes diversity, and fosters inclusion within its organizational structure. These indicators help companies identify factors such as employee satisfaction, turnover rates, and community investments, underscoring the importance of social performance.

Social KPIs focus primarily on aspects of work culture and employee treatment, as well as adapting work to achieve maximum job satisfaction. Governance KPIs primarily encompass issues related to leadership, ethics, and legal aspects. Governance KPIs ensure that the organization operates in compliance with ethical and legal principles, and that all information is transparent. These KPIs are essential for building trust with stakeholders within the organization.

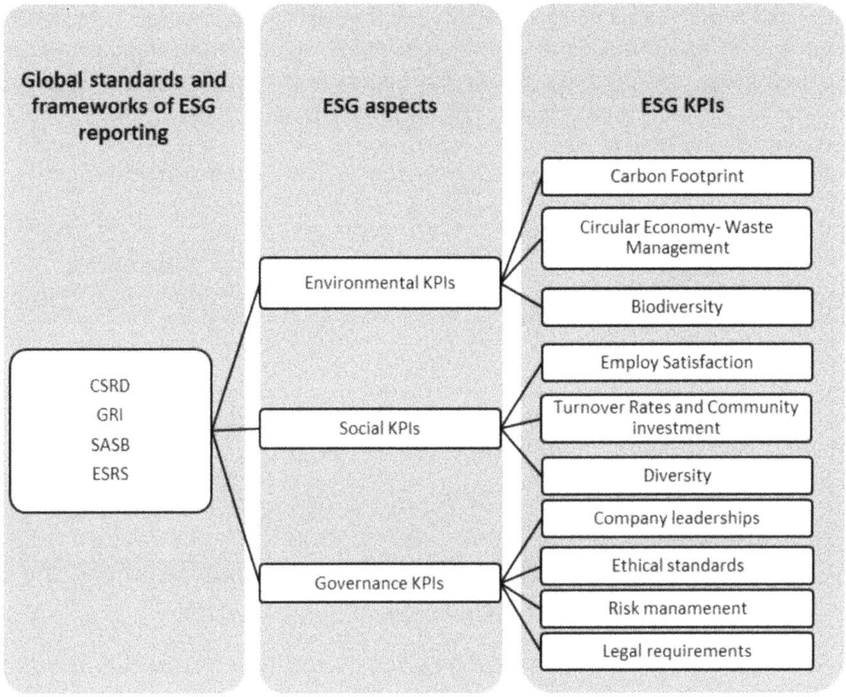

Fig. 1. ESG KPIs for organization.

The CSRD and ESRS guidelines establish standards for developing ESG reporting, making these reports more transparent, useful, and comparable across organizations in different industries. Therefore, it is crucial to develop best practices and develop Key Performance Indicators for organizations working in specific business environments and for stakeholders. This applies particularly to new aspects of ESG reporting, such as environmental indicators and, in particular carbon footprint in accordance with the GHG Protocol. Carbon footprint is one of the important environmental impact categories that must be included in ESG reporting [11]. Carbon footprint is expressed as equivalent of carbon dioxide and includes gases emissions such as carbon dioxide, methane, nitrous oxide etc.

3 Sustainability Strategy at Škoda Auto

The sustainability strategy at Škoda Auto is built on social, economic, and environmental pillars [12]. These pillars include the Corporate Social Responsibility (CSR) strategy, the Green Future environmental strategy, and principles of ethical and transparent conduct. The company's current Sustainability Strategy is based on three main ESG pillars, encompassing eight strategic areas identified as Top ESG KPIs (see Fig. 2.), each with specific targets set for 2025 or 2030. Environmental responsibility is an integral part of Škoda Auto's strategy. The company takes actions aimed at minimizing fuel consumption and exhaust emissions, utilizing the latest available technologies. The company's strategy and actions in social sustainability are inspired by the principles of Corporate Shared Value (CSV), focusing on education, mobility, volunteerism, employee engagement, well-being, and diversity. Škoda Auto operates in compliance with applicable laws, ethical standards, and principles of integrity. Figure 8 illustrates the ESG KPIs achieved by Škoda Auto in 2023.

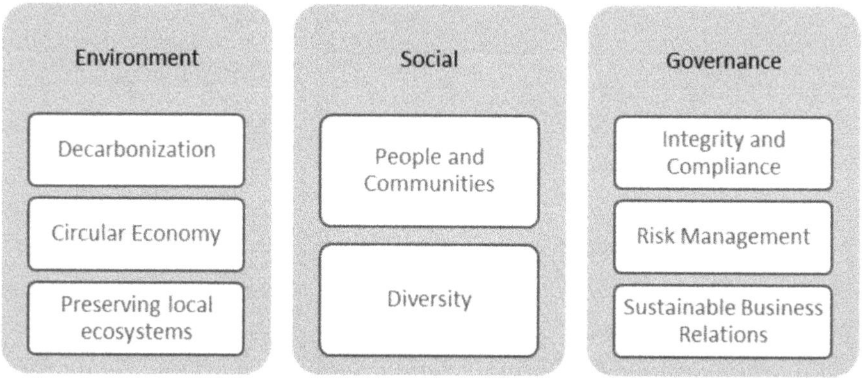

Fig. 2. Top ESG KPIs strategies in Skoda Auto [12].

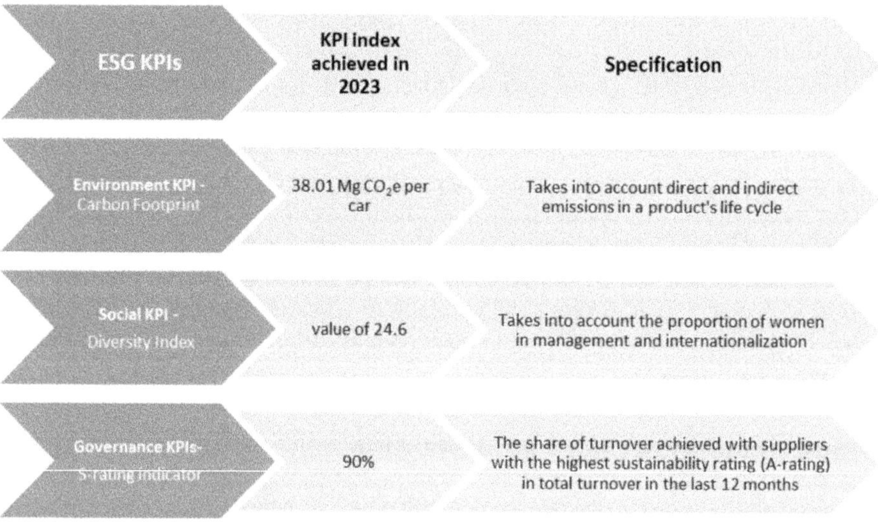

Fig. 3. ESG KPIs indexes achieved by Skoda Auto [12].

Environmental responsibility is a key component of Škoda Auto's strategy. Within the environmental pillar, Škoda Auto aims to become an innovator in the reuse of natural resources through sustainable Simply Clever solutions. The company has developed a strategy for reducing its environmental impact, taking into account a life-cycle approach based on environmental standards and the company's voluntary commitments. A key element in a company's environmental assessment is the consideration of various aspects, including life-cycle management. Strategic areas related to the environmental aspect include the circular economy, decarbonization, and the protection of local ecosystems.

The social aspect of ESG reporting at Škoda Auto considers people and community management, supporting education, mobility, well-being, volunteering, and diversity. The organization actively engages in local communities, supports employees, and provides humanitarian aid. The governance pillar of ESG reporting encompasses ethical aspects and resilience management. Strategic areas within this ESG pillar include risk management, compliance with legal regulations and ethics, and sustainable business relationships.

4 Environmental Key Indicators Analysis in Sustainability ESG Reporting

One of the main and most complex pillars of ESG in Škoda Auto's Sustainability Strategy is the environmental pillar. According to the new European Union guidelines, there are numerous new indicators and standards that need to be considered for ESG reporting. The Environmental KPIs in Škoda Auto are presented in Fig. 4., while the strategies for reducing GHG emissions are illustrated in Fig. 5.

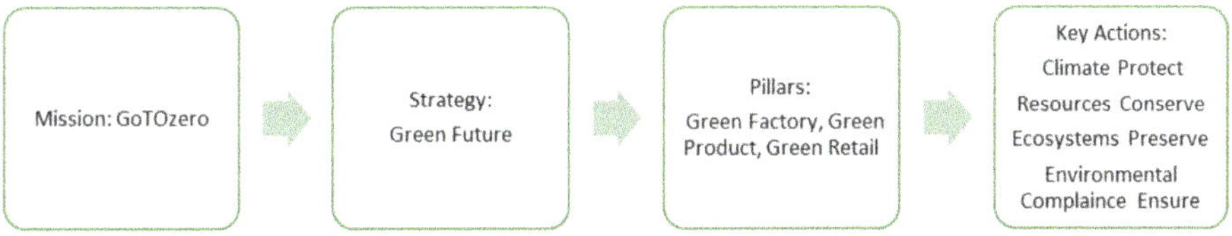

Fig. 4. Environmental KPIs in Škoda Auto [12].

According to the Green Factory goals, Škoda Auto efficiently uses resources and promotes a circular economy management to the use of energy, by-product and water. By 2025, the company plans to reduce external production-related costs by 52.5% per vehicle compared to 2010. In line with the Green Product strategy, the core principle of Škoda Auto's strategy to reduce negative environmental impacts is the sustainable and responsible production of energy-efficient vehicles. The company plans to increase the share of electric vehicle sales in Europe to 70% by 2030, depending on market developments. Škoda Auto dealers participate in Green Retail activities, implementing eco-friendly procedures and processes for efficient resource and material management. The main focus is on maximizing waste utilization, reducing energy and material consumption, and using rainwater. At the same time, the company continuously seeks opportunities to improve environmental conditions.

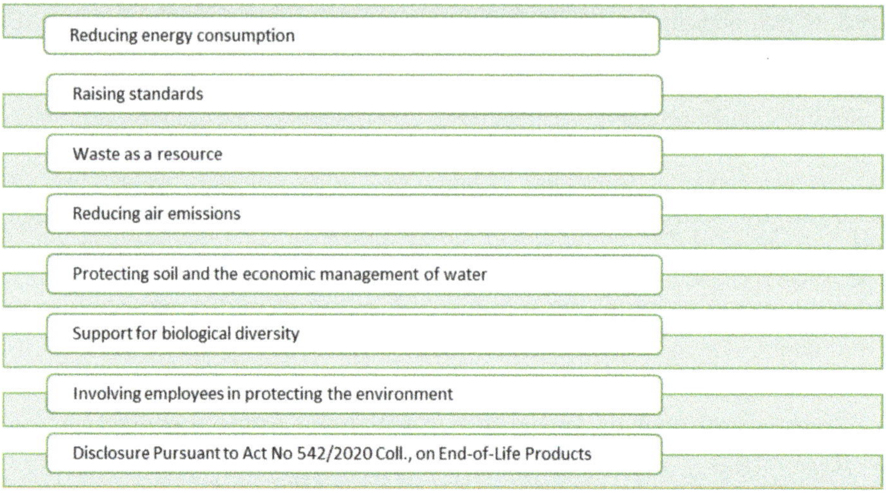

Fig. 5. Strategies of carbon neutrality in Škoda Auto [12].

5 Discussion and Conclusions

ESG reporting is the new standard for non-financial reporting for organizations across various sectors, including the transport sector. For this reason, organizations must become familiar with new principles, procedures, and standards related to the implementation of sustainability ESG reporting, including the consideration of economic, environmental, and social factors. Defining ESG KPIs is a key element for any organization required to prepare an ESG report, serving as a foundation for integrating sustainability aspects into the company's strategy. The selection of KPIs lays the groundwork for organizations aiming to achieve business success in a sustainable and ethical manner. Subsequent stages involve continuous review and enhancement of the ESG strategy accordance to legal guidelines and standards. Using the example of key indicators in ESG sustainability reporting by Škoda Auto, it has been demonstrated that preparations for non-financial reporting are already underway in the automotive industry, and the topic of introducing sustainability principles is highly significant.

In the European Union, companies have mainly focused on the environmental aspects of ESG reporting, presenting initiatives in carbon reduction, waste management, and renewable energy adoption. However, regarding the ESRS, the accounting entities are now expected to balance environmental efforts with strong social standards. This shift encourages companies to highlight their commitment to employee well-being and fair treatment. Many EU companies are already prepared to meet these social standards, boasting numerous employee-focused benefits. For instance, many offer vacation days that exceed statutory requirements, additional paid sick leave, and a wide range of corporate benefits, including health and wellness programs. These initiatives not only align with ESRS social criteria but also help to support a motivated and satisfied workforce.

References

1. European Commission: The European Green Deal. https://commission.europa.eu/strategy-and-policy/priorities-2019–2024/european-green-deal_en. Last accessed 24 Oct 2024
2. European Union: Directive (EU) 2022/2464 of the European Parliament and of the Council Amending Regulation (EU) No 537/2014, Directive 2004/109/EC, Directive 2006/43/EC and Directive 2013/34/EU, as regards Corporate Sustainability Reporting. https://eur-lex.europa.eu/legal-content/EN/TXT/?uri=CELEX:32022L2464. Last accessed 24 Oct 2024
3. Tsang, A., Frost, T., Cao, H.: Environmental, social, and governance (ESG) disclosure: A literature review. Br. Account. Rev. **55**(1), 101149 (2023)
4. Pinheiro, A.B., Panza, G.B., Berhorst, N.L., Toaldo, A.M., Segatto, A.P.: Exploring the relationship among ESG, innovation, and economic and financial performance: evidence from the energy sector. Int. J. Energy Sect. Manag. **18**(3), 500–516 (2023)
5. Nenavani, J., Prasuna, A., Siva Kumar, S., Kasturi, A.: ESG measures and financial performance of logistics companies. Lett. Spat. Resour. Sci. **17**(1), 5 (2024)
6. Khaw, T.Y., Azlan, A., Teoh, A.P.: Factors influencing ESG performance: A bibliometric analysis, systematic literature review, and future research directions. J. Clean. Prod. **448**, 141430 (2024)
7. Senadheera, S.S. et al.: Scoring environment pillar in environmental, social, and governance (ESG) assessment. Sustain. Environ. **7**(1), 1960097 (2021)
8. Ahmad, H., Yaqub, M., Lee, S.H.: Environmental, social, and governance related factors for business investment and sustainability: A scientometric review of global trends. Environ. Dev. Sustain. **26**(2), 2965–2987 (2024)
9. Hristov, I., Chirico, A.: The role of sustainability key performance indicators (KPIs) in implementing sustainable strategies. Sustainability. **11**(20), 5742 (2019)
10. Llanos, C.G., A.F., Vijaya, A., Wicaksono, H.: Rating ESG key performance indicators in the airline industry. Environ. Dev. Sustain. **26**, 27629–27653 (2023)
11. Jursova, S., Burchart-Korol, D., Pustejovska, P.: Carbon footprint and water footprint of electric vehicles and batteries charging in view of various sources of power supply in the Czech Republic. Environments. **6**, 1–11 (2019)
12. Auto, Š.: Sustainability Report (2023). https://www.skoda-auto.com/company/sustainability. Last accessed 24 Oct 2024

Environmental and Economic Impacts of Introducing Hydrogen Buses—A Case Study of Krakow's Public Transport

Katarzyna Gdowska(✉) and Karolina Osielczak

AGH University of Krakow, Krakow, Poland
kgdowska@agh.ed.pl, osielczak@student.agh.edu.pl

Abstract. The need to reduce emissions from the transport sector and urbanization is critical, as climate change intensifies. The EU's Fit for 55 initiative aims to decarbonize the energy sector, promoting hydrogen as a clean energy solution for urban mobility. In Poland, a major emitter of CO2 from transport, Kraków faces severe air quality issues and plans to implement a Low Emission Zone by 2024, including expanding its fleet of hydrogen buses. Hydrogen-powered buses offer a zero-emission alternative to fossil fuels, overcoming the limitations of electric vehicles. The adoption of hydrogen buses necessitates careful planning, substantial investment, and the development of suitable infrastructure. This study examines the environmental and economic implications of introducing hydrogen buses on two Kraków bus lines, namely lines 128 and 304. A techno-economic framework compares hydrogen and diesel buses, assessing fuel consumption, emissions, and operational costs. The analysis finds that hydrogen buses reduce carbon emissions compared to diesel buses, with cost savings for both lines; however, the environmental benefits are impacted by the emissions associated with hydrogen delivery. The SWOT analysis highlights the advantages of hydrogen buses, including alignment with sustainability goals and operational efficiency, but also identifies challenges such as high costs and the need for infrastructure development. Despite these hurdles, strategic investments, technological advancements, and public support could help Kraków lead in sustainable urban mobility, making hydrogen buses a promising solution for the future.

Keywords: hydrogen fuel · public transport · Fit for 55 · urban mobility · transport management

1 Introduction

The increasing greenhouse gas emissions from the transport sector and urbanization underscore the need for strategies to reduce pollution and enhance the urban quality of life. This urgency is driven by growing evidence of climate change. The EU's Fit for 55 initiative aims to decarbonize the energy sector, which is responsible for 75% of the EU's emissions, by supporting cleaner energy sources, such as hydrogen, promoting energy system integration, and connecting energy infrastructure across member states

J. Feliks et al. (Eds.): CLC2024 2024, 2026.
https://doi.org/10.1007/978-3-032-28931-5_9

[1, 2]. Hydrogen, produced from water and reverted to water upon oxidation, is a promising clean energy solution for decarbonizing urban mobility, particularly in cities. The transition to hydrogen buses is critical for meeting environmental targets and improving urban efficiency. Global momentum for hydrogen is rising due to decreasing production costs and the need to reduce greenhouse gas emissions [3]. Collaboration between the hydrogen and renewable energy sectors enhances market growth and increases energy demand flexibility, making hydrogen a vital tool for reducing emissions.

As of 2022, Poland ranked 62nd for air quality, the 6th worst in the EU, and leads in CO_2 emissions from transport, contributing 62 million tons in 2021 [4, 5]. Poland also emitted over 11,000 tons of PM2.5, more than double Germany's emissions [6]. Kraków, the 16th most polluted city in Poland [7], faces severe air quality challenges, exacerbated by coal burning for heating and high vehicle congestion. The city plans to implement a Low Emission Zone by July 2024, restricting vehicles that fail to meet specific EURO standards [8, 9], while expanding environmentally friendly transport, such as hydrogen buses.

Hydrogen-powered vehicles, particularly buses, provide a zero-emission alternative to fossil fuels, overcoming the limitations of electric vehicles, such as slow refueling times and battery degradation. Technological advancements enhance hydrogen's role in the transport market [10, 11]. The adoption of hydrogen buses requires careful planning, investment, and simulation analyses to ensure smooth integration into existing systems. Initiatives such as the European Industrial Strategy 1.7.5 are making renewable energy sources more cost-effective, promoting sustainable practices for a viable, environmentally responsible future [12].

This paper contributes with an analysis of the environmental and economic implications of introducing hydrogen buses on two Krakow municipal bus lines, 128 and 304. It addresses the following research questions:

- RQ1: To what extent can the implementation of hydrogen buses in Krakow's public transportation system quantitatively reduce emissions?
- RQ2: What quantitative evaluations can be conducted to assess the operational costs associated with introducing hydrogen buses into Krakow's public transportation network?
- RQ3: How can a qualitative SWOT analysis systematically evaluate the strengths, weaknesses, opportunities, and threats of integrating hydrogen buses into Krakow's transport system, providing insights into the expected benefits and challenges for the local transport industry?

2 Introduction of hydrogen buses in Kraków

This research analyzes the environmental and economic implications of introducing hydrogen buses on two bus lines in Kraków. Since 2022, Krakow has been testing three hydrogen buses, with the latest trial in February 2023. The city serves 804,200 residents and an estimated 127,600 students, both of whom contribute significantly to transportation demand. Public transport is contracted to MPK S.A. and Mobilis Sp. z o.o. by the City of Krakow, with oversight from the Public Transport Authority in Kraków (ZTP). Kraków operates 22 tram lines, 166 bus lines, 121 electric buses, and 34 hybrid buses, with one hydrogen-powered bus recently added to the fleet. Krakow plans to acquire 40 hydrogen buses by 2025, including 30 buses of 12 m and 10 of 18 m. By 2025, zero-emission vehicles are expected to account for 20% of the city's operational fleet under the Electromobility Act [13].

The study uses a techno-economic analysis framework that combines greedy algorithms and cost analysis [14] to assess the feasibility of hydrogen buses in Krakow. It compares the economic performance of hydrogen buses with that of diesel buses, examining environmental factors such as fuel consumption and greenhouse gas emissions. The goal is to provide decision-makers with a comprehensive understanding of the implications of introducing hydrogen buses on specific lines in Krakow, offering valuable insights for integrating eco-friendly buses into public transportation systems globally, supporting efforts to reduce carbon footprints and improve air quality.

3 Techno-Economic Analysis of Introducing Hydrogen Buses at Selected Bus Routes in Krakow

This analysis examines the potential to replace diesel-powered buses with hydrogen buses on two bus routes in Krakow and its surrounding agglomeration. The study considers the existing hydrogen refueling infrastructure, primarily the station in the town of Trzebinia, which supplies fuel to Krakow. A collaboration between the municipal public transport services provider Miejskie Przedsiębiorstwo Komunikacyjne S.A. (MPK S.A.) and PKN ORLEN S.A. in 2022 established Krakow's first hydrogen mobile refueling station in the Płaszów region. Hydrogen fuel is delivered by a diesel-powered truck from a biorefinery in the town of Trzebinia, 36 km from Krakow.

The operational details for two bus lines in Krakow are as follows:

- Bus Line 128 runs from the Płaszów depot (where hydrogen buses were included in the testing period) to Prądnik Czerwony. The route covers 11.3 km and takes about 28 min. On weekdays, the line operates 64 courses (128 total), with buses running every 17 min. On weekends, 52 courses operate, with departures every 40 min [15].
- Bus Line 304 connects Wieliczka to the Main Railway Station in Krakow, covering 18.4 km in about 40 min. It has 18 stops and operates as a faster line. Unlike Line 128, Line 304 does not have direct access to the hydrogen refueling station, requiring an additional 11.3 km trip to the Płaszów depot for refueling. On weekdays, the line runs 95 courses (190 total), with departures every 10 min during peak hours and every 20 min during non-peak hours. On weekends, the line operates 48 courses, with departures every 20 min until 6 pm, then every 30 min [15].

The analysis compares hydrogen buses (ARTHUR H2 Bus, Solaris Urbino 12 hydrogen, Autosan Sancity 12LFH) to the Solaris Urbino 12 LE Lite hybrid bus, which features a diesel engine and an AdBlue tank. AdBlue helps reduce nitrogen oxide emissions by breaking them down into ammonia and carbon dioxide, which are expelled through the exhaust system. The Solaris Urbino 12 LE Lite has a 200-l diesel fuel tank and an additional 40-l AdBlue tank. The hybrid bus, powered by a 145 kW generator, consumes less than 30 l of diesel per 100 km, providing a range of approximately 667 km per refueling. The estimated cost of a hybrid bus in 2018 was 1.5 million zloty [14].

In the comparison, the hydrogen bus is assumed to have a fuel consumption of 8.5 kg per 100 km, with a 36 kg hydrogen tank, and a refueling range of 425 km. The mobile refueling station, which carries 400 kg of hydrogen, can refuel up to 11 buses simultaneously. The cost of a hydrogen bus is estimated at € 3.06 million. The average carbon dioxide equivalent (CO2e) emission from diesel fuel is 2.67 kg per liter. The production carbon footprint for diesel is 0.46 kg of CO_2 per kilogram of fuel, which translates to 0.383 kg of CO_2 per liter of diesel [14].

For hydrogen delivery, a diesel-powered truck transports hydrogen from the town of Trzebinia, located 60 km away from Krakow. The truck consumes about 26 l of diesel per 100 km, resulting in a CO_2 footprint of 41.65 kg for a single delivery, calculated as: 60 km × 26 l/100 km × 2.67 kg/l = 41.65 kg of CO_2

The parameters used in the analysis are:

- Diesel bus: fuel consumption = 30 l/100 km, CO_2e consumption = 2.67 kg/l, CO_2eq production = 0.383 kg/l
- Hydrogen bus: fuel consumption = 8.5 kg/100 km, CO2e production = 0.364 kg CO_2e/kg, fuel delivery footprint = 41.65 kg CO_2eq.

In the comparative analysis, three scenarios were adopted:

- Scenario 1: Line 128 is examined with three hydrogen buses and three diesel-powered hybrids, reflecting a real-life operational line. This scenario serves as a baseline for comparing the two bus types under actual operational conditions.
- Scenario 2: Exclusively hydrogen buses are compared with diesel-powered buses on Line 128. This scenario evaluates the potential impact of fully transitioning to hydrogen-powered vehicles on this specific route, with a focus on both environmental and economic implications.
- Scenario 3: Line 304 is examined with a focus on hydrogen-powered buses versus conventional diesel buses. This scenario explores the feasibility and benefits of introducing hydrogen buses on longer routes, evaluating their performance and impact in comparison to traditional buses.

3.1 Scenario 1: Line 128—three hybrid Diesel-powered vs. three hydrogen buses

For bus line 128, serviced by six vehicles, the analysis considers a mixed fleet of 3 Diesel-powered and three hydrogen-powered buses. Each bus covers 241.07 km daily, totaling 723.20 km for three buses. Diesel buses consume 219.69 l of fuel daily, resulting in 579.28 kg of CO_2 emissions and an additional 83.10 kg of CO_2 emissions from production. Hydrogen buses, consuming 61.47 kg of fuel, produce only water vapor, with a carbon footprint of 28.78 kg CO_2—comprising 6.40 kg from delivery and 22.38 kg from production.

Refueling intervals are approximately every two days for hydrogen buses and every three days for Diesel buses under economic driving conditions. Weekend operations were excluded from the analysis due to the lack of confirmed data and the redundancy of comparing equal fleet distributions (three vs. three buses).

3.2 Scenario 2: Line 128—Six Hybrid Diesel-Powered Vs. Six Hydrogen

Bus line 128 is serviced by six buses on weekdays, covering 241.07 km per bus for a total of 1446.40 km. Diesel buses require refueling every 3 days, consuming 433.92 l of fuel daily and emitting 1158.57 kg of CO_2, plus an additional 166.19 kg from fuel production. In contrast, hydrogen buses consume 122.94 kg of hydrogen fuel (about 30.74% of the Płaszów depot's capacity) daily, with a total carbon footprint of 57.55 kg CO_2, including 12.80 kg from delivery and 44.75 kg from production.

On weekends, three buses cover a total of 195.87 km each, totaling 587.60 km. Diesel buses consume 176.28 l of fuel, emitting 470.67 kg of CO_2, of which 67.52 kg are from production. Refueling occurs approximately every 3.5 days. Hydrogen buses use 49.95 kg of fuel (12.49% of Płaszów's capacity), with a carbon footprint of 23.38 kg CO_2 (5.20 kg from delivery, 18.18 kg from production), and require refueling every 2 days.

3.3 Scenario 3: Line 304—Hybrid Diesel-Powered Fleet Versus Hydrogen Fleet

For bus line 304, each bus travels 218.5 km daily on weekdays, with a total daily fleet distance of 3496 km. Diesel buses consume 1048.80 l of fuel daily, emitting 2800.30 kg of CO_2 and an additional 401.69 kg from production. Refueling is required every three days. Hydrogen buses, operating under similar conditions but requiring daily refueling for 8 of the 16 buses due to the distance to the Płaszów depot, cover an adjusted total of 3676.80 km daily. Hydrogen consumption totals 312.53 kg (78% of storage capacity), with delivery and production emissions amounting to 32.54 kg and 113.76 kg of CO_2, respectively, yielding a total of 146.30 kg.

During weekends and holidays, with a reduced fleet of 6 buses, each bus travels 294.40 km, maintaining a total daily distance of 1902.00 km. Diesel buses consume 529.92 l of fuel, emitting 192.89 kg of CO_2 and an additional 202.96 kg from production. Hydrogen buses require 150.1 kg of fuel, plus an additional 11.57 kg for refueling, for a total consumption of 161.67 kg. Emissions from hydrogen delivery are 16.83 kg CO_2, with production adding 112.36 kg, for a total of 129.19 kg.

3.4 Lines 128 and 304—30-Day Operational Analysis

Over a 30-day period, a comparative analysis of fuel consumption, emissions, and costs was conducted for full Diesel and hydrogen fleets operating on bus lines 128 and 304. The evaluation considered 22 weekdays and 8 weekend days, using fuel costs reported by Eurostat for Germany in the first quarter of 2022 (€9.05/100 km for Diesel and €7.60/100 km for hydrogen) [16].

For Line 128, the monthly total distance traveled was 36,521.60 km. A Diesel fleet would consume 10,956.50 l of fuel, resulting in 4,196.33 kg of carbon dioxide emissions during production and 29,253.80 kg during operation. The total fuel cost for Diesel buses was estimated at €3,305.20. In comparison, a hydrogen fleet on the same line would consume 3,104.34 kg of hydrogen fuel, producing 1,129.98 kg of carbon dioxide during production and 666.40 kg from the eight required delivery trips. The total cost of hydrogen fuel for the month was calculated at €2,775.64.

For Line 304, which covered 91,043.20 km in the same period (76,912.00 km on weekdays and 14,131.20 km on weekends), the Diesel fleet was estimated to consume 27,501.56 l of fuel. This resulted in emissions of 10,535.35 kg from production and 73,481.09 kg during operation, totaling €8,239.91. The hydrogen fleet on this route consumed 8,186.98 kg of fuel, resulting in 2,924.74 kg of emissions from production and 1,749.30 kg from 21 required delivery trips (equivalent to 42 one-way trips). The total fuel cost for hydrogen buses was estimated at €7,228.45.

Overall, hydrogen fleets demonstrated lower carbon emissions in both production and operation compared to diesel, with cost savings observed across both lines. Line 128 achieved a monthly savings of €529.56, while Line 304 saved €1,011.46. However, the logistical demands of hydrogen delivery, especially for Line 304, introduced additional emissions that must be considered.

3.5 Key Findings of SWOT Analysis

The SWOT analysis highlights the potential and challenges associated with integrating hydrogen buses into Krakow's public transportation system. This initiative supports sustainability goals, aligns with policies like the European Green Deal and the Polish Hydrogen Strategy, and reduces reliance on fossil fuels. Strategic investments by companies like ORLEN ensure a stable hydrogen supply chain, while the buses' long-distance range and quick refueling enhance operational efficiency—these advantages position Krakow as a leader in sustainable urban mobility.

However, the transition involves high costs, including the purchase price of approximately 3 million zlotys per bus and the development of refueling infrastructure. The constrained production capacity and high demand for hydrogen buses limit accessibility. Long-term feasibility requires careful consideration of total ownership costs, including maintenance and operational expenses.

Despite these challenges, Krakow has opportunities to secure funding and subsidies, benefit from ongoing technological advancements, and foster collaborations with stakeholders to optimize resources. However, risks like hydrogen supply chain disruptions and competition from alternative technologies must be mitigated to ensure the initiative's success. Public awareness campaigns will also play a critical role in achieving community support for this environmentally friendly solution.

4 Conclusion

The analysis of integrating hydrogen buses into Krakow's public transport system highlights both opportunities and challenges. Hydrogen buses offer significant environmental benefits, including reduced greenhouse gas emissions, which align with EU sustainability goals, such as the European Green Deal. However, financial and logistical barriers, including high initial investment and the need for refueling infrastructure, must be addressed. The economic analysis revealed cost efficiency on some routes, particularly those with nearby refueling hubs, but higher costs on routes lacking such access. Public awareness campaigns are also essential for societal acceptance. A SWOT analysis identified strategic partnerships and technological advancements as potential solutions;

however, threats such as supply chain issues and competition from other technologies remain. Overall, while promising, adopting hydrogen buses requires careful planning, investment, and community engagement for success.

Acknowledgements. The research was supported by Europe's Rail Flagship Project 4—Sustainable and green rail systems, which has received co-funding from the Horizon Europe Framework Program (HORIZON) under grant agreement no. 101101917 and the Polish Ministry of Education and Science under grant agreement no. MEiN 5337/HE/2023/2 under the program titled International Co-Financed Projects. The project is implemented at the AGH University of Krakow in 2022–2026.

References

1. European Council: Fit for 55. https://www.consilium.europa.eu/en/policies/green-deal/fit-for-55-the-eu-plan-for-a-green-transition/ (2023). Last Accessed 05 Nov 2024
2. European Commision: Key Actions of the EU Hydrogen Strategy. https://energy.ec.europa.eu/topics/energy-systems-integration/hydrogen/key-actions-eu-hydrogen-strategy_en (2020). Last Accessed 05 Nov 2024
3. Gielen, D., Taibi, E., Miranda, R.: Hydrogen: a renewable energy perspective. Int. Renew. Energy Agency. https://www.irena.org/Publications/2019/Sep/Hydrogen-A-renewable-energy-perspective (2019). Last Accessed 05 Nov 2024
4. IQAir: World's most polluted countries & regions. World air quality report. Switzerland (2022). https://www.iqair.com/pl/world-most-polluted-countries. Last Accessed 05 Nov 2024
5. EUROSTAT: Air emissions accounts related to road transport by NACE Rev. 2 activity (2023). https://ec.europa.eu/eurostat/databrowser/view/env_ac_ainah_rd__custom_9142352/default/table?lang=en. Last Accessed 05 Nov 2024
6. Metabolic Institute: Instytut Innowacji i Odpowiedzialnego Rozwoju. Circular Cities Program Krakow. https://www.metabolic.nl/publications/circular-cities-program%2D%2Dkrakow/ (2021). Last Accessed 05 Nov 2024
7. Sejmik Województwa Małopolskiego. Uchwała Nr XXXII/452/17 Sejmiku Województwa Małopolskiego z dnia 23 stycznia 2017 r. w sprawie wprowadzenia na obszarze województwa małopolskiego ograniczeń i zakazów w zakresie eksploatacji instalacji, w których następuje spalanie paliw (Dz. Urz. 2017 poz. 787). https://edziennik.malopolska.uw.gov.pl/WDU_K/2017/787/akt.pdf (2017). Last Accessed 05 Nov 2024
8. Kraków.pl. Strefa Czystego Transportu w Krakowie. https://www.krakow.pl/strefa_czystego_transportu/276724,2353,komunikat,strefa_czystego_transportu_w_krakowie.html (31 Oct 2023). Last Accessed 05 Nov 2024
9. Miejska Platforma Internetowa ‚Magiczny Kraków". Strefa Czystego Transportu – kogo zapisy nie będą dotyczyć?. https://www.krakow.pl/get_pdf.php?dok_id=267746 (27 Oct 2023). Last Accessed 05 Nov 2024
10. IQAir: World's most polluted cities. World Air Quality Report, Switzerland (2022). https://www.iqair.com/pl/world-most-polluted-cities. Last Accessed 05 Nov 2024
11. Addcomposites Oy: How Hydrogen Fuel Cell Technology is Revolutionizing Transport. https://www.addcomposites.com/post/how-hydrogen-fuel-cell-technology-is-revolutionizing-transport (2023). Last Accessed 05 Nov 2024
12. McKinsey & Company. The clean hydrogen opportunity for hydrocarbon-rich countries. https://www.mckinsey.com/industries/oil-and-gas/our-insights/the-clean-hydrogen-opportunity-for-hydrocarbon-rich-countries#/ (2022). Last Accessed 05 Nov 2024

13. Osielczak, K.: Hydrogen Fuel Production and Supply Chain Management for Sustainable Transportation: a Production Engineering Perspective. Bachelor's Thesis. AGH University of Krakow, Krakow (2024)
14. Stecuła, K., Olczak, P., Kamiński, P., Matuszewska, D., Duong Duc, H.: Towards sustainable transport: techno-economic analysis of investing in hydrogen buses 57 in public transport in the selected City of Poland. Energies. **12**(4), 9456 (2022)
15. Public Transport Authority in Kraków. Linie Komunikacji Miejskiej w Krakowie. https://ztp.krakow.pl/transport-publiczny/komunikacja-miejska/linie-kmk (2023). Last Accessed 05 Nov 2024
16. European Alternative Fuels Observatory: Fuel price comparison. https://alternative-fuels-observatory.ec.europa.eu/consumer-portal/fuel-price-comparison. Last Accessed 05 Nov 2024

Management's Green Motivation in Supply Chain Management—A Case Study of Green Supply Chain

Agnieszka Jagoda(✉), Anna Baraniecka, and Joanna Lyszkiewicz–Potaczek

Wroclaw University of Economics and Business, Komandorska 118/120, 53–345, Wroclaw, Poland

{agnieszka.jagoda,anna.baraniecka, joanna.lyszkiewicz-potaczek}@ue.wroc.pl

Abstract. This article examines the motivation of management in making pro-environmental decisions within supply chain management, using Sikla Poland as a case study. The analysis explores both internal (intrinsic) and external (extrinsic) drivers influencing managerial decisions. Internal motivation stems from personal values, environmental awareness, and a genuine commitment to sustainability, while external motivation is primarily determined by regulatory compliance, customer expectations, and competitive pressures. The findings indicate that, although both sources of motivation play a significant role, internal motivation, rooted in environmental concern constitutes the main driver of managerial engagement in green practices. The study also reveals that managerial attitudes towards the environment shape the ecological behaviors of employees and strengthen the organization's ability to implement sustainable solutions across the supply chain. By identifying key motivational factors, this article contributes to the broader understanding of how managerial values and environmental responsibility accelerate the greening of supply chains. The results emphasize that effective implementation of green supply chain management requires not only compliance with external pressures but also genuine internal commitment from leadership.

Keywords: green motivation · green supply chain management · management's motivation

1 Introduction

The concept of the green supply chain (GSC) was first introduced in 1996 by the Manufacturing Research Consortium at Michigan State University as an attempt to combine process optimization in the supply chain with minimizing the negative impact on the natural environment [1]. Since then, the concept of the green supply chain has developed and evolved. According to S.K. Srivastava [2], green supply chain management (GSCM) is a comprehensive process encompassing the full cycle of design, production, packaging, sales, use, and recycling. It includes storage, transportation, and information exchange processes that meet appropriate environmental standards. Greening supply

J. Feliks et al. (Eds.): CLC2024 2024, 2026.
https://doi.org/10.1007/978-3-032-28931-5_10

chains incorporates environmental concerns in a supply chain and aids organizations to improve their footprint on sustainability [3].

Green supply chain management differs from classic supply chain management by incorporating aspects such as green production, waste management, green design, and reverse logistics into the decision-making process [2]. According to some researchers, green supply chain management also involves integrating a pro-environmental approach at every stage – from product design, material selection, production and work organization, delivery, to the collection of used products from the market, while considering external aspects such as legal regulations and stakeholders [4]. Among the factors contributing to the success of green supply chains are offering green products, cooperation with suppliers, process measurement, training, information management, and importantly, from the perspective of achieving the goal of this article, the involvement of management and employees [5]. Significantly, despite numerous scientific studies on motivation theory and motivators stimulating various actions undertaken by employees in enterprises, as well as an increasing number of publications focused on the issue of green supply chain management, the perspective of staff motivation as a factor in greening the supply chain has not been clearly emphasized so far. Therefore, the aim of this article is to identify the green motives of managerial staff when making pro-environmental decisions in supply chain management, in light of the identified gap. The considerations aimed at filling this gap are both literary and empirical in nature – the article presents a case study of the Sikla Poland supply chain.

2 Executive's Green Motivation as an Accelerator of Greening Supply Chains

Green motivation refers to the internal driving force that encourages individuals to engage in environmentally friendly behaviors and pursue environmental goals. It encompasses two main aspects [6]:

- Internal motivation - this type of motivation arises from an individual's intrinsic interests and values. It includes a sense of responsibility and a desire for personal fulfillment in environmental protection efforts. Individuals motivated internally are likely to engage in green behaviors, such as conservation and sustainable practices, due to their love for the environment rather than external rewards.
- External motivation - this type of motivation is driven by external factors, such as incentives or recognition from others. It involves behaviors that are influenced by rewards or approval related to environmental actions, such as participating in company green programs for potential benefits.

Both, internal and external motivations play a crucial role in fostering green behaviors. Internal motivations are primarily driven by personal values and emotional satisfaction, while external motivations are influenced by social dynamics and organizational factors. Understanding these motivations can help in developing effective strategies to promote sustainable practices in the area relating to business management. For example Li, Yan et al. [7] showed intrinsic and extrinsic green motivation as mediating variables between transformational leadership and employees' green creativity. Another research indicate that the ecological behavior or environmental preferences of certain individuals can be predictors of the ecological behaviors of others [8]. It means that green preferences of managerial staff are helpful in promoting and maintaining the behaviors of other employees, which is very important for pro-environmental decisions regarding the management of the organization and the supply chain. This assumption correlates with research results emphasizing the importance of top management support and involvement for the development of green actions [9] and green supply chain strategies [10]. Employees generally follow their leaders in whatever they ask and in whatever direction they push, thereby highlighting the importance of trust in leadership [11, 12] Consequently, employees will be able to understand the importance of environmental protection, promote the formation of a sense of environmental responsibility, and take environmentally responsible actions [13].

There are thus various green external or internal factors that motivate an executive's to adopt green supply chain management practices. For example, Liu et al. [13] consider three external motivational factors (legal regulations, customer requirements, and competitive pressure) and two internal motivational factors (cost savings and environmental concern). The motivations of executives to implement green supply chain management practices can come from individual pro-environmental beliefs of the management, analysis of potential costs and benefits that GSCM practices can bring to the enterprise, or from external sources such as customer pressure, legal regulations, and industry standards. Therefore, to encourage companies to adopt a strategic orientation towards GSCM, it is important to understand specific motivations to strengthen them.

3 Methods and Characteristics of the Surveyed Company

The research method used for the preparation of this article is a case study analysis utilizing a focused interview. The interview was conducted with the Managing Director of Sikla Poland. The interview was direct and conducted using a questionnaire containing standardized open-ended questions. The aim of the interview was to determine the motives behind making pro-environmental decisions in the Sikla Poland supply chain.

Sikla Poland was established in 2001 as a branch of the international Sikla Group. The Sikla Group has branches and representatives in many countries across Europe, as well as in Africa, North America, Central and South America, the Middle East, the Far East, and Africa. Sikla Poland is one of the leading suppliers of innovative fastening systems in the domestic market. Since its inception, the company's priority has been the high quality and safety of the products offered while simultaneously reducing their negative environmental impact. Sikla Poland builds its competitive advantage on high standards of logistical customer service. The company guarantees nearly 100% completeness, accuracy, and timeliness of its deliveries, with very short delivery times (1–2 days). For the company's customers, this certainty constitutes added value, which, in the context of product shortages in the market, translates into an increasing scale of orders and customer loyalty.

The supply chain of Sikla Poland is highly centralized in the purchasing area. Supplier selection, price negotiations, purchase planning, and setting delivery conditions are conducted centrally. Sikla Poland mainly influences operational procurement (execution of goods receipt and delivery). During the COVID-19 pandemic, Sikla Poland, like most enterprises, had to reconfigure its supply chain. The actions taken allowed for increased product availability and also contributed to reducing the negative environmental impact of the company's supply chain. The company's development strategy for the years 2022–2032 includes environmental goals. These primarily involve reducing the company's negative impact on the natural environment, which is specified in the partial goals of the logistics strategy related to reducing the carbon footprint by at least 20% in 2024 compared to 2022 and achieving climate neutrality in transportation by 2030. These goals have a very high priority and are correlated with marketing, sales, and CSR goals.

Environmental protection is also a very important element of the value system manifested in Sikla Poland. Research on values and organizational culture has shown that it is one of the most important values for the company's employees, with its level of saturation (manifestation) being close to 100%. The fact that Sikla Poland, with the support of the headquarters, is "greening" its logistics and supply chain means that its partners are obliged to plan and undertake pro-environmental actions. Suppliers sign a declaration of compliance of their actions with the principles of sustainable development proposed by the Sikla Group, including Sikla Poland.

Based on the analysis of data and information obtained during the interview it can be unequivocally stated that the Sikla Poland supply chain is green. This is evidenced by the large scale and frequency of "green" initiatives. The company collaborates with entities in its supply chain to realize the vision of a green future, simultaneously providing external information about its environmental impact (so far mainly in the scope of mandatory reports) and, most importantly, planning continuous improvement actions in this area, seeking and implementing eco-innovations, including within the internal innovation development program called Si-Innovations.

4 Results—Factors Motivating pro-Environmental Decisions in the SIKLA POLAND Supply Chain

Similar to Liu et al. [13], Sikla Poland considers five significant factors motivating pro-environmental decisions and actions in the supply chain. The external motivational factors at Sikla Poland include legal regulations, customer requirements, and competitive pressure, while the internal factors include cost savings and environmental concern. Below (Table 1) lists and characterizes the individual factors motivating pro-environmental decisions in the company's supply chain.

Sikla Poland has established levels of importance for the criteria for evaluating "green" projects. Each pro-environmental project and initiative is assessed according to

Table 1 Factors motivating pro-environmental decisions in the Sikla Poland supply chain.

Source of green motivation	Factor	Company's level of knowledge (1–low, 3–high)	Importance for the company	Importance for the decision-maker (personal)	Decision-maker's goal regarding the level of compliance	Current level of implementation	Top management's level of motivation to act
External motivation	Legal regulations	3 (support from an external company)	Very large legal and financial consequences	Very large element of the Managing Director's performance evaluation by the Board	Compliance with all regulations–100%	100%	High "We must comply with all guidelines"
	Customer requirements	2 (customer satisfaction surveys, but without questions related to sustainability; information obtained from the market based on non-standardized and not centrally coordinated customer interviews)	Very large (potentially) Determines the choice made by the customer (affects the achievement of strategic goals—revenue growth)	Very large Little-known element, but considered potentially important	Meeting all customer-reported expectations—100%	20% (meeting the most frequently reported expectations, e.g., ISO 14001)	High "We must respond positively to all customer expectations"
	Pressure from competitors	1 (limited information on competitors' activities; no benchmarking)	Medium there is no data on the importance of pro-environmental actions for the customer and thus on the impact on the competitiveness of individual players in the market	Medium reluctance to compare with competitors, need to pursue pioneering and individualized development paths	Achieving better results than competitors in pro-environmental actions. Due to the lack of agreements on the level of advancement of competitors' actions—incomplete data	Due to the lack of agreements on the level of advancement of competitors' actions—incomplete data	Medium-low

(*continued*)

Table 1 (*continued*)

Source of green motivation	Factor	Company's level of knowledge (1–low, 3–high)	Importance for the company	Importance for the decision-maker (personal)	Decision-maker's goal regarding the level of compliance	Current level of implementation	Top management's level of motivation to act
Internal motivation	Cost savings	2 (moderate level of knowledge about the scale of costs and financial benefits associated with pro-environmental actions)	Large the branch is accountable for cost levels, works with limits, and is required to achieve an appropriate level of profits	Very large element of the Managing Director's performance evaluation by the Board	High efficiency of pro-environmental investments. Target value has not been defined due to lack of data on costs and benefits	Achieved value has not been defined due to lack of data on costs and benefits	High
	Environmental concern	3 (a position for quality and environmental management was established, environmental audits were conducted, and reporting on the company's environmental impact was initiated)	Very large the company's strategy includes goals related to reducing the company's negative impact on the environment, an environmental policy and strategy have been created, a new transport policy related to environmental goals has been prepared, ISO 14001 has been implemented. Respect for people, animals, and the environment is part of the company's values, demonstrated by over 80% of employees. Employee pro-environmental actions are planned	Very large the personal attitude and involvement of the Managing Director in environmental protection activities contribute to the dynamism of actions in this area (including the scientific factor in research on the company's environmental impact, acceptance of pro-environmental expenditures, introduction of a project generating eco-innovations, etc.)	The company should be climate neutral–100%	Currently, there is no complete data. The reporting process has relatively recently begun, preceded by the development of indicators and data collection	Very large

criteria related to individual sources of motivation, to which weights are assigned, and which are rated in the evaluation process. In the evaluation of pro-environmental projects and initiatives at Sikla Poland, it is very important that a project can be implemented even if it only reduces the negative impact on the environment (i.e., it does not meet other criteria). Therefore, the criterion of reducing the negative impact on the environment is the leading criterion that conditions the implementation of the project. The motivation for such decisions is the personal involvement of the Board, including the Managing Director, in environmental protection issues, through financial support for such initiatives (in the social area) as well as pro-environmental consumer behaviors.

5 Conclusion

The analysis shows that both external and internal motives of the management staff are significant in making pro-environmental decisions in the supply chain management of Sikla Poland. However, the main source of motivation is internal and is the concern for the environment. Since responsibility towards the environment and concern for it are considered one of the predictors of ecological behavior intention (Kaiser and Shimoda 1999), in this case, it is a significant factor in the implementation of green solutions in the supply chain. The pro-environmental attitude of the management is also important in informing other company employees about the importance of ecological issues in developing the company's operational standards and the entire supply chain. As indicated in the literature part of the article, the environmental preferences of the management, their promotion, and maintenance, strengthen the ecological behaviors of other employees both in shaping corporate ecological practices and in implementing green supply chain management practices. The example of Sikla Poland shows that there are companies that, by managing their supply chains, engage in greening them not to meet mandatory or customer-required standards but mainly due to the managers' concern for the environment, which, considering the pro-environmental projects presented in the article, goes beyond mere compliance with legal regulations or cost savings.

References

1. Beamon, B.M.: Designing the green supply chain. Logist. Inf. Manag. **12**(4), 332–342 (1999)
2. Srivastava, S.K.: Green supply chain management: a state-of-the-art literature review. Int. J. Manag. Rev. **9**(1), 53–80 (2007)
3. Bhatia, M.S., Gangwani, K.K.: Green supply chain management: scientometric review and analysis of empirical research. J. Clean. Prod. **284**, 124722 (2021). https://doi.org/10.1016/j.jclepro.2020.124722
4. Walker, H., Preuss, L.: Fostering sustainability through sourcing from small businesses: public sector perspectives. J. Clean. Prod. **16**(15), 1600–1609 (2008)
5. Zaid, A.A., Jaaron, A.A., Bon, A.T.: The impact of green human resource management and green supply chain management practices on sustainable performance: an empirical study. J. Clean. Prod. **204**, 965–979 (2018)
6. Liao, B., Li, L., Yang, Z.: Perceived social green preference: the motivation mechanism of inducing green behaviour. Curr. Psychol. **41**, 1–17 (2022)

7. Li, W., Yan, L., Bhutto, T.A., Li, Y.: The influence of green transformational leadership on Employees' green creativity. Manag. Sci. Lett. **33**(02), 87–101 (2020)
8. Masri, H.A., Jaaron, A.A.M.: Assessing green human resources management practices in Palestinian manufacturing context: an empirical study. J. Clean. Prod. **143**, 474–489 (2017)
9. Zhao, H., Zhou, Q.: Exploring the impact of responsible leadership on organizational citizenship behaviour for the environment: a leadership identity perspective. Sustainability. **11**(4), 1–15 (2019)
10. Hoejmose, S., Brammer, S., Millington, A.: Green supply chain management: the role of trust and top management in B2B and B2C markets. Ind. Mark. Manag. **41**(4), 609–620 (2012)
11. Islam, M.N., Furuoka, F., Idris, A.: Employee engagement and organizational change initiatives: does transformational leadership, valence, and trust make a difference? Glob. Bus. Organ. Excell. **40**(3), 50–62 (2021)
12. Islam, T., Khan, M.M., Ahmed, I., Mahmood, K.: Promoting in-role and extra-role green behavior through ethical leadership: mediating role of green HRM and moderating role of individual green values. Int. J. Manpow. **42**(6), 1102–1123 (2021)
13. Liu, S., Eweje, G., He, Q., Lin, Z.: Turning motivation into action: a strategic orientation model for green supply chain management. Bus. Strateg. Environ. **29**(7), 2908–2918 (2020)

Pro-Environmental Behaviour Factors in Choosing the Last Mile Delivery Method. A Gender Perspective

Tomasz Kołakowski[1](✉), Agnieszka Jagoda[1], Katarzyna Cheba[2], and Magdalena Mucowska[1]

[1] Wroclaw University of Economics and Business, Komandorska 118-120, 53-345 Wroclaw, Poland
{tomasz.kolakowski,agnieszka.jagoda, magdalena.mucowska}@ue.wroc.pl

[2] West Pomeranian University of Technology, Al. Piastów 17, 70-310 Szczecin, Poland
katarzyna.cheba@zut.edu.pl

Abstract. In recent years, the e-commerce market has grown significantly in terms of its population share. However, the situation has changed dramatically since the onset of the COVID-19 pandemic. Electronic marketplaces have undergone substantial transformation due to rapid digitalisation and the evolution of shopping behaviours. Online services offer customers various delivery methods, such as home or out-of-home delivery. This aspect of the e-commerce market has garnered increased attention from both practitioners and academics, particularly in the context of sustainable last-mile deliveries. Interestingly, the existing literature primarily focuses on analysing the impact of e-commerce on last-mile delivery. However, identifying factors influencing the choice of delivery methods and the motivations that drive e-customers to select eco-friendly delivery options remains an underexplored area of research. Therefore, the primary objective of this paper is to compare the factors across genders that motivate e-customers to choose a delivery method. The research was conducted in 2021 among 1,110 e-customers in Poland. The paper addresses the following research question: What factors motivate each gender to choose a specific delivery method?

Keywords: E-Commerce · E-Customer · Sustainable Last-Mile Delivery · Delivery Method · Gender

1 Introduction

One of modern societies' most significant challenges is combating climate change, which increasingly disrupts socio-economic processes. Air pollution is a particularly concerning area contributing to global environmental threats, including those affecting human health. Its consequences include phenomena such as acid rain, smog, the depletion of the ozone layer, the greenhouse effect, and unpleasant odours. A sizeable portion of this pollution originates from the transportation sector. According to a 2015 report by the WHO

J. Feliks et al. (Eds.): CLC2024 2024, 2026.
https://doi.org/10.1007/978-3-032-28931-5_11

Regional Office for Europe and the OECD [1], air pollution in Europe imposes socio-economic costs of $1.6 trillion annually. Of this, approximately $1.4 trillion is linked to about 600,000 premature deaths, while the remaining $200 billion relates to diseases caused by polluted air and the associated treatment costs. Similarly, a 2018 report by the Centre for Research on Energy and Clean Air and Greenpeace [2] estimated the global economic cost of air pollution from fossil fuels at $2.9 trillion, equivalent to 3.3% of global GDP. The report highlighted that 4.5 million deaths in 2018 were attributable to exposure to air pollution from fossil fuels, including those used in transportation. Moreover, in its Energy Technology Perspectives report [3], the International Energy Agency predicts that global transportation activity will double by 2070, with car ownership expected to rise by 60%, leading to a significant increase in total transportation emissions.

A significant portion of freight transportation is now associated with orders placed through the e-commerce market, often involving delivering products directly to customers' homes. The rapid growth of e-commerce has introduced substantial challenges for the sustainable transportation of goods to customers [4, 5]. On the one hand, there is an urgent need to reduce CO_2 emissions in transportation; on the other hand, the steady increase in online retail purchases is driving a rise in the volume of goods requiring delivery. The global share of online retail sales increased from 13.6% in 2019 to 19.5% in 2023 and is expected to rise to 23% in 2027 [6].

For this reason, it is essential to develop appropriate infrastructure to promote sustainable transportation and implement so-called green last-mile delivery solutions. These include green modes of transportation, IT systems to support environmentally friendly delivery and other measures. Equally important is accurately identifying customer preferences and motivations that encourage environmentally friendly behaviour, including e-customers' selection of sustainable delivery methods for their purchases. Although scientific publications often analyse the impact of e-commerce on last-mile delivery [5, 7], relatively few studies focus on identifying the factors that motivate e-customers to choose environmentally friendly delivery options. Thus, the primary aim of this article is to examine the factors that can influence consumer behaviour toward more environmentally sustainable choices, with particular attention to one characteristic of the respondents: their gender. This focus on gender is supported by ample evidence in the literature suggesting that gender—understood as the socially constructed roles, behaviours, actions, and attributes considered appropriate for men and women—influences pro-environmental behaviour. This influence operates through mediating factors such as pro-environmental values and attitudes [8, 9].

2 Gender Perspective on Pro-Environmental Behaviour

As noted, climate change, pollution, and environmental contamination have increasingly affected lives worldwide in recent years. Although awareness of these issues is rising among the general population, many individuals still fail to make sustainable choices. This gap is particularly evident in consumer purchasing behaviour - despite widespread expressions of environmental concern and a declared desire to engage in greener consumption practices, these intentions often do not translate into corresponding actions [10].

A pro-environmental behaviour (PEB) is defined as a deliberately chosen action that uses fewer resources or emits less greenhouse gas compared to the default, more common or convenient behaviour [11]. Such actions may either reduce ecological harm or actively contribute to environmental benefits. The definition of PEB is context-dependent and often described in terms of its relative environmental impact rather than as an absolute measure [12]. Numerous factors that motivate individuals to adopt more sustainable practices shape customer engagement in pro-environmental behaviour. People who engage in PEB often share a combination of sociocultural, psychological, and economic characteristics that influence their environmental decisions and actions. Understanding the drivers behind these behaviours is critical for designing effective interventions and policies to reduce carbon emissions and promote sustainable consumption practices. The literature and existing research indicate that gender - understood as the socially constructed roles, behaviours, and attributes deemed appropriate for women and men - influences pro-environmental behaviour through mediating factors such as values and attitudes [13, 14]. A substantial body of scholarship across various countries, including Canada, the United States, Sweden, China, and Great Britain, highlights that women generally exhibit more ecological behaviours and express greater concern for the environment than men [9]. Gender differences in pro-environmental behaviour extend beyond perceptions of environmental problems; women are more likely than men to engage in pro-environmental actions. This distinction is particularly noticeable in everyday behaviours, such as sustainable consumer practices [8, 15]. In the context of this study, Schniederjans and Starkey [16] found that a positive attitude toward environmentally friendly consumption increased intentions to choose green transportation. Their research also demonstrated that women and younger adults (up to 44 years) were more willing to pay for green transportation than their counterparts.

3 Research Methodology

The analyses presented in this paper are based on research conducted between July and August 2021 on a sample of over 1,100 adult residents of voivodship cities in Poland (602 women and 508 men) who had made online purchases within the year preceding the survey. The study employed quasi-representative sampling, with the sample structured according to place of residence, gender, and age [17]. In previous research in this area [18], the authors analysed preferences for last-mile delivery in the e-commerce market, considering the respondents' age. In this study, the focus shifts to differences in the preferences and behaviours of market participants based on gender (female/male). Particular attention is given to the pro-environmental behaviour of respondents concerning their choice of an environmentally friendly parcel collection method. Table 1 presents the demographic structure of the survey respondents, categorised by age and gender.

The table shows that the survey included 54.23% women (602 participants) and 45.76% men (508 participants) representing different age groups. Women predominated across all age categories, and the sample structure closely mirrored that of the population in provincial towns in Poland. A two-stage research procedure was employed to compare the preferences of residents of Poland's provincial towns concerning last-mile deliveries in the e-commerce market, focusing on gender distribution. In the first stage,

Table 1. Structure of the sample according to age and gender.

Age	Gender				Total	
	Woman		Man			
	N	%	N	%	N	%
19–34	143	51.25	136	48.75	279	100.00
35–44	123	50.83	119	49.17	242	100.00
45–64	172	53.75	148	46.25	320	100.00
65 and more	164	60.97	105	39.03	269	100.00
Total	602	54.23	508	45.76	1110	100.00

the respondents' answers to individual survey questions were analysed by gender and age. In the second stage, cluster analysis was applied to investigate how women and men participate in this market based on various types of behaviour. Cluster analysis encompasses several classification algorithms and is an exploratory data analysis tool designed to organise data into meaningful structures or groupings. Its objective is to group objects so that the degree of association within the same group is maximised while the association between objects in separate groups is minimised. This study used the generalised k-means method to classify respondents according to their behaviour in the e-commerce market. The V-test implemented in STATISTICA was employed to determine the optimal number of clusters. This test is designed to identify patterns in the data [19, 20].

4 Study Results

Following the described methodology, the first step involved identifying factors that differentiate the behaviour of e-commerce market participants in terms of their pro-environmental behaviour, considering their gender. Table 2 presents the results of this analysis. Participants in the study were tasked with identifying factors that could influence them to choose a more environmentally friendly method of receiving a parcel outside their place of residence. The analysed factors included: lower delivery price (P), the possibility of a free return (R), access to a green delivery point - within walking distance (E), possibility of choosing a green delivery method offered by a courier company (C); shorter delivery time (T); proximity to parcel collection points (DP); low weight and small size of the parcel (S); greater flexibility in selecting the delivery time and day (CN); the need to pay an additional environmental charge if an environmentally unfriendly delivery method is chosen (AK); social pressure to engage in environmentally friendly behaviour (SP); personal rules and environmentally friendly behaviour (EB); weather conditions, such as good weather or pleasant views on the way to the parcel collection point(W); well-developed infrastructure for cyclists, and pedestrians - (BI); the awareness of the health benefits associated with this type of delivery (HL); greater safety of deliveries (SD).

Table 2. Study results—ecological preferences

Factors	Women				Men				Sum			
	Yes		No		Yes		No		Yes		No	
	N	%	N	%	N	%	N	%	N	%	N	%
P	416	69.10	186	30.90	383	75.39	125	24.61	799	71.98	311	28.02
R	319	52.99	283	47.01	255	50.20	253	49.80	574	51.71	536	48.29
E	211	35.05	391	64.95	158	31.10	350	68.90	369	33.24	741	66.76
C	76	12.62	526	87.38	73	14.37	435	85.63	149	13.42	961	86.58
T	294	48.48	308	51.16	274	53.94	234	46.06	568	51.17	542	48.83
DP	772	73.42	160	26.58	332	65.35	176	34.65	1104	99.46	336	30.27
S	265	44.02	337	55.98	190	37.40	318	62.60	455	40.99	655	59.01
CN	227	37.71	375	62.29	180	35.43	328	64.57	407	36.67	703	63.33
AK	49	8.14	553	91.86	46	9.06	462	90.94	95	8.56	1015	91.44
SP	37	6.15	565	93.85	42	8.27	466	91.73	79	7.12	1031	92.88
EB	88	14.62	514	85.38	88	17.32	420	82.68	176	15.86	934	84.14
W	143	23.75	459	76.25	109	21.46	399	78.54	252	22.70	858	77.30
BI	120	19.93	482	80.07	104	20.47	404	79.53	224	20.18	886	79.82
HL	98	16.28	504	83.72	102	20.08	406	79.92	200	18.02	910	81.98
SD	149	24.75	453	75.25	126	24.80	382	75.20	275	24.77	835	75.23

From the data presented in the table, it is evident that three factors are particularly influential in the decision to choose an environmentally friendly method of collecting parcels outside the home: proximity of collection points (DP), lower delivery price (P), and the possibility of free return (R). These are the only factors for which the percentage of respondents confirming their importance exceeds 50%. Two of these factors are directly associated with the cost of environmentally friendly parcel collection. At the same time, the third - the most important for respondents - relates to the convenience of such a collection method. It is noteworthy that factors involving external pressure to adopt environmentally friendly behaviours - such as the requirement to pay an additional environmental charge for choosing an environmentally unfriendly delivery method (AK) or the influence of social pressure to engage in environmentally friendly behaviour (SP)—are not perceived by respondents as significantly important when selecting an environmentally friendly delivery method. Additionally, the analysis of the collected data indicates that the differences in preferences between male and female respondents are smaller than anticipated, suggesting a relatively uniform set of motivations across genders.

In conclusion, the analyses conducted represent an effort to uncover structures or patterns of behaviour within the data, considering all identified factors as well as the gender of the respondents. Table 3 shows the results of the analyses, which were performed using the generalised k-means method.

Table 3. Study results—cluster analysis

Cluster 1	Cluster 2
Gender—mainly women Number of respondents: 923; 83.15%	Gender—mainly men Number of respondents: 187; 16.85%
Lower delivery price (P); The possibility of a free return (R); Proximity to parcel collection points (DP)	The possibility of a free return (R); Shorter delivery time (T)

The information presented in the table confirms earlier observations that only a limited number of the analysed factors significantly influence respondents' decisions to choose an environmentally friendly delivery method. Differences between male and female respondents are minimal. Women are more likely than men to opt for this type of delivery when it offers a lower price, free returns, or proximity to parcel collection points. Conversely, men are more inclined to prioritise free returns and shorter delivery times as key factors influencing their choice of environmentally friendly delivery options.

5 Conclusion

While numerous studies support the view that women are generally more pro-environmental than men and exhibit a slightly higher willingness to pay for green transportation [16], our research suggests that gender may not be the most significant predictor of environmental behaviour. Analysis of the collected data reveals that the differences between respondents' choices by gender are smaller than anticipated based on previous empirical studies. Our survey provides evidence that both male and female Polish online shoppers are unwilling to pay for last-mile delivery services or to pay extra to reduce greenhouse gas emissions. They also do not change their behaviour in response to social pressure to take environmentally friendly actions. Among the factors of most significant importance in the decision to choose an environmentally friendly method for picking up parcels outside their residence are proximity to parcel pick-up points, lower delivery price, and the possibility of free returns, with the latter two factors being related to the cost of environmentally friendly parcel collection.

Our findings are inconsistent with the observed consumer awareness of environmentally friendly consumption. These results could inform policymakers in Poland to develop new policies to implement carbon pricing for freight and subsidies for e-stores offering green delivery methods. Such measures would provide incentives for consumers to choose climate-friendly delivery options and contribute to reducing greenhouse gas emissions in the e-commerce sector.

Acknowledgements. The paper is a part of the research project, which is being financed by the National Science Centre in Poland, granted based on decision number 2018/31/B/HS4/03711.

References

1. WHO, Air pollution costs European economies US$ 1.6 trillion a year in diseases and deaths, a new WHO study says, World Health Organization. (2015). https://www.who.int/europe/news/item/28-04-2015-air-pollution-costs-european-economies-us-1-6-trillion-a-year-in-diseases-and-deaths-new-who-study-says. Accessed 19 Nov 2024
2. Myllyvirta, L.: Quantifying the Economic Costs of Air Pollution from Fossil Fuels, Centre for Research on Energy and Clean Air, Finland. (2020). https://energyandcleanair.org/wp/wp-content/uploads/2020/02/Cost-of-fossil-fuels-briefing.pdf. Accessed 19 Nov 2024
3. EEA, Trends and projections in Europe 2023, European Environment Agency Report 07/2023, Copenhagen, Denmark. (2023). https://www.eea.europa.eu/publications/trends-and-projections-in-europe-2023. Accessed 19 Nov 2024
4. Van Loon, P., Deketele, L., Dewaele, J., Mckinnon, A., Rutherford, C.: A comparative analysis of carbon emissions from online retailing of fast moving consumer goods. J. Clean. Prod. **106**, 478–486 (2015)
5. Allen, J., et al.: Understanding the impact of e-commerce on last-mile light goods vehicle activity in urban areas: the case of London. Transp. Res. Part D: Transp. Environ. **61**, 325–338 (2018)
6. Statista. Global e-commerce share of retail sales 2027. (2024). https://www.statista.com/statistics/534123/e-commerce-share-of-retail-sales-worldwide/. Accessed 20 Nov 2024
7. Cárdenas, I., Beckers, J., Vanelslander, T.: E-commerce last-mile in Belgium: developing an external cost delivery index. Res. Transp. Bus. Manag. **24**, 123–129 (2017)
8. Hunter, L.M., Hatch, A., Johnson, A.: Cross-national gender variation in environmental behaviors. Soc. Sci. Q. **85**(3), 677–694 (2004)
9. Granda, L., Moya-Fernández, P.J., Soriano-Miras, R.M. et al.: Pro-environmental behaviour in household water use. A gender perspective. Sustain. Water Resour. Manag. **10**, 49 (2024)
10. Young, W., Hwang, K., Mcdonald, S., Oates, C.J.: Sustainable consumption: green consumer behaviour when purchasing products. Sustain. Dev. **18**(1), 20–31 (2010)
11. Stern, P.C.: New environmental theories: toward a coherent theory of environmentally significant behavior. J. Soc. Issues **56**, 407–424 (2000)
12. Topf, S., Speekenbrink, M.: Evidence of 'Green'behaviours: exploring behavioural traces of pro-and anti-environmental behaviors. J. Environ. Psychol. **84** (2022)
13. Elert, N., Lundin, E.: Gender and climate action. Popul. Environ. **43**(1), 470–499 (2022)
14. Ergas, C., Greiner, P.T., Mcgee, J.A., Clement, M.T.: Does gender climate influence climate change? The multidimensionality of gender equality and its countervailing effects on the carbon intensity of well-being. Sustainability **13**(7) (2021)
15. Zelezny, L., Chua P-P, Aldrich Ch.: New ways of thinking about environmentalism: elaborating on gender differences in environmentalism. J. Soc. Issues **56**(3), 443–457 (2000)
16. Schniederjans, D.G., Starkey, C.M.: Intention and willingness to pay for green freight transportation: an empirical examination. Transp. Res. Part D: Transp. Environ. **31**, 116–125 (2014)
17. Kiba-Janiak, M., Cheba, K., Mucowska, M., De Oliveira, L.K.: Segmentation of e-customers in terms of sustainable last-mile delivery. Oeconomia Copernicana **13**(4), 1117–1142 (2022)
18. Jagoda, A., Kołakowski, T., Marcinkowski, J., Cheba, K., Hajdas, M.: E-customer preferences on sustainable last mile deliveries in the e-commerce market: a cross-generational perspective. Equilibrium. Q. J. Econ. Econ. Policy **18**(3), 853–882 (2023)
19. Sinaga, K.P., Yang, M-S.: Unsupervised K-means clustering algorithm. IEEE Access **8**, 80716–80727 (2020)
20. Na, S., Xumin, L., Yong, G.: Research on k-means clustering algorithm: an improved k-means clustering algorithm. In: 2010 Third International Symposium on Intelligent Information Technology and Security Informatics. IEEE, Jian, China, pp. 63–67 (2010)

The Management of the Technological Process According to the Product Lifecycle and Circular Economy Concept

Olena Stryhunivska[1](✉), Rafał Rumin[1], Jerzy Feliks[1], and Olena Paliienko[2]

[1] AGH University of Krakow, Krakow, Poland
{olestr,rumin,jfeliks}@agh.edu.pl
[2] State University of Trade and Economics, Kyiv, Ukraine
o.paliyenko@knute.edu.ua

Abstract. The purpose of this study is to identify and analyze strategies for effective management of technological processes in the context of sustainable product development. Particular emphasis is placed on the principles of the product life cycle and the circular economy. The research problem focuses on methods of integrating life cycle management with the concept of the circular economy to minimize waste while maintaining high product quality. The study analyzes the key stages of the product life cycle, including design, production, use, and recycling. It also evaluates management approaches that support resource efficiency and improve production quality.The findings indicate the growing importance of technological innovations and process improvements that promote the closing of material and energy loops. The conclusions confirm that integrating life cycle management with circular economy principles is an effective tool for supporting sustainable industrial development, improving resource efficiency and enhancing the long-term competitiveness of enterprises

Keywords: Sustainable product development · Life cycle management · Circular economy · Waste management · Circular economy model

1 Introduction

The management of technological processes according to the product lifecycle (PLC) and circular economy (CE) concept has become one of the core priorities in the sustainable transformation of modern industry. The development of new-generation, high-efficiency technologies has created new requirements for both the socio-economic system and the natural environment. Moreover, contemporary conditions of hypercompetition emphasize the importance of implementing circular mechanisms within economic systems, which is both timely and practically relevant. The circular economy contributes to the modernization of production and the implementation of innovative technologies by enhancing environmental awareness, protecting natural ecosystems and reducing harmful emissions. At the same time, it is necessary to differentiate approaches to implementing circular economy methods and tools across various socio-economic systems, considering their structural, technological and institutional characteristics [1].

J. Feliks et al. (Eds.): CLC2024 2024, 2026.
https://doi.org/10.1007/978-3-032-28931-5_12

The circular economy model is based on interconnected and complementary economic concepts that combine economic growth with environmental security. The key distinction between the sustainable development concept and the circular economy lies in the latter's broader scope — encompassing economic, ecological, social and restorative dimensions of development. Recent research indicates a deepening scientific understanding of circular economy mechanisms and a gradual formation of theoretical and methodological frameworks for their implementation [2]. The circular economy is defined in multiple interrelated ways as a model, activity, system, strategy, process, instrument and philosophy, each reflecting different aspects of circular transition [3, 4].

Within this broader theoretical context, technological process management plays a crucial role in ensuring the practical realization of CE principles throughout the product lifecycle. Effective control of technological processes enables optimization of resource flows, reduction of production waste and maintenance of high product quality. Integrating PLC management with circular economy principles helps to design closed-loop production systems, where materials and energy are continuously reused through repair, refurbishment and recycling, thereby reducing environmental impact and improving competitiveness [5, 6].

The ongoing industrial transition is closely tied to the digitalization of production and the introduction of Industry 4.0 technologies such as IoT, digital twins and advanced data analytics, which facilitate the monitoring of process efficiency and product condition across life cycle stages. These technologies support the practical implementation of circular strategies by enabling predictive maintenance, material traceability and real-time quality control [7–10].

However, despite the rapid expansion of CE-related research, there is still a lack of standardized methods for integrating life cycle management with circularity assessment and readiness frameworks. Many enterprises face difficulties in quantifying circular performance, balancing environmental and economic goals and aligning process optimization with circular principles [11, 12].

The purpose of this study is to identify and analyze strategies for managing technological processes in alignment with the product life cycle and the circular economy concept. The research focuses on the relationship between process management, product quality and resource efficiency, emphasizing how life cycle-based approaches can support the transition toward circular industrial models. The central hypothesis assumes that effective technological process management, when integrated with circular design and life cycle assessment, minimizes waste generation, optimizes resource use and enhances production sustainability [13–15].

The study examines technological process management within life cycle and circular economy frameworks. It reviews current theoretical and practical approaches. The research identifies innovations that enhance circularity. It also provides recommendations for implementing circular process management in manufacturing systems. This approach provides a conceptual and methodological foundation for supporting sustainable industrial development, improving resource efficiency and strengthening the competitiveness of enterprises in the circular economy context [16].

2 Food Waste Management within the Framework of the Circular Economy

Food waste represents one of the most pressing environmental and economic challenges of the modern world. According to the Food and Agriculture Organization of the United Nations (FAO), approximately one-third of all food produced globally is lost or wasted each year. This phenomenon not only results in considerable economic losses but also generates serious environmental consequences, including the excessive use of natural resources and significant greenhouse gas emissions. The concept of the circular economy provides effective approaches aimed at reducing food waste and promoting its reuse within sustainable production and consumption systems [17].

Food waste can be classified into two main categories: losses occurring at the production stage and waste generated at the consumption level. The principal causes of food waste include overproduction and inefficient inventory management, insufficient infrastructure for storage and transportation, non-compliance with product quality and appearance standards, as well as unconscious consumer behavior and improper food handling. The excessive generation of food waste leads to the depletion of natural resources, particularly water, soil and the accumulation of waste in landfills contributing to methane emissions. It also causes social consequences, as the loss of edible products occurs alongside persistent global hunger. Reducing food waste effectively requires the implementation of a comprehensive set of measures involving technological, behavioral, regulatory and innovative solutions. The development of modern refrigeration, packaging and logistics technologies plays a key role in extending the shelf life of food products and minimizing spoilage. Advanced refrigeration systems and freezing methods enable longer preservation of food, while innovative biodegradable and active packaging materials reduce environmental contamination and extend product usability. Optimized logistics processes and improved inventory management further decrease losses during storage and transportation [18].

Equally important is the transformation of consumer behavior through awareness-raising campaigns and educational initiatives that promote conscious and responsible consumption. Consumers can be encouraged to plan purchases carefully, store food properly and prepare meals with minimal waste. The growing use of mobile applications and digital platforms for sharing surplus food or selling near-expiry products at reduced prices contributes to waste reduction at the household level. Additionally, clear food labeling, including precise explanations of expiration and best-before dates, allows consumers to make more informed decisions and avoid premature disposal of edible products. Legislative regulation is another crucial component of food waste reduction. The development of policies and incentives for enterprises implementing circular economy principles fosters more responsible food management practices. Legal frameworks that prohibit the disposal of edible food and require its redistribution for charitable purposes have proven effective in several countries. Furthermore, financial mechanisms such as tax incentives for companies engaged in food waste recycling stimulate innovation and encourage broader participation in circular initiatives. Effective governance also entails supporting technological innovations in food waste utilization and creating specialized collection and processing centers. It should also strengthen control over compliance with safety and waste management standards [19].

Innovative solutions play a pivotal role in the transition towards a circular food system. The development of alternative products made from recycled food residues enables the conversion of organic waste into valuable resources, including bioplastics and food additives. The application of fermentation and hydrolysis technologies further enhances this process. The integration of artificial intelligence (AI) and the Internet of Things (IoT) into the food supply chain allows for more accurate demand forecasting, helping to prevent overproduction and minimize losses. Drones and automated monitoring systems improve the control of product quality and enable timely responses to potential spoilage, thereby reducing waste levels. The implementation of circular economy principles in food waste management constitutes an effective strategy that mitigates environmental impacts. It also creates new economic opportunities for enterprises and communities. The adoption of these principles contributes to sustainable development, rational resource utilization and a reduction in environmental pollution. Close cooperation among governments, businesses, and citizens is essential to establishing a resilient and efficient system for food waste management. Such collaboration ensures long-term environmental and economic sustainability [20].

3 The Basic Model of the Circular Economy

The vast majority of researchers defining the principles of the circular economy employ the prefix "re" (from Latin, meaning "again"), which characterizes the fundamental essence of this economic paradigm. The basic "3R" model of the circular economy encompasses three core principles:

- reduce – aimed at minimizing the consumption of raw materials and resources, thereby reducing waste generation,
- reuse – involves the repeated use of materials and resources in production processes, decreasing the need for new resource extraction,
- recycle – ensures the transformation of used materials into new products, contributing to the reduction of environmental pollution.

At the initial stage of the concept's development, the circular economy was predominantly examined through the lens of the "3R" system, which focused on resource recirculation. Over time, however, this approach was expanded, resulting in the formation of the extended "10R" model [21]. While preserving the fundamental principles of the original "3R" system, the "10R" framework incorporates additional dimensions that enhance its scope. Scholars emphasize that this model is not final, as the circular economy continues to evolve.

The main components of the "10R" framework include:

1. refuse—minimizing or entirely avoiding excessive resource consumption,
2. rethink—optimizing methods of material and product utilization,
3. reduce—decreasing the use of natural resources and materials,
4. reuse—employing products that have already been used,
5. repair—restoring the functionality of products for continued use,
6. refurbish—upgrading older products to extend their life cycle,
7. remanufacture—producing new goods using components from old ones,

8. repurpose—changing the intended function or application of used products,
9. recycle—reprocessing waste materials into new ones,
10. recover—obtaining energy from waste and unused materials.

The expansion of the circular economy framework promotes efficient resource utilization, waste reduction and the establishment of sustainable production and consumption models.

Consequently, the circular economy can be regarded as both a prerequisite and a catalyst for a new industrial revolution. Conceptually, it pursues two overarching objectives. On one hand, the value of used products must be restored to ensure maximum economic efficiency. On the other hand, such restoration inherently reduces the negative environmental impact, thereby fulfilling the socio-economic and ecological requirements of sustainable development.

At the micro level, the implementation of circular economy principles involves integrating environmental considerations into product and process design (eco-design), organizing low-emission and clean production, establishing waste prevention systems and increasing consumer responsibility through eco-labelling and green public procurement. At the meso level, it encompasses the development of eco-industrial parks and agro-ecosystems, supported by environmentally friendly design and efficient waste management. At the macro level, the application of circular economy principles is directed toward the creation of eco-cities, eco-societies and eco-regions. In essence, the circular economy constitutes a foundation for the implementation of environmental innovations and the enhancement of ecological responsibility, grounded in rational resource and energy consumption, mitigation of environmental impacts and the advancement of a socially integrated and sustainable society [21–24].

4 The Basic Model of the 8E–CE

The concept of the circular economy involves designing production processes that integrate the use of secondary raw materials and aim to extend the life cycle of manufactured products. According to ISO 14040, the "life cycle" is defined as a series of "consecutive and interrelated stages." This definition highlights the necessity of adopting a systemic and holistic approach to environmentally responsible design. Effective spatial planning and design play a fundamental role in implementing the principles of the CE. Conventional approaches to the design of food industry facilities primarily address the requirements of end users. From the circular economy perspective, however, this process must be approached in a more comprehensive manner. It requires the development of professionals equipped with advanced engineering and technological knowledge. They must be capable of making creative and environmentally responsible decisions concerning the design, location, and operation of technological lines, machinery, and equipment in compliance with CE principles.

Furthermore, beyond focusing on the final product intended for consumers, it is essential to consider the influence of production processes on each component of the technological chain. It is also necessary to account for their interaction with all stakeholders involved in the design process. Such an integrated approach ensures the reintegration of the final product into the circular flow. The analysis was carried out based on the 6E framework, which evolved from the original 4E concept and was subsequently expanded into the 8E–CE model. This development enabled the formulation of an effective design process model grounded in the principles of the circular economy.

The main components of the 8E–CE process model for food industry enterprises include:

1. Ecological process—integration of industrial environmental architecture with innovative technological approaches, grounded in ecological responsibility. It focuses on rational spatial design that minimizes environmental impact, incorporates eco-friendly production processes, enhances food nutritional value and supports the use of energy-efficient technologies and renewable energy sources (RES).
2. Economic process—optimization of financial and material resources through rational technological and spatial solutions across all project stages—from design and construction to operation. It is based on the 10R principles, aiming to reduce consumption, minimize waste generation, extend the life cycle of technological equipment and recover energy to lower production costs and CO_2 emissions.
3. Energy-efficient process—optimization of energy performance through the appropriate spatial arrangement of facilities and the selection of energy-saving technologies. It includes the application of materials, systems and installations that utilize non-conventional, and renewable energy sources (solar, wind, hydro, and geothermal) to conserve natural resources.
4. Ergonomic process—design of production environments that enhance worker safety, comfort and efficiency through proper workstation configuration. It also includes the analysis of human–technology interactions while ensuring compliance with spatial, technical and organizational constraints.
5. Eco-circular process—implementation of life-cycle thinking at all stages of design, construction and operation to ensure responsible resource management. In this process, waste is treated as a reusable material, and reuse scenarios for materials and products are defined at the earliest design stages.
6. Elastic process—development of adaptable, modular and functional solutions that enable quick and effective spatial or technological adjustments in production facilities, even during operation. It promotes the adoption of closed-loop systems and new sustainable consumption models.
7. Efficient process—pursuit of ecological and efficient changes in production systems through the integration of theoretical and practical knowledge. It extends the framework of "effective process time" modeling to analyze system performance, material and energy flows, and technical, organizational, and environmental conditions.
8. Eco-innovative process—implementation of innovative and comprehensive solutions is aligned with circular economy principles. It involves the integration of Building Information Modeling (BIM) and the development of BIM libraries. It also includes the application of multi-criteria optimization methods for sustainable, energy-saving and intelligent building systems. These systems are connected with renewable energy technologies.

Fig. 1 presents the integration of the 8E–CE model with the 10R framework, illustrating the interconnections between design and production processes in the context of the circular economy.

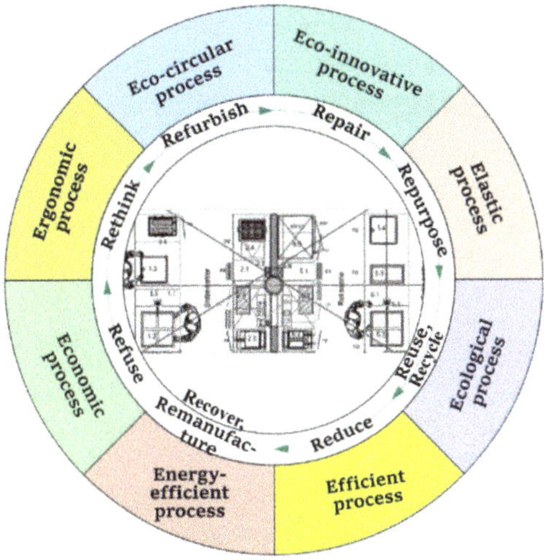

Fig. 1. Integration of the 8E–CE model with the 10R framework for industry enterprises [25].

The outer ring represents the eight key processes of the 8E–CE model. The inner ring visualizes the 10R principles which define the hierarchy of circular actions aimed at minimizing waste, extending product life cycles, and recovering value from materials and energy. The central part of the figure contains a schematic spatial layout of a food production facility, symbolizing the integration of design, technology and circular resource management. This visual model emphasizes that the combination of the 8E and 10R frameworks supports the creation of sustainable, resource-efficient and innovative industrial systems, contributing to the transition toward a fully circular economy. The integration of these two approaches enables a comprehensive strategy for the design, operation, and modernization of food industry enterprises. This strategy encompasses ecological, economic, energy-efficient, and ergonomic aspects. The advantages of integrating the 8E and 10R models include:

- Holistic life-cycle management—by incorporating the principles from refuse to recover, the model supports reducing the use of primary raw materials and minimizing waste generation at all stages of the process.
- Optimization of technological and energy processes—the 8E approach promotes the design of energy-efficient, flexible and effective production systems aligned with the principles of sustainable development.
- Improved economic efficiency—through the implementation of reuse, repair and remanufacture strategies, enterprises can reduce investment and operational costs.
- Enhanced innovation and competitiveness—integrating eco-innovative processes with the 10R principles encourages the adoption of advanced BIM technologies, environmental monitoring systems, and renewable energy solutions.
- Increased environmental responsibility—the model supports the conscious design of production and logistics spaces that minimize environmental impact.

- Flexibility and resilience of production systems—through the flexible and efficient processes, enterprises can rapidly adapt to market, technological and environmental changes while maintaining operational stability.

As a result, the integration of the 8E–CE model with the 10R framework provides a foundation for developing modern, sustainable, and resilient production structures in the food industry, fully consistent with the principles of the circular economy [25–26].

5 Conclusion

The conducted research confirmed that the integration of technological process management with the principles of the product life cycle and circular economy constitutes an effective strategy for achieving sustainable industrial development. The proposed 8E–CE model, when combined with the 10R framework, provides a comprehensive methodological foundation for optimizing material and energy flows, improving production efficiency and enhancing environmental responsibility. The model promotes systemic thinking by incorporating ecological, economic, ergonomic and energy-efficient dimensions, supporting enterprises in the transition toward circular production systems.

The results of the analysis emphasize the importance of adopting an interdisciplinary approach that connects life cycle management, eco-innovation and digital transformation. Industry 4.0 technologies such as IoT, BIM and data analytics enable real-time process monitoring and the implementation of predictive maintenance, which facilitate closed-loop material management. This integration contributes to reducing waste, extending product life cycles and improving competitiveness in the context of sustainable development. However, the study also revealed that the practical implementation of circular economy principles in production systems still faces significant challenges. These include the lack of standardized indicators for assessing circularity, limited integration of digital tools across life cycle stages and insufficient alignment of organizational strategies with circular goals. Further research should focus on developing quantitative and simulation-based models that assess the economic and environmental effects of applying the 8E–CE model in various industrial sectors. Another promising direction involves extending the proposed framework with digital twin technology for monitoring and optimizing life cycle performance in real time. Future investigations should also explore policy and management mechanisms that foster cooperation among stakeholders and support the large-scale adoption of circular process management in both manufacturing and service industries.

Acknowledgements. Research project supported/partly supported by program "Excellence initiative—research university" for the AGH University of Krakow

References

1. Amin, M.B., Asaduzzaman, M., Debnath, G.C., Rahaman, M.A., Oláh, J.: Effects of circular economy practices on sustainable firm performance of green garments. J. Clean. Prod. **456**, 140–158 (2024)

2. Brglez, K., Perc, M., Lukman, R.K.: The complexity and interconnectedness of circular cities and the circular economy for sustainability. Sustain. Dev. **32**(3), 2049–2065 (2024)
3. Abu-Bakar, H., Charnley, F.: Developing a strategic methodology for circular economy roadmapping: a theoretical framework. Sustainability. **16**(15), 6682 (2024)
4. Brogi, S., Menichini, T.: The pathway towards circular economy: measuring circular advantage of eco-innovations. Bus. Strateg. Environ. **33**(4), 3005–3038 (2024)
5. Obiuto, N.C., Ninduwezuor-Ehiobu, N., Ani, E.C., Andrew, K.: Implementing circular economy principles to enhance safety and environmental sustainability in manufacturing. Int. J. Adv. Multidiscip. Res. Studies. **4**(2), 22–29 (2024)
6. Perotti, F.A., Bargoni, A., De Bernardi, P., Rozsa, Z.: Fostering circular economy through open innovation: insights from multiple case study. Bus. Ethics, Environ. & Responsibility. **34**(2), 390–408 (2025)
7. Afshari, H., Gurtu, A., Jaber, M.Y.: Unlocking the potential of solid waste management with circular economy and industry 4.0. Comput. Ind. Eng. **195**, 110457 (2024)
8. Matarneh, S., Piprani, A.Z., Ellahi, R.M., Nguyen, D.N., Le, T.M., Nazir, S.: Industry 4.0 technologies and circular economy synergies: enhancing corporate sustainability through sustainable supply chain integration and flexibility. Environ. Technol. Innovation. **35**, 103723 (2024)
9. Shahzad, M.F., Liu, H., Zahid, H.: Industry 4.0 technologies and sustainable performance: do green supply chain collaboration, circular economy practices, technological readiness and environmental dynamism matter? J. Manuf. Technol. Manag. **36**(1), 1–22 (2025)
10. Marak, Z.R., Sharma, A., Uikey, A.A.: Exploring the interlinkages between industry 4.0, circular economy, and sustainable performance. In: Implementing ESG Frameworks through Capacity Building and Skill Development, pp. 301–328. IGI Global Scientific Publishing, Hershey, PA (2025)
11. Pakuła, N., Łapniewska, Z., Dutra, C.J.: Comparative measurements of circular economy performance among European countries: reviewing approaches and limitations. J. Environ. Manag. **375**, 124414 (2025)
12. Mora-Contreras, R., Carrillo-Hermosilla, J., Hernández-Salazar, G., Torres-Guevara, L.E., Mejia-Villa, A., Ormazabal, M.: Eco-innovation for circular economy and sustainability performance: insights and evidence from manufacturing firms. Bus. Strateg. Environ. **34**(1), 1231–1256 (2025)
13. Vera-Guerrero, D. et al.: Sustainable kombucha production: a conceptual framework integrating life cycle assessment and circular economy principles. Trends Food Sci. Technol. **104996** (2025)
14. Carlos, R.L., De Souza, E.B., Mattos, C.A.: Enhancing circular economy practices through the adoption of digital technologies. Bus. Strat. & Development. **7**(1), e330 (2024)
15. Das, S.K., Bressanelli, G., Saccani, N.: Clustering the research at the intersection of industry 4.0 technologies, environmental sustainability and circular economy: evidence from literature and future research directions. Circ. Econ. Sustain. **4**(4), 2473–2504 (2024)
16. Obradović, V., Todorović, M., Cvijović, J.: Integrating circular economy and project management: a conceptual framework. In: Research Handbook on Sustainable Project Management, pp. 203–223. Edward Elgar Publishing, Cheltenham (2024)
17. Canton, H.: Food and agriculture organization of the United Nations – FAO. In: The Europa Directory of International Organizations 2021, pp. 297–305. Routledge, London (2021)
18. Adaryani, R.L. et al.: Predicting household food waste behavior: bringing food literacy and purchasing power into the theory of planned behavior. J. Retail. Consum. Serv. **82**, 104119 (2025)
19. Acquier, A., Carbone, V., Ezvan, C.: Scaling circular economy business models: a capability perspective. Bus. Ethics, Environ. & Responsibility. **34**(2), 377–389 (2025)

20. Hernandez, J.C.R., Villa-Enciso, E., Cardona-Acevedo, S., Valencia, J., Velasquez Salas, S.: Smart innovation for a circular economy: a systematic review of emerging trends and the future of AI in the sustainable economy. Sustainability. **17**(13), 5793 (2025)
21. Favaro, R. et al.: Teaching the "10 Rs" of circular economy in technological-level undergraduate courses offered in Brazil: an exploratory analysis via fuzzy TOPSIS class. Int. J. Sustain. High. Educ. (2024)
22. Stryhunivska, O., Gdowska, K., Rumin, R.: Implementation of synergetic planning for factory. Zeszyty Naukowe. Organizacja i Zarządzanie/Politechnika Śląska. No. **155**, 503–512 (2022)
23. Otasowie, O., Aigbavboa, C.O., Oke, A.E.: The principles of circular economy. In: Circular Economy Business Model for Construction Organisations, pp. 13–32. Springer Nature Switzerland, Cham (2025)
24. Tran, G.T., Le Dinh, T., Labelle, F., Chouchane, R.: Circular economy transformation through intelligent systems: a systematic review. In: International Conference on Resources and Environmental Research, pp. 104–116. Springer Nature Switzerland, Cham (2024)
25. Stryhunivska, O., Dzyundzya, O., Rasulov, R.: Harmonizacja podejścia do przeplanowania układów przestrzennych w zakładach przemysłu spożywczego według zasad GOZ. In: Innowacje w naukach inżynieryjno-technicznych – wybrane aspekty: Monografia naukowa, pp. 107–119. Wydawnictwo Naukowe TYGIEL, Lublin (2023)
26. Lai, L.W., Kwong, J.W.: Principles of Town Planning: Translating ConceptstoApplications. Edward Elgar Publishing, Cheltenham (2025)

Advanced Logistics Technologies

Lean Logistics Methods and Sustainable Growth: Case Study of the Assembly Workplace

Helena Fidlerova, Jarmila Blahová, and Helena Makyšová(✉)

Faculty of Materials Science and Technology in Trnava, Slovak University of Technology in Bratislava, Bratislava, Slovakia
helena.fidlerova@stuba.sk

Abstract. The implementation of lean management principles and innovative practices within a company has been shown to significantly enhance process efficiency, optimise logistics flows, and strengthen competitiveness. This paper presents the case study focused on process standardisation at an inverter assembly workplace, providing a practical demonstration of the effective integration of lean production methods. The objective is to demonstrate the effective use of the selected lean production methods and tools in order to present broader possibilities for improving logistics. The study analyses the use of standard work, 5S, and visual management to streamline material flow, reduce variability, and improve workplace. The research applies the PDCA methodology to design, implement, and evaluate process standardisation, focusing on the reduction of time losses, non-value-added activities, and training inefficiencies. A comparative time analysis of four process stages demonstrates that value-added activities account for over 80% of total cycle time, confirming a substantial improvement in workflow efficiency and labour utilisation. The findings indicate that standardisation not only stabilises processes and enhances logistics flow but also supports the company's transition towards sustainable manufacturing in line with SDG 12.

Keywords: Industry Innovation · Lean Logistics · PDCA · Sustainable Manufacturing · SDG 12

1 Introduction

Innovations combined with lean manufacturing practices, significantly improve operational performance. Nevertheless, many organizations encounter difficulties in the effective integration of these approaches. Implementing these new principles can be highly advantageous, offering the potential to save money, time, energy, as well as material and human resources. Standardization can help scale innovation across the logistics network ensuring the consistent implementation of new technologies and processes. However, limited research has examined how process standardisation in logistics directly contributes to sustainable growth in manufacturing enterprises.

The objective of the paper is to provide an example of good practice based on the presented case study of the application of standardization in logistics and to point out other possibilities of elimination of time and effort in the wider context of sustainability.

J. Feliks et al. (Eds.): CLC2024 2024, 2026.
https://doi.org/10.1007/978-3-032-28931-5_13

1.1 Sustainable Innovation

For lean logistics, new technologies and innovation present an opportunity to empower employees to drive continuous improvement. By integrating lean methodologies with Industry 4.0 tools, manufacturers can achieve more than they would by pursuing these initiatives independently. Real-time data and communication between people, machines, and systems offer a comprehensive view of production, enabling frontline workers to make real-time adjustments [1].

Authors [2] reported sustainable innovation specifically from an open innovation perspective, some of which are green process innovation capacity, green product innovation capacity, product life-cycle management capability, sustainable innovation orientation development, sustainable partnership and collaboration. Sustainable innovation includes green logistics and reverse logistics, which support green process innovation [3]. It also involves developing products, services, and business models that align with long-term sustainability goals [4, 5].

The value of lean logistics and lean manufacturing is defined in minimising operational complexity, eliminating waste, and improving productivity by empowering workers to make continuous adjustments and to facilitate the performance of work activities in logistics. That is why, there is needed to define Muda as a crucial element of Lean. By Walter [6] „Muda" is a Japanese term that means waste or inefficiency. In the context of Lean Muda refers to all activities and processes that do not add value to the customer. This means that if an activity is considered Muda, it should either be eliminated or minimized because it only adds cost without benefit to the product. Identifying and eliminating Muda is the foundation stone of lean manufacturing. Waste in the production and logistics process not only increases costs, but also slows down the entire process, reduces quality and negatively affects customer satisfaction. By eliminating Muda, we can achieve greater efficiency, faster lead times, better quality and overall lower costs. That's why it's important to constantly monitor and analyse processes, identify potential areas of waste and take the necessary steps to eliminate them [7].

During the introduction of the lean method the PDCA (Plan-Do-Check-Act) methodology was used. The first phase "PLAN" included the creation of an implementation plan respecting time and space limitations and the human factor. The goals were defined that might be achieved in terms of the efficiency of the logistics process. Then in phase "DO" specific measures for improvement were implemented (standardisation) in logistics. In "CHECK" phase we monitored and measure processes, compare with the policies, objectives and requirements of the customer requirements. In last phase "ACT" we provide an action if required for continuous improvement of process performance in selected workplace to achieve the desired goals.

2 Lean Logistics—Theoretical Background

Several studies have explored the integration of Lean Manufacturing, innovations and Industry 4.0 from various perspectives to clarify the relationship between these paradigms. Ding [5] indicates in the systematic review of the literature that links Industry 4.0 with lean and agile manufacturing, identified specific areas for improvement within both strategies. The integration enhances cost-competitiveness in terms of performance, while its integration with agile manufacturing mainly boosts flexibility.

According to [8], applying lean manufacturing principles helps reduce waste, standardise processes, and organize work, thus facilitating the implementation of various Industry 4.0 technologies in companies.

2.1 Lean Principles and Industry 4.0

Conversely, the technological advancements offered by innovation help in achieving significant results and an exceptional level of maturity. Important are lean management concepts such as standardisation, organization, and transparency. Standardisation of processes, where it is possible to standardize and continuously improve every process to avoid waste. Minimizing of unnecessary movement of people—eliminate any movement that does not add value. Minimizing of material movement helps to eliminate any material movement in logistics flow that does not add value to the final product or service. Applying lean thinking approaches plays a pivotal role in accelerating sustainable development by systematically eliminating inefficiencies and integrating process improvements across the entire value chain [9].

Manufacturers using 4.0 technology can minimise downtime through predictive and prescriptive maintenance using, e.g., sensors and other IIoT devices [10]. This approach maximises utilization, identifies and addresses issues, enables real-time decisions—such as automatically halting a machine in case of a safety concern, and supports more informed decision-making through data visualisation. Marinelli [11] conducted a comprehensive study of industry practices regarding the combined use of Industry 4.0 and lean tools in the manufacturing sector whereby the results of study highlighted that, although there is consensus on the positive impact of lean practices, the perceived benefits of Industry 4.0 tools are less evident, with their potential to enhance efficiency and business profitability not yet fully acknowledged. Kassem [12]. highlight a consensus on the synergistic relationship among the nine pillars of Industry 4.0, showing their positive impact on operational performance. The research reveals that the strongest synergy is found between just-in-time practices and the Industrial Internet of Things (IIoT). This mean that prediction that investments in IoT and cyber-physical systems will have the greatest influence on interactions with the pillars of the House of Lean (HoL).

Although all studies confirm the positive correlation between lean implementation and operational performance, their approaches differ in scope: highlights flexibility and agility [5], emphasises digital integration [12], on the other hand the study [13] underlines organisational maturity as a prerequisite for Lean 4.0. This comparison reveals that the success of lean–digital integration depends primarily on the stability of standardised processes. In parallel, Knapčíková et al. [14] demonstrated measurable gains in production process efficiency using lean tools in Slovak manufacturing, confirming the practical relevance of standardized workflows and continuous improvement methodologies. The critical role of logistics outsourcing in enhancing efficiency and flexibility within Slovenian enterprises, aligning with lean logistics objectives [15].

Fidlerova et al. [16] highlight that optimizing logistics processes not only reduces waste and improves resource efficiency, but also enhances the sustainability profile of operations.

In fact, lean manufacturing is often considered a prerequisite for Industry 4.0. Qualitative survey relationship within small and medium-sized enterprises confirmed that lean manufacturing is essential for Industry 4.0, arguing that a production process cannot be efficient, even with advanced digitalisation, if it contains waste [17].

Most studies in lean manufacturing analyse production or technological dimensions, whereas the logistics perspective—linking lean tools with sustainability outcomes—remains underexplored. This gap provides the rationale for the case study presented in the following section.

3 Analysis of the Process Standardisation in Assembly Operations

In this part of paper, the findings will be presented in the selected industrial enterprise. First, based on the study of internal documents, observation, and a controlled interview with the company's employees, we found out to what extent the selected methods of lean production are applied (Kaizen, Kanban, Teamwork, Value Stream Mapping, and 5S methods). Based on positive results from the previous implementation of mentioned methods of lean management in the company the focus was established on standardization of logistics processes.

3.1 Improving Efficiency through Standardized Work

Among the expected goals of standardised work are stabilization of the work process, increase in productivity of the workplace, increase in the quality of the final product in the workplace, easy training, and implementation of innovations. Standardised work serves the efficient realisation of work activities, is intended to eliminate nonconformity, increase quality and supports the process of continuous improvement. It is a tool for synchronising the sequence of employee movements and helps to improve all processes. The standardised procedure serves as documentation of the currently best-known and reproducible process. Important prerequisites for the introduction and functioning of standardized work are cyclical, stable processes, continuous training of workers, creation and maintenance of workplace standards. For the implementation, a workplace was chosen where was identified a waste of time during the training of new workers and during job rotation. The job rotation is important for the worker to alternate different activities, and thus avoid fatigue, frustration, and the stereotype of everyday work. It is necessary to optimise times.

The workplace for the assembly of hydrodynamic converters for truck transmissions, especially buses, was selected for analysis. The workplace can be described as semi-automated, closed without dust, the room is air-conditioned with a constant temperature. In the following, we will briefly characterize the main activities at this workplace with two employees, where it is divided into four stages (takts). For clarity, the term *takt* is used in this study to denote a single work cycle within the assembly process.

First, the worker prepares the material, he scans the lid and the torsion damper with a reader. Using unambiguous light signalling by reading the code makes it possible to reduce the number of errors and simplify activities at the given workplace. If the code changes to green, it is possible to continue working, otherwise, if the code is red, the material needs to be replaced. A piston with two rubber bands is inserted into the lid, and the bearing is pressed. The torsion damper, together with the disc and the locking ring, are inserted into the bearing and pressed. For better grip, six slats (three outer and three inner) and twelve pins are added to centre the slats. To assure the quality a pressure test is performed, and the dimensions are checked. If everything is in order, a despatch note is added to the bottom of the converter and follows the second takt. Subsequently, in second stage the other worker has the task of assembling the upper part of the converter. A sliding sleeve is pressed into the turbine. The assembled timing wheel is inserted, and the bearing is pressed. A cover with a pump can be placed on top. The entire assembly is centred and moves to the third takt together with the accompanying documentation. The third and fourth measures are automated without the intervention of a worker. First, the machine works in this section, which takes the necessary number of screws and tightens them. Then the machine controls the required dimensions and axial play of product, a pressure test is performed. Finally, the completely assembled inverter is preserved, and the final control follows. During it, accompanying documentation is scanned, gearing, height, draft, diameters are measured. If all parameters are within limits, the converter is completely cleaned and packed in cartons.

3.2 Designing and Documenting Standardized Work Procedures

For the implementation of a standardized work, it was necessary to define and describe all activities, outline the layout of the workplace and visualise the worker's movements (to plan and to set goals. Subsequently, each activity in the entire work process was monitored. Based on the analysis of collected data, a standardised worksheet was created for the first and second stage, detailing the key features and description for each activity with photos. The key feature outlines how the activity should be performed, while the justification explains why it should be carried out in that specific manner. The critical feature regarding the quality and/or safety of the worker was marked with a specific symbol <C> and warns the worker to pay attention. The cyclic times are the measured values for realisation of the individual activities. Symbols for processes can represent the work performed by a person (hand symbol), the work performed by a machine (gear wheel symbol), and the movement of person (walking stick figure symbol).

As presented, Fig. 1 is visualisation of the assembly workplace during the first takt. The orange lines represent the worker's paths, while the blue dashed line indicates the logistics flow of material. The letter "I" in the triangle represents the place of supply and the symbol exclamation mark "!" in the red triangle represents a place to dispose of waste. The orange diamond represents the quality control. The letter **"X"** in the oval indicates the location of the workflow. The numbers in circles correspond to the individual activities on the standardized worksheet. The worker received only simple instructions without unnecessary text, supplemented by a photo visualisation of the process.

After creating a standardised worksheet, the sheet was placed in eye level by a worker at the workplace of the load converters. The worker was properly trained to perform standardised work. It was expected that standardized workflow and visualization of activities improve performance at the workplace because the worker performs activities for which he is trained. The focus of the measures was to improve the quality of production and the

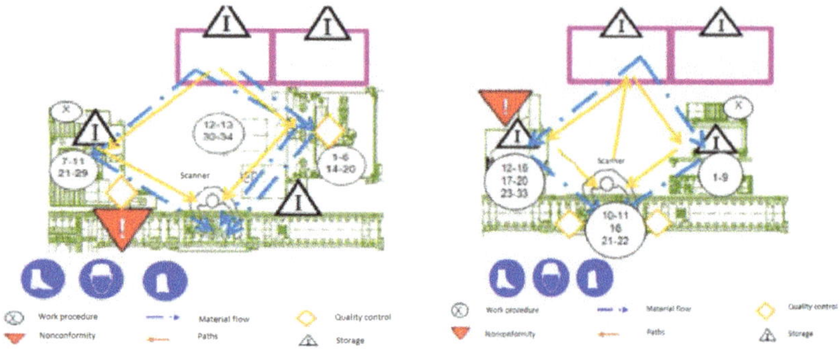

Fig. 1. The layout of the assembly workplace *Source* Authors, based on [17]

well-being of the worker. Another advantage of standardised workflow is that it considers both types of activities: cyclical (those that are repeated regularly) and periodic (those that are performed once in a certain time). Therefore, there is no waste of time during work rotation. The procedure for standardised work is applicable even if the company receives an order for a different type of parts.

A time study at the assembly workplace was performed for all four takts. For the first two takts, the times of individual activities performed by the workers were measured. Since the third and fourth takt are automated, the activities of the machines were measured. The time analysis is shown at Fig. 2. The collected data were recorded and processed using Microsoft Excel 365, while descriptive time analysis and graphical visualisation of process cycles were performed in Minitab 21. These tools enabled statistical comparison of time categories, identification of non-value-added activities, and visual representation of takt time distribution across all stages.

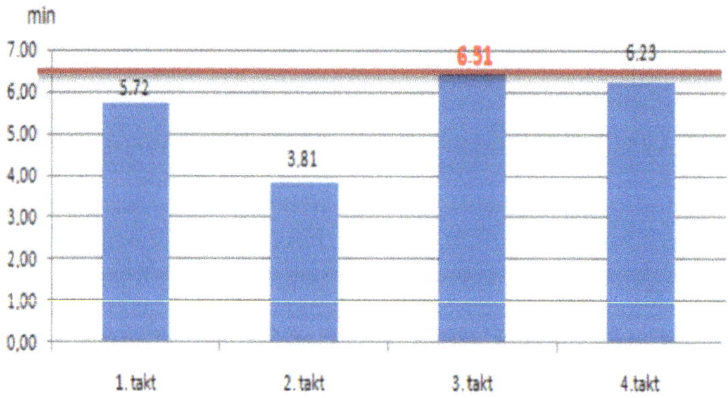

Fig. 2. The time distribution analysis in minutes at workplace. *Source* Authors, based on [17]

4 Results and Discussion

As shown in the bar graph (Fig. 3) the largest share of measured times is represented by activities, which are marked with green and represent the times of activities that represent added value for the overall process (in percentage 84.8% first takt, 73.9% the second takt and total time 100% third and fourth takt) representing machine work and assembly operations. The colour with a lower proportion is orange (11.1% first takt and 21.3% second takt. Necessary activities that need to be performed in order to be able to carry out value-added work such as grasping, inserting, storing, and transferring are indicated in orange. Work without added value, marked in red, represents the smallest share and is mainly about the transfer of the worker (4.1% and 4.8%).

It can be concluded that there has been a notable advancement in development of standardisation at the workplace of the assembly of load converters. From the beginning of the project, no standard was introduced in the workplace. The implementation of a standardised work procedure at the selected workplace will facilitate continuous improvement. Following the implementation of the standards and subsequent evaluation of the results, it can be stated that value-added work represents the highest share in the workplace for assembly of load converters.

As previously discussed, activities lacking added value represent in selected workplace the smallest share, which is beneficial for the company. The main benefit is the optimisation of time, because time was wasted in the assembly workplace during the induction of new workers and during work rotation.

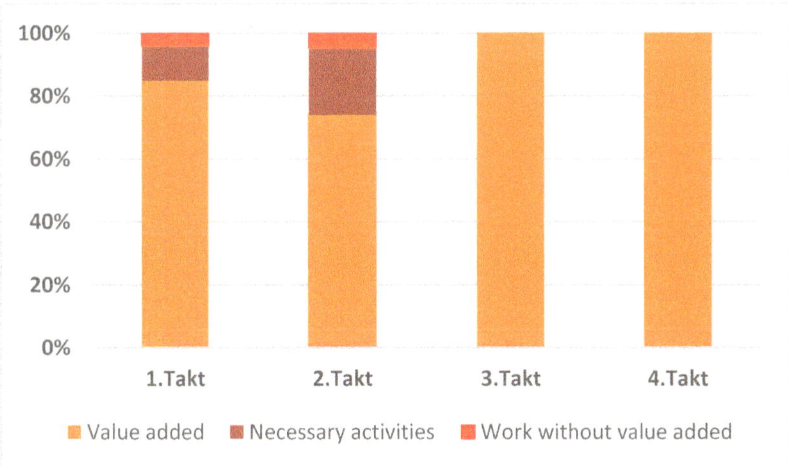

Fig. 3. Standardizatio—share of activity types in %. *Source* Authors

The expected benefits of the described solution are better visualisation and clarity and better comprehensibility. The implementation of standardised work practices serves to stabilise the workflow and, in turn, contribute to an increase in productivity within the workplace. The solution facilitates rapid comprehension of the requisite competencies and the training of new workers to perform tasks in the assembly workplace as part of the training and development of workers.

It is essential that management demonstrate their support for the introduction of standardised work [18]. The implementation of Sustainable Innovation 4.0 can be achieved through the appropriate utilisation of Lean methodologies, including standardisation, with a particular focus on the attainment of selected sustainable development goals from a triple bottom line perspective. As stated by [19, 20] lean practices and sustainability innovation positively influence sustainability (economic and social).

Suitable for the implementation of standardised work are cyclically repeating activities, stable processes, continuous training of workers, creation, and maintenance of standards on site. The expected advantage of standardised workflow is that it prioritises safety and quality, and it facilitates introduction of changes. The observed reduction of time losses, waste elimination, and improved training efficiency directly support the objectives of SDG 12 – Responsible Consumption and Production, particularly in the area of resource efficiency and sustainable industrial processes.

5 Conclusion

Finally, we can conclude based on results that standards support innovation and that standards are important for achieving business strategy goals. Innovation encourages flexibility and adaptability, which is essential in a rapidly changing market, while standardization ensures that operations run smoothly and reliably. The challenge for enterprises is to find a balance. The implementation of standard operating procedures (SOP) is an effective method of reducing complexity and cost.

Standardisation can mitigate risks associated with implementing innovations. For example, when a new logistics technology is introduced, SOPs can define protocols to handle exceptions, manage data, or ensure safety. Conversely, standard operating procedures (SOPs) – defined as a set of step-by-step instructions outlining the correct method for completing a routine task – can potentially lead to a lack of engagement and motivation among workers. This emphasises the importance of providing workers with opportunities for job rotation and other forms of work enrichment to maintain their interest and wellbeing.

Meanwhile, researchers in Sweden have initiated a shift in focus towards Industry 5.0, which is centred on human-centric manufacturing [21]. The study suggests that, while Lean production and Industry 4.0 practices should be integrated to complement each other. The research revealed that manufacturing organisations encounter difficulties when attempting to integrate Lean principles, which emphasise production flow efficiency and waste reduction, with the focus of Industry 4.0 on digital technology, particularly in a socially sustainable manner within the context of Industry 5.0. Furthermore, it emphasises the ambiguities and challenges organisations face when implementing both paradigms simultaneously, particularly in terms of managing organisational change and new work practices. Interest in industrial humanization, sustainability and resilience is rapidly increased in recent years [22]. Authors highlight that implementing circular business models promotes both resource efficiency and long-term competitiveness, reinforcing the synergy between lean logistics and sustainability objectives [23].

Similar conclusions regarding the link between logistics management and environmental sustainability were drawn by authors, who demonstrated that inefficient material flows and waste logistics can directly impact water quality and regional sustainability performance [24]. Their findings further confirm that lean and standardized logistics systems play a key role in preventing resource loss and supporting responsible environmental management.

Acknowledgements. This article is part of the research project VEGA no. 2/0013/24" Acceptance and use of innovations 4.0 in relation to cognitive gains and load in the context of sustainable development goals."

References

1. Angel, M.: How industry 4.0 is transforming lean manufacturing. Available from: https://tulip.co/blog/how-industry-4-0-is-transforming-lean-manufacturing/. Last accessed 29 Sept 2024
2. Mubarak, M.F., Tiwari, S., Petraite, M., Mubarik, M. and Raja Mohd Rasi, R.Z. How industry 4.0 technologies and open innovation can improve green innovation performance? Manag. Environ. Qual., Vol. 32 5, pp. 1007–1022.2021. doi:https://doi.org/10.1108/MEQ-11-2020-0266 (2021).
3. Richnák, P., Gubová, K.: Green and reverse logistics in conditions of sustainable development in Enterprises in Slovakia. Sustainability. **13**(2), 581 (2021). https://doi.org/10.3390/su1302 0581
4. Tavanti, M.: Understanding Sustainability Innovation. In: Developing Sustainability in Organizations: A Values-Based Approach. Sustainable Development Goals Series, pp. 341–361. Springer (2023). https://doi.org/10.1007/978-3-031-36907-0_11
5. Ding, B., Ferràs Hernández, X., Agell Jané, N.: Combining lean and agile manufacturing competitive advantages through Industry 4.0 technologies: an integrative approach. Production Planning & Control. **34**(5), 442–458 (2021). https://doi.org/10.1080/09537287.2021.193 4587
6. Walter, J.: What is Muda and how to avoid it in the production process, https://www.beewatec.com/en/blog/what-is-muda-and-how-to-avoid-it-in-the-production-process. Last accessed 29 Sept 2024
7. Goodhue, A.: Muda: The daily logistics challenges, https://www.linkedin.com/pulse/muda-daily-logistics-challenges-albert-goodhue/. Last accessed 29 Sept 2024
8. Naziri, L., Mouhib, Z., Gallab, M., Nali, M., Abbou, R., Kebe, A.: Lean and industry 4.0: a leading harmony. Procedia Comput. Science. **200**, 394–406 (2022)
9. Sinton, J.: Using lean thinking approaches to speed sustainable development. In: Sustainable Development: Principles, Frameworks and Case Studies, p. 242. CRC Press, New York (2010)
10. Westrom, D.: Lean Manufacturing and Industry 4.0, https://www.iiot-world.com/smart-manufacturing/lean-manufacturing-and-industry-4–0/. Last accessed 29 Sept 2024
11. Marinelli, M., Deshmukh, A.A., Janardhanan, M., Nielsen, I.: Lean manufacturing and industry 4.0 combinative application: practices and perceived benefits. IFAC-PapersOnLine. **54**(1), 288–293 (2021)
12. Kassem, B., Callupe, M., Rossi, M., Rossini, M., Portioli-Staudacher, A.: Lean 4.0: a systematic literature review on the interaction between lean production and industry 4.0 pillars. J. Manuf. Technol. Manag. **35**(4), 821–847 (2024). https://doi.org/10.1108/JMTM-04-2022-0144

13. Tsukada, O., Ibusuki, U., Kuchii, S., De Almeida, A.T.: Is lean manufacturing maturity a prerequisite for industry 4.0? Survey of SMEs in Japan and Brazil. Int. J. Lean Six Sigma. **15**(5), 1102–1126 (2024). https://doi.org/10.1108/IJLSS-08-2023-0149
14. Knapčíková, L., Martiček, M., Husár, J., Kaščák, J.: An increase in the efficiency of selected production processes using lean tools. Appl. Mech. Mater. **830**, 388–393 (2023)
15. Bajec, P.: The possibility of developing intelligent logistics outsourcing in Slovenia. Transp. Telecommun. J. **14**(4), 344–353 (2013). https://doi.org/10.3846/16484142.2013.829520
16. Fidlerova, H., Makyšová, H., Sklenárová, L., Bajdor, P.: Streamlining packaging as part of sustainable reverse logistics processes. Acta Logistica. **8**(4), 423–433 (2021). https://doi.org/10.22306/al.v8i4.252
17. Klačanová, M.: The proposal for the use of the selected lean production method at workplace of freight converters assembly in ZF Slovakia, a.s. Graduate Thesis, MTF STU 75 (2019)
18. Opoku, R.K., Issifu, R., Ofori, D., Wafa, S., Asiedu, A.: Lean sustainability and triple bottom line performance of manufacturing industries in a developing economy: does top management commitment matter? Int. J. Qual. Reliab. Manage. **41, Iss. 8** (2024). https://doi.org/10.1108/IJQRM-05-2024-0164
19. Jum'a, L., Zimon, D., Ikram, M., Madzík, P.: Towards a sustainability paradigm; the nexus between lean green practices, sustainability-oriented innovation and triple bottom line. Int. J. Prod. Econ. **245**, 2022, 108393 (2021). https://doi.org/10.1016/j.ijpe.2021.108393
20. Khan, I.S., Ahmad, M.O., Majava, J.: Industry 4.0 innovations and their implications: an evaluation from sustainable development perspective. J. Clean. Prod. **405**, 137006 (2023). https://doi.org/10.1016/j.jclepro.2023.137006
21. Eriksson, K.M., Olsson, A.K., Carlsson, L.: Beyond lean production practices and industry 4.0 technologies toward the human-centric industry 5.0. Technol. Sustain. **3**(3), 286–308 (2024). https://doi.org/10.1108/TECHS-11-2023-0049
22. Grabowska, S., Saniuk, S., Gajdzik, B.: Industry 5.0: improving humanization and sustainability of industry 4.0. Scientometrics. **127**(6), 3117–3144 (2022). https://doi.org/10.1007/s11192-022-04370-1
23. Ingaldi, M., Ulewicz, R.: The business model of a circular economy in the innovation and improvement of metal processing. Sustainability. **16**(13), 5513 (2024). https://doi.org/10.3390/su16135513
24. Glova Végsöová, O., Straka, M.: Illegal waste dumps and water quality: environmental and logistical challenges for sustainable development—a case study of the Ružín reservoir (Slovakia). Environments. **12**(8), 251 (2025). https://doi.org/10.3390/environments12080251

Risk Management in Logistics: Digital Twins and Their Role in Operational Threat Assessment

Sylwia Werbinska-Wojciechowska(✉), Robert Giel, and Klaudia Winiarska

Wroclaw University of Science and Technology, Wroclaw, Poland
sylwia.werbinska@pwr.edu.pl

Abstract. Modern logistics systems, integral to global supply chains, face growing complexity and vulnerability to disruptions such as equipment failures, delivery delays, and demand fluctuations. To combat these challenges, digital twins have emerged as essential tools for real-time risk management. This paper explores the role of digital twins in logistics risk management, focusing on operational risk assessment and the application of the Fuzzy Analytical Hierarchy Process method. By leveraging fuzzy logic, Fuzzy AHP enables the prioritization of key digital twin parameters, incorporating expert judgment and uncertainty. The study highlights the benefits of combining digital twins with Fuzzy AHP to improve risk identification, evaluation, and mitigation strategies in dynamic logistics environments.

Keywords: Risk Management · Logistic Systems · Digital Twins

1 Introduction

Modern logistics systems, forming the backbone of global supply chains, face increasing complexity and vulnerability to disruptions such as equipment failures, delivery delays, and cost fluctuations [1]. Managing these risks has become essential, prompting the adoption of advanced technologies like digital twins. These virtual replicas of logistics assets enable real-time monitoring of vehicles, warehouses, and equipment, allowing early detection of issues and reducing downtime [2]. By simulating scenarios like transportation delays or equipment failures, DTs enhance operational efficiency, improve system resilience, and support proactive decision-making. This makes them vital tools for addressing the challenges of modern supply chains [3].

The aim of this paper is to analyze the role of digital twins in risk management in logistics, with a particular focus on their application in operational risk assessment. Indeed, the paper will focus on the application of the Fuzzy Analytical Hierarchy Process method as a tool for selecting the most critical digital twin parameters that should be considered in assessing the risks of logistics systems. The Fuzzy AHP method, by utilizing fuzzy logic, allows for incorporating uncertainty and subjective expert evaluations in decision-making, which is particularly valuable in the dynamic and unpredictable logistics environment.

J. Feliks et al. (Eds.): CLC2024 2024, 2026.
https://doi.org/10.1007/978-3-032-28931-5_14

2 Digital Twin for Risk Assessment in Logistics

Digital twins (DTs) have emerged as transformative tools for managing risks in logistics, offering unparalleled insights into the functioning and vulnerabilities of complex supply chain systems. In logistics, risk can stem from numerous factors, including equipment failures, delays in delivery schedules, fluctuating demand, or adverse environmental conditions. Traditional risk management tools often fall short in their ability to provide real-time insights and predictive capabilities [4]. Digital twins, however, bridge this gap by offering [3]: real-time monitoring, scenario simulation, and predictive analytics. By employing DTs, logistics managers can shift from reactive responses to proactive risk management, thereby minimizing downtime, reducing financial losses, and enhancing overall system resilience.

2.1 Key Parameters for Risk Assessment in Digital Twins

In the context of logistics, digital twins (DTs) are invaluable tools for assessing and managing risks within complex systems. Their ability to monitor, analyze, and simulate various aspects of logistics operations enables proactive decision-making and effective risk mitigation. However, the effectiveness of digital twins in risk assessment depends largely on the key parameters they monitor and analyze. Identifying and prioritizing these parameters is essential for maximizing the value of digital twin technology in identifying potential threats and vulnerabilities within logistics networks [3].

Key parameters for risk assessment in digital twins include equipment performance, equipment health monitoring, delivery delays, inventory levels, temperature control, external factors, and cost efficiency. **Equipment performance** focuses on tracking KPIs such as speed, fuel consumption, and load capacity to detect inefficiencies or potential failures. Early identification of anomalies enables preventive maintenance, reducing downtime. **Equipment health monitoring** involves using sensors to measure variables like vibration and temperature, predicting maintenance needs, and avoiding unexpected breakdowns.

Delivery delays are critical in supply chains. Digital twins analyze real-time data on routes and conditions, forecasting disruptions and suggesting alternatives. This is essential for time-sensitive shipments, minimizing financial losses and customer dissatisfaction.

Inventory levels and resource allocation are optimized through real-time monitoring, allowing prediction of demand patterns and avoiding stockouts or overstocking. Temperature control, vital in industries like food and pharmaceuticals, ensures products remain within safe conditions using sensor data to prevent spoilage.

External factors, such as weather or geopolitical issues, are assessed through data integration, enabling contingency planning for disruptions. Finally, **cost efficiency** evaluates the financial impact of risks and mitigation strategies, guiding resource allocation and operational improvements. These parameters collectively support robust risk management in logistics through proactive decision-making and optimization.

2.2 Linking Parameters to Risk Identification

In logistics, the relationship between key parameters monitored by digital twins and the identification of risks is crucial for proactive risk management. By analyzing and linking specific parameters to potential risks, organizations can anticipate disruptions, mitigate threats, and optimize their response strategies. Understanding how each parameter influences the likelihood of risks allows logistics managers to implement more precise and effective risk mitigation measures. Table 1 illustrates how specific parameters monitored by digital twins contribute to identifying various risks within logistics systems.

Table 1 Key parameters for risk identification in logistics

Parameter	Exemplary KPIs	Risk identified	Impact on logistics
Equipment performance	speed, fuel consumption, load capacity, operating temperature	Equipment failure, inefficiency	Potential breakdowns, operational delays, increased maintenance costs
Equipment health monitoring	Vibration, pressure, temperature, usage hours	Mechanical failure, wear and tear	Unexpected downtime, need for unplanned maintenance, reduced asset lifespan
Delivery delays	On-time delivery rate, transportation time, route efficiency	Shipment delays, late deliveries	Customer dissatisfaction, penalties, missed deadlines
Inventory levels	Stock quantity, inventory turnover, order fulfillment rate	Stockouts, overstocking, demand–supply imbalance	Supply chain disruptions, financial losses, storage inefficiencies
Resource allocation	Labor allocation, machine utilization, production schedules	Resource shortages, under/overutilization	Operational inefficiency, increased labor costs, delays in delivery
Temperature control	Temperature stability, humidity levels, refrigeration status	Product spoilage, non-compliance with safety standards	Loss of goods, quality degradation, increased waste or spoilage
Environmental and external factors	Weather conditions, traffic data	Disruptions due to external conditions (e.g., weather, strikes)	Route delays, increased transportation costs, cargo damage
Cost efficiency	Operational cost per unit, resource cost, total logistics cost	Rising operational costs, poor cost management	Reduced profitability, unsustainable operations, inefficiency

Monitoring these parameters with digital twins provides real-time insights into potential disruptions, enabling logistics managers to take corrective actions before issues escalate into significant risks.

3 Selection of Key Parameters for Monitoring in Digital Twin Systems to Mitigate Operational Risks Based on the Fuzzy AHP Method Use

In logistics systems based on Digital Twin technology, various parameters significantly impact the identification and management of risks. However, due to the dynamic nature of logistics operations, it can be challenging to determine which parameters are most crucial to monitor at any given moment. Furthermore, data-related uncertainties and experts' subjective judgment complicate the decision-making process. Fuzzy AHP allows decision-makers to incorporate uncertainty and subjective expert evaluations, which is essential in the logistics environment where data is often incomplete or subject to significant variability. FAHP is widely applied in many areas. For recent survey, see, e.g., [5].

In the presented paper, the authors introduce a fuzzy AHP method (FAHP) based on fuzzy triangular numbers to select the key parameters for monitoring in DTs, considering the three main criteria: **scalability, predictability,** and **management costs**. The fuzzy AHP allows users to structure complex problems in the form of a hierarchy or a set of integrated levels. Its implementation is usually based on three main steps [6]. The adopted methodology is presented in Fig. 1.

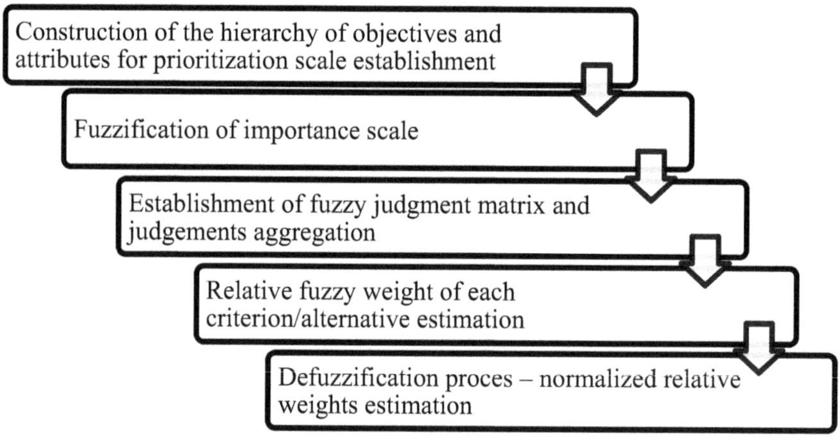

Fig. 1 Adopted fuzzy AHP methodology

In order to develop a fuzzy pairwise comparison matrix, the evaluation of linguistic terms has to be established. For the proposed approach, fuzzy triangular numbers represent all elements in the judgment matrix and weight factors. Moreover, the arithmetic operation laws for any two triangular fuzzy numbers are described, e.g., in [7]. Table 2 shows the fuzzification process for the linguistic variables.

Based on the scale defined in Table 2, pairwise comparisons of the elements of a particular level are performed. Usually, an element that receives a higher rating is viewed as superior compared to another one that receives a lower rating. The comparisons are used to form a matrix of pairwise comparisons called the judgment matrix $\boldsymbol{A^d}$ and is obtained for every expert [8]:

$$A^d = \begin{bmatrix} a_{11}^d & \cdots & a_{1j}^d \\ \vdots & \ddots & \vdots \\ a_{i1}^d & \cdots & a_{ij}^d \end{bmatrix} \tag{1}$$

Following [9], matrix A is a positive and reciprocal matrix.

Table 2 Fuzzy triangle scale (based on [8])

Linguistic scale	Fuzzy scale	Reciprocal fuzzy scale
Equally important (EI)	(1, 1, 1)	(1, 1, 1)
Weakly more important (WI)	(1, 3/2, 2)	(1/2, 2/3, 1)
Strongly more important (SI)	(2, 5/2, 3)	(1/3, 2/5, 1/2)
Very strongly more important (VI)	(3, 7/2, 4)	(1/4, 2/7, 1/3)
Absolutely more important (AI)	(4, 9/2, 9/2)	(2/9, 2/9, 1/4)

Later, in order to obtain a compromise between the k decision-makers, the matrices $\boldsymbol{A}$ have to be aggregated into a single matrix, which is the aggregated fuzzy matrix for pairwise comparisons. For this purpose, the fuzzy numbers a_{ij}^d expressed by each decision-maker will be aggregated using the geometric mean.

The next step (according to Fig. 1) is connected with the relative fuzzy weight of each criterion or alternative estimation. Thus first, the geometric mean of fuzzy comparison values of each criterion/alternative is calculated, according to:

$$r_i = \left(\prod_{j=1}^{n} a_{ij}\right)^{1/n}, i = 1, 2, \ldots, n \tag{2}$$

Based on this estimation, the fuzzy weight of criterion/alternative i (w_i) may be estimated:

$$w_i = r_i \otimes (r_1 \oplus r_2 \oplus \ldots \oplus r_n)^{-1} \tag{3}$$

where w_i are still fuzzy triangular numbers.

The last step regards the defuzzification process. For the purposes of this work, the Centre of area method is applied [10]:

$$M_i = \frac{lw_i + mw_i + uw_i}{3} \tag{4}$$

The M_i is a non-fuzzy number, but it needs to be normalized according to the equation:

$$N_i = \frac{M_i}{\sum_{i=1}^{n} M_i} \tag{5}$$

The proposed methodology is to be performed to find the normalized weights of both criteria and the alternatives. Then, the scores for each alternative are calculated by multiplying each alternative weight with related criteria. Finally, based on the obtained results, the best solutions for the decision-maker can be defined.

4 Case Study

- The Fuzzy AHP method can be applied to risk management across different levels of the supply chain, specific links, or subsystems within an organization. This approach enables the prioritization and selection of key parameters for monitoring and mitigating risks at various stages of the logistics process. In our case, we focus on assessing the risks associated with Autonomous Mobile Robots (AMR) operating within a warehouse environment. By using Fuzzy AHP, we can identify and evaluate the most critical factors impacting the AMR's operation, helping to improve risk management. Following the expert opinions, there have been preliminary selected three key parameters, represented by selected five KPIs. A short overview is provided in Table 3. The last important issue is connected with the selection criteria definition. Again, the experts preliminarily defined three basic criteria for the selection process. These criteria are:
- ***Scalability***—refers to the ability of the system to adapt and scale with changing operational conditions,
- ***Predictability***—evaluates how well a parameter allows for the prediction of future risks,
- ***Risk management costs***—refers to the costs involved in monitoring and managing a particular parameter.

As a result, it was possible to develop a hierarchy structure for the analyzed problem (Fig. 2). Finally, the output of the FAHP is developed to suggest appropriate settings in prioritizing of KPIs to be monitored based on the agreed three criteria.

After completing the first four steps of the methodology, the geometric mean of fuzzy comparison values of each criterion and alternative is calculated based on Eq. (2). An example of calculations for the "Scalability" criterion (r_1) is shown below:

$$r_i = \left[(1*1*2)^{\frac{1}{3}}; \left(1*\frac{3}{2}*\frac{5}{2}\right)^{\frac{1}{3}}; (1*2*3)^{\frac{1}{3}} \right] = [1.26; 1.55; 1.82] \tag{6}$$

This gives the possibility to estimate the fuzzy weight of each criterion and alternative (following Eq. (3)). An example of calculations for the "Scalability" criterion (w_1) is shown below:

$$w_1 = [(1.26*0.365); (1.55*0.305); (1.82*0.257)] = [0.324; 0.473; 0.663] \tag{7}$$

Table 3 Selected key parameters with KPIs

Parameter	Selected KPIs	Description	No
Equipment performance	Task completion rate	Measures the percentage of tasks successfully completed by the AMR compared to planned tasks. A low completion rate may indicate operational inefficiencies, breakdowns, or issues with the robot's capability to handle specific tasks, thus posing a risk to overall system performance	K1
	Mean time between failures (MTBF)	Measures the average time between consecutive failures of an AMR. This KPI is essential for assessing how reliable the equipment is over time	K2
Equipment health monitoring	Condition monitoring index	A composite index or score that aggregates data from various sensors to provide an overall health status of the AMR. This includes metrics such as temperature, vibration, and pressure to monitor the robot's condition	K3
	Component wear and tear rate	Measures the rate at which key components of the AMR, such as wheels, motors, and sensors, wear down over time. It helps to track the aging process of individual parts	K4
External factors	Traffic and obstacle density	The amount of traffic and the density of obstacles (such as other AMRs, humans, or machinery) in the AMR's path can directly impact its ability to perform tasks efficiently. High traffic or obstacle density increases the likelihood of delays, collisions, and operational disruptions	K5

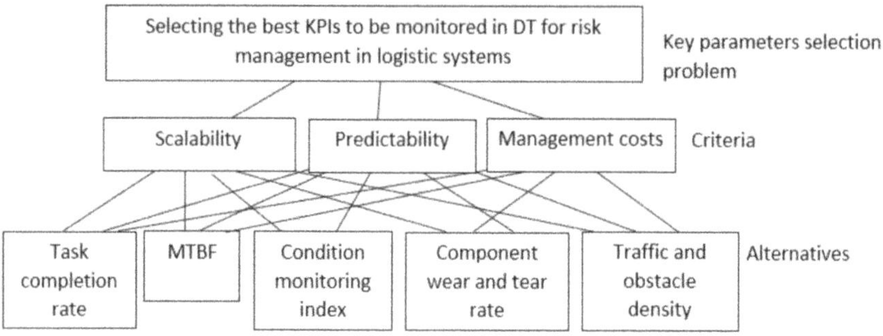

Fig. 2 Hierarchy for key parameters for monitoring selection problem

Hence, the w_i is still a fuzzy number; the defuzzification and normalization processes are needed (according to Eqs. (4–5)). The normalized weights of each criterion and alternative are calculated and tabulated in Table 4. The final score is also estimated.

Table 4 Aggregated results for each alternative according to each criterion

Criteria		Scores of alternatives with respect to related criterion				
	Weights	K1	K2	K3	K4	K5
Scalability	0,4675	0,1555	0,1917	0,3780	0,1466	0,1282
Predictability	0,3664	0,1334	0,0930	0,4277	0,2313	0,1147
Management costs	0,1661	0,1548	0,1912	0,3780	0,1477	0,1283
Total		**0,147**	**0,155**	**0,396**	**0,178**	**0,123**

The highest score of the alternatives gives the idea about the appropriate maintenance resource selection for its sharing within the factories. The order of ranking the alternatives using the fuzzy AHP method results as follows:

$$K3 > K4 > K2 > K1 > K5$$

According to the FAHP results, the most important KPI for risk management in AMR operations is the **Condition Monitoring Index.** The **Condition Monitoring Index** ranks highest because it directly reflects the real-time health of the AMR's key components, making it essential for the early identification of potential failures. This aligns well with **Digital Twin (DT)** technology, which excels at continuous monitoring and providing actionable insights into the asset's status. A comprehensive condition monitoring system allows for more accurate risk assessment and proactive maintenance, minimizing unexpected breakdowns.

5 Conclusions

This paper proposes the Fuzzy AHP method for selecting Key Performance Indicators (KPIs) in DT monitoring and risk management. The aim is to identify critical KPIs that support proactive risk mitigation and enhance logistics efficiency. Proper KPI selection is essential for the success of DT-based systems, ensuring effective monitoring of assets and processes.

The FAHP method effectively prioritizes KPIs by incorporating uncertainty and expert judgment. However, limitations such as defuzzification techniques and external factor integration suggest areas for improvement. Future research will focus on enhancing FAHP's accuracy and relevance for DT-driven logistics risk management.

References

1. Tubis, A.A., Werbińska-Wojciechowska, S.: Risk management maturity model for logistic processes. Sustainability (Switzerland) **13**(2), 1–19 (2021)
2. Sharma, A., Kosasih, E., Zhang, J., Brintrup, A., Calinescu, A.: Digital twins: state of the art theory and practice, challenges, and open research questions. J. Indus. Inform. Integrat. 100383 (2022)
3. Ivanov, D., Dolgui, A.: A digital supply chain twin for managing the disruption risks and resilience in the era of industry 4.0. Product. Planning Control **32**(9), 775–788 (2021)
4. Pournader, M., Kach, A., Talluri, S.: A review of the existing and emerging topics in the supply chain risk management literature. Decis. Sci. **51**(4), 867–919 (2020)
5. Kubler, S., Robert, J., Derigent, W., Voisin, A., Le Traon, Y.: A state-of the-art survey and testbed of fuzzy AHP (FAHP) applications. Expert Syst. Appl. **65**, 398–422 (2016)
6. Durán A.O.: Spare parts criticality analysis using a fuzzy AHP approach. Tehnicki Vjesnik **22**(4), 899–905 (2015)
7. Wang, Z.J.: Consistency analysis and priority derivation of triangular fuzzy preference relations based on modal value and geometric mean. Inform. Sci. **314**, 169–183 (2015)
8. Chang, D.Y.: Applications of the extent analysis method on fuzzy AHP. Europ. J. Operat. Res. **95**(3), 649–655 (1996)
9. Shiraishi, S., Obata, T., Daigo, M.: Properties of a positive reciprocal matrix and their application to AHP. J. Operat. Res. Soc. Japan **41**(3), 404–414 (1998)
10. Ayhan, M.B.: A fuzzy Ahp approach for supplier selection problem: a case study in a gear motor company. Int. J. Managing Value Supply Chains **4**(3), 11–23 (2013)

Enhancing Manufacturing Logistics Efficiency through 3D Simulation in Visual Components

Simona Spirkova, Martin Straka, and Jakub Kovalcik(✉)

Technical University of Kosice, Institute of Logistics and Transportation, Košice, Slovakia
{simona.spirkova,martin.straka,jakub.kovalcik}@tuke.sk

Abstract. The increasing complexity of manufacturing logistics demands innovative solutions to optimize workflows, reduce costs, and improve efficiency. Visual Components, a 3D simulation software, has emerged as a powerful tool for visualizing and refining logistics operations within manufacturing environments. This article explores the use of Visual Components for simulating logistics workflows, focusing on its capacity to model complex systems like conveyor networks and robotic handling. Through practical applications and case studies, we examine how this software enables manufacturers to virtually test and optimize layouts, streamline material flows, and predict bottlenecks before physical implementation. The benefits of 3D simulation, including enhanced collaboration, reduced design errors, and data-driven decision-making, underscore its value in logistics optimization. We also discuss future trends, such as the integration of AI and digital twins, which are set to further transform the role of 3D simulation in manufacturing logistics.

Keywords: Robotic workplace · Designing · Simulation · Manufacturing logistics · Efficiency

1 Introduction

The progression of the Industrial revolution has driven transformative changes across all areas of society. However, as new social priorities like sustainability, human-centered approaches, and carbon reduction emerge, the traditional manufacturing system has struggled to meet the needs of an innovative, forward-thinking society [1]. Recently, forward-looking concepts such as Industry 5.0 and Society 5.0 have gained significant attention. The European Commission introduced Industry 5.0 as a vision for a sustainable, human-centered, and resilient European industry. This paradigm looks beyond productivity, aiming for a manufacturing system that is adaptable, inclusive, and responsive to human needs. Industry 5.0 emphasizes agility and resilience by harnessing flexible technologies that respect environmental limits and foster diversity, talent, and empowerment [2, 3]. The innovative Industry 5.0 vision seeks to bridge the gap between manufacturing practices and societal needs. Unlike previous industrial revolutions, which

J. Feliks et al. (Eds.): CLC2024 2024, 2026.
https://doi.org/10.1007/978-3-032-28931-5_15

prioritized economic gains, Industry 5.0 focused on aligning industry with human-centered values and societal well-being [1]. Advances in human-centric smart manufacturing (HSM) reflect a growing focus on integrating humans into technological processes to address the complexities of human-machine interaction. In this context, human-cyber-physical systems (HCPS) offer a promising human-centered framework that can inform the development and application of HSM. Regardless of perspective—whether humans are actively involved (human-in-the-loop) or indirectly engaged (human-out-of-the-loop)—advanced manufacturing technologies are designed by humans, serve human needs, and operate in collaboration with humans. Therefore, it is essential to incorporate both human and cyber elements within smart manufacturing systems, rather than positioning humans as external to these systems [4].

2 Literature Review

Manufacturing, Manufacturing, much like society itself, is experiencing a rapid transformation driven by widespread digitization. Routine, low-skill activities are increasingly delegated to intelligent production systems capable of automating processes that once required human adaptability and decision-making. A major milestone in this evolution is human–robot collaboration (HRC), where people and robots operate side by side to accomplish shared objectives. In more complex industrial settings, this interaction is supported by advanced methods of human–robot interaction (HRI). Automation and robotics have therefore become fundamental to the concept of smart manufacturing, with HRC recognized as a critical enabler of higher productivity and efficiency in both present and future production environments. According to [5], HRC can be classified into three principal forms:

- coexistence, in which humans and robots share or partially share a workspace without pursuing a common goal;
- cooperation, where both work toward the same objective within a shared or partially shared area;
- collaboration, when humans and robots simultaneously perform tasks together in the same space.

The need to redesign production systems has grown as industries adjust to shifting external conditions such as fluctuating demand, market competition, and evolving performance targets [6]. Careful factory-layout planning can lower operational costs by as much as 20–50%, a decisive advantage for maintaining competitiveness. A frequent challenge in facility design involves expanding existing layouts to integrate new equipment or personnel. The quality of this planning directly affects overall efficiency and the plant's ability to respond to changing requirements [7]. Including real-time human performance data in production planning enables rapid adjustments to meet specific goals. Present-day simulation tools, however, still focus primarily on technical aspects such as large-scale system handling, scene modelling, data analysis, and visualization [8]. Such tools let engineers test multiple scenarios without interrupting live operations. Their primary purpose is to evaluate system performance or to assess the implications of process modifications and the integration of new equipment [7]. Digital manufacturing technologies play a pivotal role in cutting product-development time and costs while supporting a broader range of customization. Simulation-based approaches form the core of these digital strategies because they allow experimentation and validation of different product, process, and manufacturing configurations [9]. Within the Industry 4.0 framework, computer simulations combined with virtual-reality systems and real operational data underpin the creation of digital-twin solutions. Over recent decades, computer simulation has become one of the most effective decision-support tools in logistics, with 3D dynamic simulations proving especially valuable for detecting and preventing costly design or operational errors [10].

Three-dimensional modelling itself draws on fields such as computer graphics, intelligent information processing, computer vision, and artificial intelligence. By capturing 3D data from target objects and processing it through digital technologies, these methods enable accurate reproduction and analysis, making 3D modelling an essential component of contemporary logistics engineering [11].

3 Methodology

Global interconnectivity and the complexity of modern supply chains bring numerous challenges. Key among these is limited visibility, which hinders accurate monitoring and optimization; unpredictability, which requires strong contingency planning and risk management; and relentless consumer-driven pressures, underscoring the need for flexibility and responsiveness. To remain competitive, businesses must make swift and decisive decisions, highlighting the growing importance of modelling logistics systems and related processes [12]. As computer technology continues to mature and advance, there is growing demand for experiences beyond the limitations of two-dimensional interfaces, leading to the increasingly widespread application of three-dimensional modelling technology [11]. Today, simulation systems are seen as essential tools not only in production logistics but throughout the entire logistics process. They are especially valuable for harmonizing workflows, analyzing and closely monitoring production, and supporting effective planning and management. Simulation systems can generally be categorized into three types:

- live (these involve employees operating in real time within both the simulation process and the actual system being studied),
- virtual (the most used today, virtual simulations model both personnel and systems within a software-based simulation, rather than in the real environment),
- constructive (in this type, simulated individuals operate within the model but do not influence outcomes, as they are not actively involved in determining the results) [13].

3.1 Visual Components

Visual Components is a leading developer of 3D manufacturing simulation software, widely regarded as a global leader in the industry. The software includes a library of over 4000 components from more than 30 major brands in industrial automation, and users can import CAD files into the 3D environment. Visual Components also allows seamless integration with control systems, thanks to its open architecture, which supports platform customization. Built on.NET, it provides a familiar framework for developers, along with a Phyton API to enable extensive customization—from the user interface to simulation behaviors. In essence, Visual Components is designed for flexibility, allowing simulations to be tailored to specific needs [14]. As one of the few simulation software, it also offers the option of VR simulation, which is a huge advantage these days. In recent years, the integration of VR (virtual reality) technology and 3D simulations has caused a paradigm shift and transformed traditional approaches into dynamic, immersive experiences. Today VR, AR (augmented reality) and ER (extended reality), are present in different places and in different situations. They change the world around us with a click, a tap or even a glance.

4 Case Study

The study compares simulation outcomes of a specific workplace using two different simulation software platforms, ExtendSim and Visual Components, assessing the effectiveness of 3D simulation versus a 2D simulation model. The workplace under study involves milling aluminium blocks into aluminium cylinders, which then move to the next station as a semi-finished product. The workflow is illustrated in Fig. 1, detailing each step: input material is transported via conveyor no. 1 and robot to the milling machine, where processing occurs, same robot then transfers the finished cylinders onto another conveyor, from which human operator moves them onto pallets destinated for the next workstation.

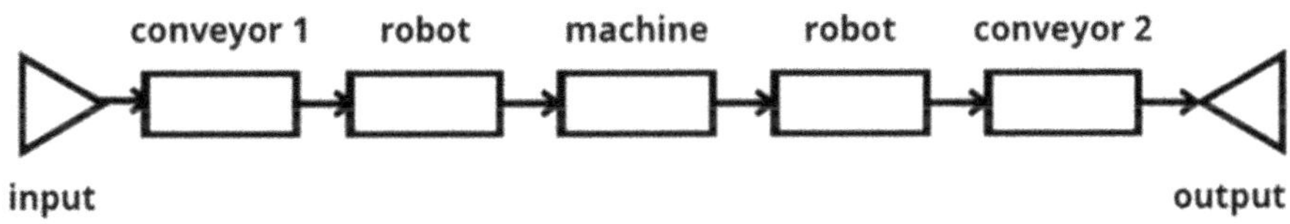

Fig. 1. Scheme of processes and operations

4.1 2D Simulation Model

The 2D simulation model was developed in ExtendSim, with slight modifications from the original scheme by adding cues before each activity. In this model, block Create (Input) represents the input generator for the production process, while Activity (Conveyor 1, Robot, Machine, Robot, Conveyor 2) blocks define handling operations such as material transport via conveyors, component repositioning with a robot, and the milling process itself. Additionally, Queue block (unnamed) simulates potential queues that may form during the process. We also encounter the Exit block (Output), which represents the manufactured products. The model also includes and Executive block (clock in the left corner), essential for running simulation, and a Plotter DE block (colored table at the end of process), which records selected data – specifically, the number of processed products and queue values before the robot picks up input material. The simulation ran for and equivalent of one 8-h work shift, showing that the system can process 1307 products within this time interval (Figs. 2 and 3).

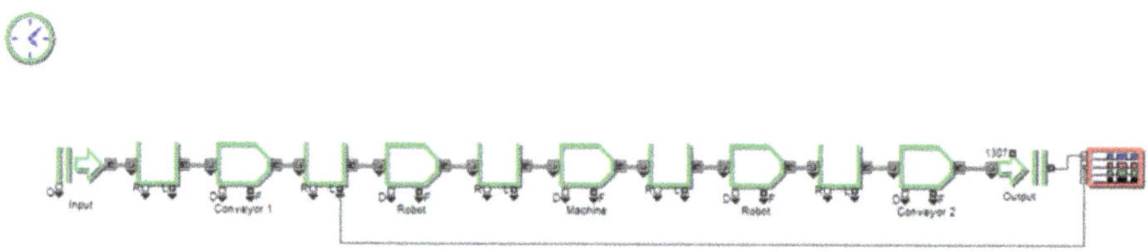

Fig. 2. 2D simulation model

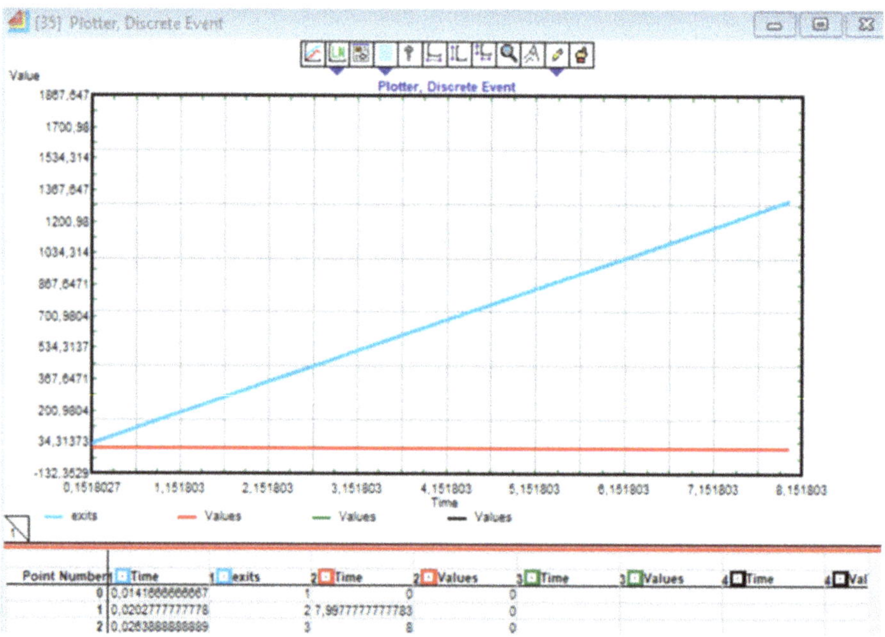

Fig. 3. 2D simulation model results

4.2 3D Simulation Model

The process of creating a 3D simulation model for the production process differs significantly from that of 2D model. For the 3D setup, Belt Conveyor components from the library were placed on the workspace to represent the workplace conveyors. The ARC-2000i component was selected as a robot, and a Robot Transport Controller was added to control its movements. A Generic 3-Jaw Gripper tool was attached to the robot's arm to handle and move materials. For the milling process, the ProMill component was chosen to represent the milling process. A human operator, labelled as Human Operator Otto, was assigned, along with a Human Transport Controller to configure his movements. Additionally, a Feeder component was used as the input generator for the production process, with aluminium blocks and cylinders incorporated to reflect material changes graphically. The final component, a Europallet, represents the pallet for stacking finished products. The simulation ran for 8-h work shift, as with the 2D model. This 3D simulation showed a slight increase in output, processing 1 312 products within the observed time, which falls within expected variation (Figs. 4 and 5).

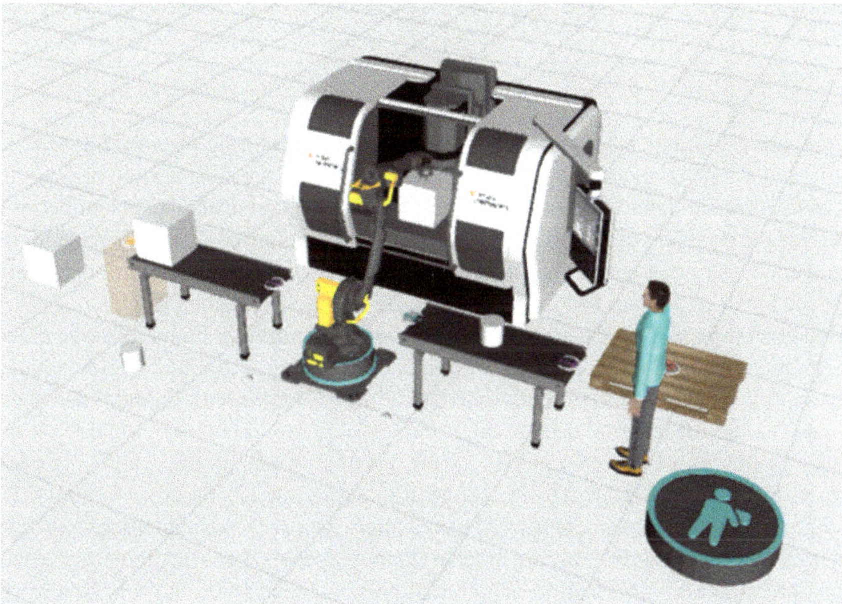

Fig. 4. 3D simulation model

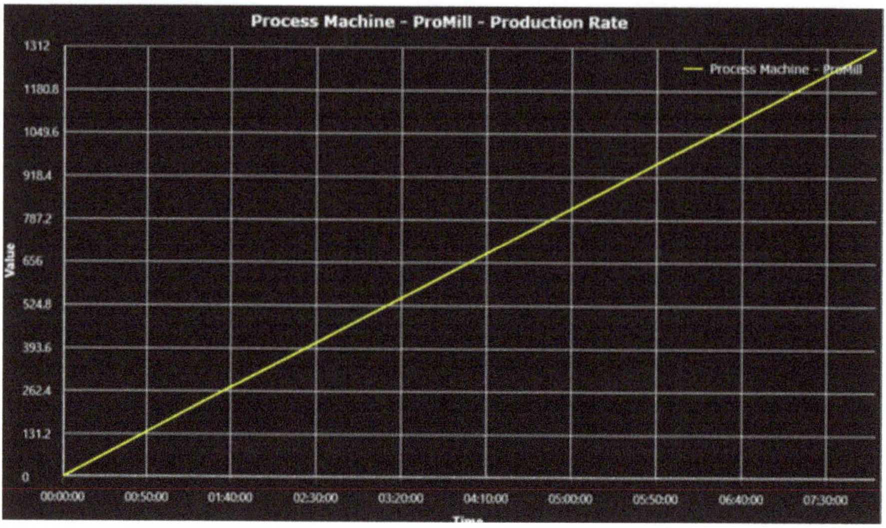

Fig. 5. 3D simulation model results

5 Conclusion

In comparing 3D and 2D simulation models for manufacturing logistics, 3D models clearly stand out for their ability to offer a comprehensive, realistic view of logistical processes, making them far more effective in identifying and resolving potential bottlenecks, spatial conflicts, and inefficiencies. Unlike 2D simulations, which provide limited insight due to their flat representation, 3D models in Visual Components allow stakeholders to see logistics operations in an immersive environment, enhancing decision-making and enabling more accurate planning. This level of detail supports better resource utilization, improved workflow accuracy, and minimized errors prior to actual implementation. The adoption of 3D simulation is also paving the way for the rise of digital twins, which represents a significant trend in manufacturing logistics. Digital twin—virtual replicas of physical systems—offer real-time monitoring and analysis capabilities, allowing manufacturers to optimize and adapt logistics operations continuously based on live data. This trend aligns with the movement toward increasingly data-driven and predictive systems, which are key in a rapidly evolving industrial landscape. As manufacturers seek greater agility and efficiency, the shift to 3D simulation and digital twins is set to become the new standard, driving the next wave of innovation in manufacturing logistics.

Acknowledgement. The submitted paper is a part of the project 010TUKE-4/2023 Application of educational robots in the process of teaching the study program industrial logistics.

References

1. Leng, J., Sha, W., Wang, B., Zheng, P., Zhuang, C., Liu, Q., Wuest, T., Mourtzis, D., Wang, L.: Industry 5.0: Prospects and retrospect. J. Manuf. Syst. **65**, 279–295 (2022)
2. Breque, M., De Nul, L., Petridis, A.: Industry 5.0: towards a sustainable, human-centric and resilient European Industry. European Commission, Brussels (2021)
3. Huang, S., Wang, B., Li, X., Zheng, P., Mourtzis, D., Wang, L.: Industry 5.0 and Society 5.0—Comparison, complementation and co-evolution. J. Manuf. Syst. **64**, 424–428 (2022)
4. Wang, B., Zheng, P., Yin, Y., Shin, A., Wang, L.: Toward human-centric smart manufacturing: a human-cyber-physical systems (HCPS) perspective. J. Manuf. Syst. **63**, 471–490 (2022)
5. Arents, J., Abolins, V., Judvaitis, J., Vismanis, O., Oraby, A., Ozols, K.: Human-robot collaboration trends and safety aspects: a systematic review. JJ. Sens. Actuator Netw. **10** (2021)
6. Herr, D., Reinhard, J., Reina, G., Krüger, R., Ferrari, R.V., Ertl, T.: Immersive modular factory layout planning using augmented reality. Procedia CIRP **72**, 1112–1117 (2018)
7. Baroroh, D.K., Chu, C.: Human-centric production system simulation in mixed reality: an exemplary case of logistic facility design. J. Manuf. Syst. **65**, 146–157 (2022)
8. Attajer, A., Darmoul, S., Chaabane, S., Riane, F.: Benchmarking simulation software capabilities against distributed control requirements: FlexSim versus AnyLogic. In: Proceeding of the International Workshop on Service Orientation in Holonic and Multi-Agent Manufacturing, pp. 520–531. Springer, Cham (2021)
9. Mourtzis, D., Papakostas, N., Mavrikios, D., Makris, S., Alexopoulos, K.: The role of simulation in digital manufacturing: applications and outlook. Int. J. Comput. Integr. Manuf. **28**, 3–24 (2015)
10. Ashrafian, A., Pettersen, O., Kuntze, K., Franke, J., Alfnes, E., Henriksen, K., Spone, J.: Full-scale discrete event simulation of an automated modular conveyor system for warehouse logistics. In: Advances in Production Management Systems. Towards Smart Production Management System, pp. 39–48. Springer, Cham (2019)
11. Liu, D., Pan, Y., Li, L.: Logistics engineering simulation using computer 3D modeling technology. J. Phys. Conf. Ser. **2143**, 012018 (2021)
12. Jackson, I., Jesus Saenz, M., Ivanov, D.: From natural language to simulations: applying AI to automate simulation modelling of logistics systems. Int. J. Prod. Res. **64**, 1434–1457 (2024)
13. AcqNotes: Modeling & Simulation. Simulator types, https://acqnotes.com/acqnote/tasks/simulator-types, last accessed 2024/11/05
14. Halldor, A.: Thesis Work: Digital twin simulation with Visual Components. The Arctic University of Norway. Faculty of Engineering Science and Technology, Department of Industrial Engineering. https://munin.uit.no/handle/10037/18109. Accessed 05 Nov 2024

Cause-Effect Analyses in the Operation of AGV Systems

Honorata Poturaj(✉)

Department of Technical Systems Operation and Maintenance, Wroclaw University of Science and Technology, Faculty of Mechanical Engineering, Wrocław, Poland
honorata.potruaj@pwr.edu.pl

Abstract. This research applied cause-effect analysis to enhance the operation of an Automated Guided Vehicles (AGV) system. AGVs, widely used for internal transport, follow preprogrammed routes and perform repetitive tasks. However, AGV still require continuous monitoring due to potential adverse events. Adverse events in AGV systems can be caused by unexpected conditions, failures, or a lack of information. Response to only these events is insufficient. It is crucial to analyze both causes and effects to improve the system and mitigate future adverse events. The research focused on analyzing the causes and effects of an adverse event in an AGV system used in an automotive company. Two methods were applied: the Ishikawa diagram and the Bowtie analysis. Data were collected from employee interviews, the AGV system database, and technical reports. The Ishikawa diagram helped identify the detailed causes of the adverse event and highlighted potential risks that could lead to its recurrence. The Bowtie analysis also provided insights into the event's effects. The Bowtie analysis, combined with the Ishikawa diagram, gave a thorough understanding of the consequences of the event, with a focus on AGV operation. The study identified controls and actions that could be implemented to prevent similar issues.

This approach allows for both preventive and recovery actions and can be applied to other logistics systems, which use automated transport solutions, leading to improved efficiency and reduced risk.

Keywords: Automated Guided Vehicles · AGV · logistics

1 Introduction

Automated guided vehicles (AGV) have become a commonly used transport solution in internal logistics in recent years. Under the term AGV, you can find many types of vehicles adapted to specific operational tasks. The popularization of their use results both from technological progress over the last few decades and from the growing costs of personnel employment and recruitment requirements for operators. The observed progress in the advancement of automatic solutions is a result of the fourth industrial revolution and the concept of Logistics 4.0 enabling the development of cyber-physical systems [1, 2]. Implementation of automated guided vehicles is an investment, and proper installation and efficient response to occurring inconsistencies allows to avoid costs related to maintenance and repairs, downtime costs, or costs of emergency scenarios.

J. Feliks et al. (Eds.): CLC2024 2024, 2026.
https://doi.org/10.1007/978-3-032-28931-5_16

AGVs have become a key component of automated material handling systems. Proper design and implementation allow AGVs to be tailored to the specific conditions and needs of a company [1, 2]. However, the long-term success of an AGV system depends on regular maintenance, adequate monitoring, and willingness to adjust when necessary. The ongoing monitoring and analysis of AGV performance are crucial to ensuring reliability, reducing downtime, and optimizing workflows [2, 3].

The operational phase of AGV systems presents unique challenges due to their interactions with the production environment. Issues such as equipment wear, the complexity of other systems, and unforeseen situations all impact performance and require continuous evaluation. Understanding how AGVs interact with each other and their environment is essential to achieving optimal performance levels. A detailed overview of issues related to risk related to AGV operation is presented in [4].

The purpose of this paper is to present a procedure to analyze the causes and effects of an adverse event in an AGV system used in an automotive company. Presented analysis are possible by accessing the data entered by employees into the AGV management system. Equally important were the interviews with employees. In the first part of the article, general information about the AGV system was presented. The next part contains a description of the cause-and-effect analysis performed. The last chapter of the work is a summary of the research described.

2 Analyzed System

Transport tasks are carried out by AGVs classified as unit load or platform with custom superstructures for compatibility with the production line, shown in Fig. 1. Vehicles transport engine components between stations in the foundry, machining, and assembly areas. Vehicles move using LIDAR navigation supported by magnetic tape or line following navigation with QR code markers. LIDAR provides localization within 1 cm, while magnetic tape or a line following with QR codes are used for tasks that require greater precision of positioning. AGVs are powered by batteries with a runtime of 10–12 h, which are manually replaced or charged at designated stations. The company uses several vehicles that form a fleet that works based on automatically generated tasks assigned to a selected AGV. AGVs are equipped with sensors to detect obstacles and maintain safety. However, the operating environment poses challenges, such as dirt and oil residue, that affect the functionality and longevity of vehicles.

The analyzed period of operation was 7 months and 11 days. Historical data was collected by the company's employees and entered the IT system that supports the service process. The information included only the name of the adverse event and the date of occurrence.

3 Cause-Effect Analyses

Reported adverse events were grouped by number of occurrences. "Driving fault" was reported 1992 times in the analyzed period and has been selected for a thorough analysis. Identification of causes was possible due to the use of the Ishikawa diagram, named after Prof. Kaoru Ishikawa of the University of Tokyo, who introduced this methodology

Fig. 1. AGVs with superstructures operating in the described system

in 1940. The diagram is often called a fishbone diagram because it resembles a fish skeleton when drawn. The effect is marked in the head of the fish, while the bones represent potential causes [3]. For the purposes of the analysis, the causes' categories are consistent with the concept of 5 M + E [4]. In the case under review, the following categories are understood as follows:

- Manpower, employees of the company operating the AGV system,
- Machine—Automated Guided Vehicle used in company,
- Material—AGV components and elements necessary for proper operation of the system,
- Method—procedures and instructions for AGV system,
- Management—management and decision-making staff of the AGV system,
- Environment—AGV's workspace and its components.

Figure 2 shows the Ishikawa diagram developed for the "driving fault" event. The analysis of potential effects of the event analyzed was carried out analogously to the analysis of causes, again using the categories described according to the concept of 5 M + E. The schematic representation of the potential effects is like the Ishikawa diagram and is presented in Fig. 3. "Driving fault' event was moved to the left side and the arrows of the main axis, the category axis and the effects were reversed.

The next stage of the research was to use the Bowtie method. The origin of this method is unclear, it is based on the "Swiss cheese model" and was first mentioned in 1979 in the ICI plc HAZAN Course Notes presented at the University of Queensland, Australia. The Bowtie method allows for an understandable visualization that considers possible causes and preventive measures, as well as potential effects and recovery measures [4]. A scheme of the Bowtie method is shown in Fig. 4. Preventive and recovery measures have been proposed and are listed in Table 1.

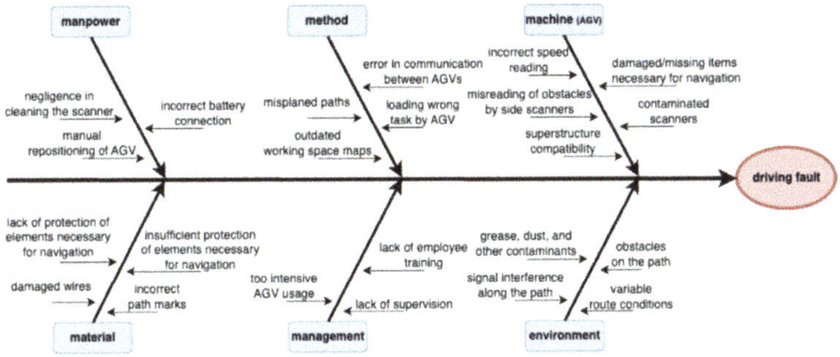

Fig. 2. Ishikawa diagram, potential causes of "driving fault" event

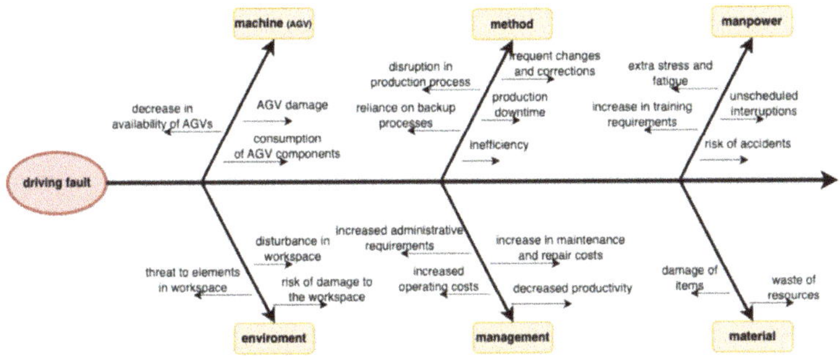

Fig. 3. Potential effects of "driving fault" event

Fig. 4. Diagram of the Bowtie method

Table 1. List of preventive and recovery measures grouped by causes and effects categories

Category	Preventive measures	Recovery measures
Manpower	Monitoring the work of personnel	An employee who monitors the AGV and is trained to intervene
	Training on AGV operation and fault intervention	Emergency procedures
	Simple instructions in the workspace	On-demand technical support
		Registration of the adverse event in the system

(continued)

Table 1. (*continued*)

Category	Preventive measures	Recovery measures
Method	Testing automated communication between AGVs	AGV temporary rerouting procedure
	Regulary updating of AGV paths and maps	Switching to manual mode
	Testing paths and maps after every change	Implementing alternative transportation methods
	Optimization of AGV paths	Analysis and updating of procedures
	Regular fleet software updates	
Machine	Inspection and maintenance	System reset in AGV
	Regular cleaning and testing of sensors	Manual localization of the AGV
		Repair of faulty components
		Procedure of emergency component replacement
Material	High-quality materials and components	Procedure of emergency component replacement
	Monitoring of battery charging process	Cleaning of AGV components
	Extra protection of fragile AGV components	Additional transportation resources
Management	Align budget and resources	Prioritization of technical or organizational activities
	Support for preventive actions	Temporary schedule changes
	Regular audits of AGV maintenance status	Determining the source of the adverse event
	Verifying transport procedures	Providing the necessary tools and resources
	Process modification management	
Environment	Providing clean operating environment	Removal of obstacles along the AGV path
	Eliminating obstacles along AGV paths	Restoring cleanliness
	Adapting AGVs to harsh conditions	Emergency adjustment of the environment

4 Conclusion

The presented procedure enables a comprehensive analysis of an adverse event that regularly disrupts the operation of the AGV system. Research conducted shows how diverse the causes and effects of such an event can be. Continuous observation of the AGV operation, reporting of adverse events that occur and analysis of their causes and effects after their occurrence are necessary. This will enable a thorough verification of the proposed procedures before their implementation. However, introducing appropriate preventive actions will reduce the frequency of the "driving fault" event. Additionally, the implementation of recovery procedures will reduce the time needed to restore the AGV to a state of correct operation. The research conducted also showed that the cause that leads to undesirable effects is not always possible to eliminate within the scope of the company activities using the AGV. In the case of system, material, or programming changes, cooperation with the manufacturer of automatically controlled vehicles is necessary. The described procedure was carried out with real data from the automated guided vehicle system implemented in the company. However, the versatility of the tools used, and their availability allow for similar performance of this analysis in any logistic system using this type of automatic method of transport.

References

1. Poturaj, H.: Identification of information needs in the process of designing safe logistics systems using AGV. In: New Advances in Dependability of Networks and Systems, pp. 217–226. Wrocław: Springer International Publishing, Cham (2022)
2. Tubis, A.A., Poturaj, H.: Risk related to AGV systems—open-access literature review. Energies (Basel). **15**(23), 8910 (2022). https://doi.org/10.3390/en15238910
3. Rubio, F., Llopis-Albert, C., Valero, F.: Multi-objective optimization of costs and energy efficiency associated with autonomous industrial processes for sustainable growth. Technol Forecast Soc Change. **173**, 121115 (2021). https://doi.org/10.1016/j.techfore.2021.121115
4. Foit, K., Gołda, G., Kampa, A.: Integration and evaluation of intra-logistics processes in flexible production systems based on OEE metrics, with the use of computer modelling and simulation of AGVs. Processes. **8**, 1648 (2020). https://doi.org/10.3390/pr8121648
5. Németh, I., Püspöki, J., Viharos, A.B., Zsóka, L., Pirka, B.: Layout configuration, maintenance planning and simulation of AGV based robotic assembly systems. IFAC-PapersOnLine. **52**, 1626–1631 (2019). https://doi.org/10.1016/J.IFACOL.2019.11.433
6. Wong, K.C., Woo, K.Z., Woo, K.H.: Ishikawa Diagram. In: Quality Improvement in Behavioral Health, pp. 119–132. Springer International Publishing, Cham (2016)
7. Gołaś, H., Mazur, A. Zasady, Metody i Techniki Wykorzystywane w Zarządzaniu Jakością, I. Poznań: Wydawnictwo Politechniki Poznańskiej (2010)
8. Alizadeh, S.S., Moshashaei, P.: The bowtie method in safety management system: a literature review. Sci. J. Rev. **4**, 133–138 (2015). https://doi.org/10.14196/sjr.v4i9.1933

Specific Methodology for Modeling a Logistic System Using an Educational Robot

Jakub Kovalčík(✉), Martin Staka, and Peter Kačmáry

Department of Logistics and Transport, Faculty of Mining, Ecology, Process Control and Geotechnologies, Technical University of Kosice, Park Komenského 14, Kosice 042 00, Slovakia
{jakub.kovalcik,martin.straka,peter.kacmary}@tuke.sk

Abstract. The article focuses on developing a specific methodology for modeling logistic systems using educational robots, supporting the design of efficient and functional models in logistics. Current approaches to designing logistic systems are not directly oriented toward the use of educational robots, creating a need for a new methodology that effectively integrates these tools. The main objective of this work is to present the possibility of creating a specific model of robotized workplace in a laboratory setting, where educational robots are used as small physical models to simulate logistic flows and workflows. This model accurately represents elements commonly found in industrial logistics operations and offers a practical demonstration of automated logistic processes. The significance of this methodology also lies in its application in accredited study programs focused on industrial logistics, where it enhances training and understanding of real-world applications in logistics and automation.

Keywords: methodology · logistics systems · educational robots · model

1 Introduction

Use In recent years, educational and industrial robots have gained increasing attention in various areas of research, analyzing their potential and benefits. Research focused on programming educational robots identifies the advantages and challenges these robots bring to the educational environment. Attention is given to their ability to imitate real-life scenarios, allowing students to gain practical experience and understand real-world applications [1]. The creation of robotic workplaces is a complex multidisciplinary task requiring expertise and experience in technical sciences. In this process, software tools play an essential role by enabling effective identification and design of robotic workplaces, thus facilitating collaboration between experts during various stages of development [2]. With increasing productivity in industrial environments, the importance of robotic Pick & Place operations is growing, especially in the food industry. The implementation of such solutions allows handling a large volume of products, thus maintaining the competitiveness of manufacturers in the market [3].

J. Feliks et al. (Eds.): CLC2024 2024, 2026.
https://doi.org/10.1007/978-3-032-28931-5_17

In addition to automation, modern workplaces are characterized by collaboration between humans and robots, particularly in industrial sectors. Advanced robotic devices equipped with computer vision and artificial intelligence enable tasks to be performed with high precision, enhancing productivity and efficiency at the workplace. These technologies are rapidly gaining popularity and becoming a standard part of everyday practices [4]. Contemporary engineering programs are increasingly incorporating robotics into their curricula. Educators employ various methods and techniques to introduce students to robotic systems and their applications, which increases students' interest and motivation in this field [5].

One of the goals in developing educational robots is to adapt them to meet the needs of the entire educational spectrum. A robot designed based on action research was evaluated using the Technology Acceptance Model with a sample of 116 pedagogical university students, yielding valuable insights into its usability and acceptance among students [6]. Finally, the widespread use of artificial intelligence and robotics is leading to the creation of a new scientific field aimed at modernizing and transforming education. These technologies have the potential to fundamentally change the way complex subjects are taught and understood, opening new possibilities for the development of innovative teaching approaches [7]

The aim of this article is to highlight the developed methodology for working with small physical models in order to create autonomous logistics systems. This body of literature demonstrates the wide range of applications and benefits that robotics brings to education and industrial sectors.

2 Definition of Methodology

Scientific research can be understood as a systematic and intentional process aimed at acquiring new knowledge and deepening the understanding of existing phenomena. It is not a random activity, but a structured effort to explore patterns, organize facts, and transform observations into verifiable conclusions. Fundamental attributes of such research include a clearly defined objective, an aspiration to uncover previously unknown relationships, adherence to logical and systematic procedures, and the ability to justify, test, and validate the outcomes and generalizations derived from the study.

A crucial distinction must be drawn between scientific knowledge and everyday experience-based knowledge. While common knowledge may arise from intuition or routine practice, scientific knowledge is built upon rigorously selected methods that guarantee reliability and replicability. This necessity leads to the constant development of new research strategies and techniques, particularly when encountering unexplored phenomena. In essence, research methods represent the concrete tools through which scientific aims are realized, and the scientific discipline that examines, classifies, and refines these tools is referred to as methodology.

Any form of human activity, whether in science or other domain depends not only on the subject (the individual or group conducting the activity) and the object (the phenomenon under investigation) but equally on the means and procedures employed to achieve the intended results. The method therefore embodies the "how" of research, acting as a guiding mechanism that increases efficiency, ensures accuracy, and reduces the likelihood of error. Choosing an appropriate method can be compared to navigating with a compass, as it directs the researcher toward the desired objective in a structured way.

Despite their frequent interchange in everyday use, the concepts of method and methodology are not identical. A method may be regarded as a specific approach or technique applied during investigation, while methodology encompasses the broader framework structured system of multiple methods and strategies applied in combination to study a given problem. For instance, in social sciences, methodologies often integrate both quantitative and qualitative approaches, forming a cohesive research design. In this sense, methodology is closely related to the research procedure itself, reflecting its sequence, structure, and algorithm.

It is important to emphasize that even the most suitable method may not yield reliable results if it is not applied through an appropriate methodology. Methodology can thus be viewed as the meta-level study of methods—their logical arrangement, their validation, and their application in practice. Examples of research methods that form the building blocks of methodology include [8]:

- Deduction—a way of constructing a specific conclusion based on a general premise.
- Analogy—a method of knowledge where the understanding of a set of common characteristics of objects allows us to draw conclusions about their similarity in other ways.
- Modeling—creating a model (copy) of the subject of study to analyze it [8].

General scientific methodology can be regarded as a universal framework that synthesizes methods and principles applicable to the study of phenomena across diverse scientific domains. It provides a broad perspective on how knowledge is systematically acquired, organized, and validated. Among the fundamental methods employed in scientific research are observation, experimentation, and modeling—each characterized by specific attributes that vary according to the nature of the discipline in which they are applied. Observation represents one of the oldest and most fundamental research strategies. It involves the deliberate selection of relevant facts, the determination of their essential characteristics, and their subsequent description. Such descriptions may take verbal form or be expressed through symbolic representations such as tables, diagrams, or statistical graphs. Observation further requires the identification of distinctive properties and features of the subject under study. In fields such as linguistics, this process often presupposes a strong command of the language and the ability to intuitively perceive subtle semantic or etymological connections—a capability sometimes referred to as "etymological instinct."

Experimentation, on the other hand, is based on the active manipulation of variables under rigorously defined conditions. Unlike observation, which is primarily passive, experimentation allows the researcher to test hypotheses in a controlled environment. Within linguistics, experiments may be carried out either with the assistance of technical tools and instruments or by relying exclusively on structured procedures without physical equipment.

Modeling constitutes yet another key methodological approach. It enables researchers to grasp the complexity of reality by creating representations, whether physical, visual, or conceptual, that act as substitutes for the original object or process. Such models may take the form of diagrams, schematic drawings, graphs, or even devices designed to replicate functional aspects of the phenomenon. Each model is grounded in a hypothesis about the internal structure or dynamics of the original system and is intended to serve as a functional analogue that allows indirect study.

In the broader methodological context, a method is generally understood as a structured set of theoretical assumptions combined with specific techniques for conducting research. Every method has a defined scope of application and is tailored to explore certain characteristics, dimensions, or properties of the object under investigation. Research methodology, in turn, refers to the way a chosen method is implemented in practice, shaped not only by the objectives of the study but also by the techniques of data collection and description, the analytical tools employed, and even the personal orientation of the researcher.

Methodology encompasses all stages of the scientific process: from initial observation and the collection of empirical material, through the selection of analytical units and methods of description, to the procedures of analysis and interpretation of findings. In disciplines such as linguistics, divergences between different schools of thought often do not arise from the choice of method itself, but rather from the preferred analytical frameworks, descriptive strategies, and the degree of formalization applied in interpreting the material. Ultimately, method, method application, and methodology form an interconnected triad that cannot be meaningfully separated. Their interplay determines the reliability and validity of research outcomes. The selection of methodological principles, the definition of the scope of a method, and the design of the overall methodology depend largely on the researcher's objectives, the epistemological orientation of the study, and the particularities of the phenomenon under investigation [8] (Fig. 1).

Fig. 1. Creation of methodology [9]

3 Creation of Model

The development of the model system was carried out within the "Logistics Innovation Laboratory of Robots" and was primarily intended to illustrate the adaptability of industrial robotic systems and their potential to substitute conventional machinery commonly employed in extraction and processing operations. During the construction phase, several modifications had to be introduced to accommodate both the constraints of the laboratory setting and the technical specifications of the robotic manipulators. A major limitation was the inability to configure and operate all simulated tasks from a single computer interface. This technical restriction led to the division of the model into two independent subsystems. Although this segmentation initially complicated the process of achieving simultaneous operation, the challenge was addressed by integrating the WLKATA STUDIO software, which enabled the activation of one subsystem with a deliberate delay. As a result, synchronization of both parts was achieved, allowing robotic manipulators and conveyor mechanisms to perform their designated tasks autonomously, if timing parameters were accurately calibrated [10]. The first subsystem simulates the extraction and initial handling of material. Here, a robotic arm equipped with a spoon-like tool removes bulk material from a fully loaded container. The extracted material is then placed onto a conveyor belt, which transports it into a secondary container before directing it to a designated processing unit. This stage reflects the early phases of industrial material handling and preparation. Following a pre-set processing interval, the second subsystem is automatically initiated. In this stage, the conveyor delivers the processed material to another robotic manipulator. The robotic arm executes its program by lifting and relocating a pair of wooden blocks, symbolizing processed material, onto a plastic pallet positioned at the system's base. This sequence models the final stage of handling and palletization, completing the simulated production cycle. The presented laboratory model serves as the basis for the formulation of a dedicated methodology, which will be discussed in the subsequent section of this work. This methodology aims to generalize the insights gained from the model, providing a structured approach to designing and evaluating robotic systems in logistics applications [10] (Figs. 2 and 3).

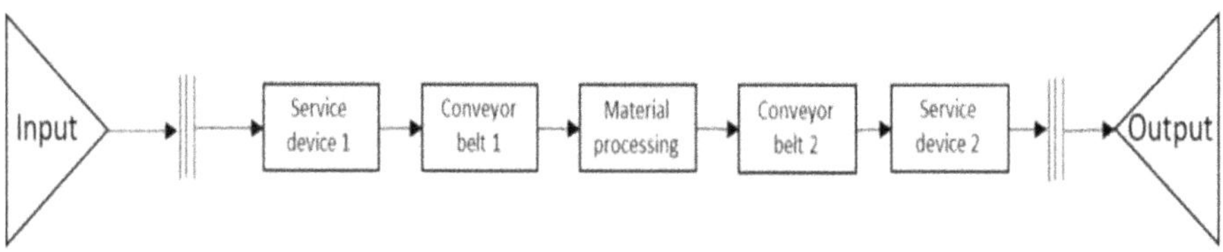

Fig. 2. Scheme of the model system [10]

This model serves as a starting point for the subsequent development of a methodology for creating autonomous logistic systems, which includes the general steps required for assembling such systems. These key elements will be illustrated in the creation of a formal scheme.

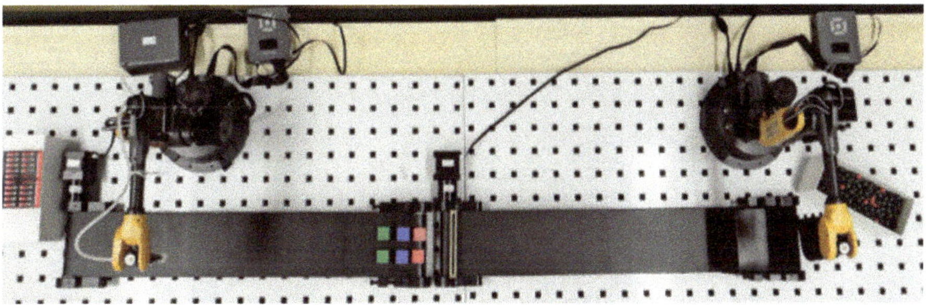

Fig. 3. Model system [10]

4 Creation of Methodology and Schemes

By creating the model system, an algorithm for building autonomous logistic systems focused on educational robots has been developed as you can see in Fig. 4. Each activity in this algorithm is essential for the proper functioning of the algorithm and its subsequent use in building logistic systems. Based on a detailed analysis of the algorithm, it can be stated that with proper use, anyone can follow the detailed steps to assemble, test, optimize, and improve logistic systems in a laboratory environment [11]. Next figure illustrates a universal algorithm that can be used to create a model of an autonomous logistic system, utilizing physical models focused on educational robots in tasks such as loading, transporting, unloading, conveying, and similar activities.

The initial phase of the project involves analyzing the market for physical components based on educational robots, which are used to create models of autonomous logistic systems. The available models that can be utilized for this purpose are assessed. The next step is to define the operations and processes that will simulate an autonomous system using small physical models. After this analysis, the necessary operations are specified to ensure they represent the correct activities within the "Logistics Innovation Robotics Laboratory" environment. Then, additional components like wooden balls, spoons, blocks, and containers are selected to represent various elements in the system. This phase also includes securing the necessary purchases and allocation of these components. Once the components are secured, the hardware assembly process begins. Software is then installed to interact with the hardware, and sensors are configured to enable the vehicle to follow a predefined path. Afterward, the system is fine-tuned, and its functionality is tested. If the system fails to meet the required criteria, the algorithm returns to analyzing the physical components, and an alternative hardware solution is considered. If everything checks out, the process moves on to the next stage.

The second step is to verify the functionality of the software. If the software is not functioning as expected, it returns to the previous step to ensure the system software is correctly installed. If the software meets the specified criteria, the process continues towards building a fully operational system. This phase involves organizing the individual stations of the model system in a logical manner. After this, the system is fine-tuned again, and its functionality is verified. If it still doesn't meet the required criteria, the entire process is repeated. Once the system is confirmed to be operational, the next

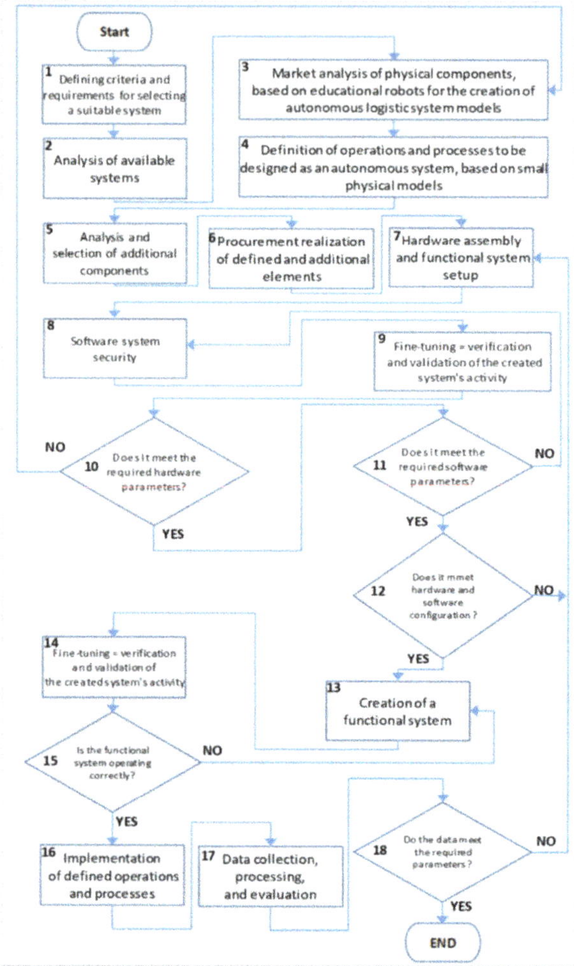

Fig. 4. Generalized formalized scheme [10]

step involves programming it to perform the defined tasks autonomously. After verifying its full functionality, the results obtained from the research are processed. Lastly, a final check is done to ensure that the collected data meets the required standards. If any discrepancies are found, the system's hardware or software is revisited for further adjustments. If the system meets all requirements, the process concludes with the system being considered fully functional [10].

5 Conclusion

The results of this research confirm that the application of educational robots in the modeling of logistic systems represents a highly effective and versatile approach. Their tangible and visual nature provides a clear advantage over purely virtual simulations, as it enables direct observation of physical interactions and workflows. Through the assembly of functional systems from readily available components, it becomes possible not only to design and test logistical structures in a relatively short time but also to identify operational risks and inefficiencies that might otherwise remain hidden in digital environments. Another important outcome lies in the adaptability of educational robotic platforms. Their modular design and increasing technological sophistication make them suitable for a wide range of practical experiments, from basic material flow demonstrations to more complex simulations of automated logistics operations. The cost-effectiveness of these solutions further supports their broader adoption, allowing institutions and laboratories to implement them without the financial burden typically associated with industrial-scale systems. The continuous advancement of educational robots also creates new opportunities for innovation in logistics education and professional training. The realistic representation of processes not only enhances the learning experience of students but also attracts the attention of skilled practitioners, who recognize the value of such environments for applied research and practical skill development. In this sense, educational robotics serves as a bridge between theoretical knowledge and industrial practice, enriching curricula and strengthening the preparation of future specialists in logistics and automation. Taken together, these findings indicate that the integration of educational robots into logistics modeling has significant long-term potential. It fosters experimental learning, reduces barriers to testing complex scenarios, and encourages interdisciplinary collaboration. Moreover, it establishes a foundation for continuous improvement and adaptation of logistics systems in line with technological progress. The insights gained from this study therefore contribute not only to academic discourse but also to the practical evolution of industrial logistics, where flexible and innovative solutions are increasingly required.

Acknowledgements. This research was funded by the grant number project KEGA 010TUKE-4/2023 Application of educational robots in the process of teaching the study program industrial logistics.

References

1. Georgieva-Trifonova, T.: Research on possibilities and benefits of programming educational robots for real-life scenarios imitation. TEM J. **13**(1), 718–725 (2024)
2. Duplak, J., et al.: Application of software tools in the identification of robotic workplace cooperation in the design phase. SAR J. **6**(4), 221–229 (2023)
3. Semjon, J., Fuhal, M., Kocan, M.: Design of the handling process at the workplace with ABB robot and modular conveyor. Tech. Sci. Technol. **21**(3), 112–119 (2020)
4. Titu, M.F.S., et al.: Experiments with cooperative robots that can detect object's shape, color and size to perform tasks in industrial workplaces. Int. J. Intell. Robot. Appl. **8**(5), 1–14 (2023)
5. Barakat, N.: The ultimate experience in learning robotics: building robots in a robotics course. In: Proceedings of the International Mechanical Engineering Congress and Exposition (2008)
6. Chatzopoulos, A., Papadakis, S., Kalogiannakis, M., Papoutsidakis, M.: A novel, modular robot for educational robotics developed using action research evaluated on technology acceptance model. Educ. Sci. **12**(4), 1–21 (2022)
7. Chatzichristofis, S.A.: Recent advances in educational robotics. Electronics. **12**(4), 1–5 (2023)

8. Bačíková, M.: Fundamentals of Pedagogical-Psychological Research Methodology [Online]. https://ik-ptz.ru/sk/dictations-on-the-russian-language%2D%2Dclass-2/chem-metodika-otlichaetsya-ot-metodologii-metody-teoreticheskogo.html (2018). Last Accessed 08 Nov 2024
9. Figure. https://www.dailyautomation.sk/wp-content/uploads/2016/05/kennzahlen.jpg. Last Accessed 08 Nov 2024
10. Kovalcik, J., Straka, M., Trojanowska, J.: Utilize educational robots to design logistic systems. Appl. Sci. **14**(13), 1–16 (2024)

Modeling Human Interaction with a Logistic System

Marek Ondov[1(✉)], Andrea Rosova[1], Marian Sofranko[2], and Lukas Horizral[1]

[1] Institute of Logistics and Transport, Technical University of Kosice, Kosice, Slovakia
{marek.ondov,andrea.rosova,lukas.horizral}@tuke.sk
[2] Institute of Earth Sources, Technical University of Kosice, Kosice, Slovakia
marian.sofranko@tuke.sk

Abstract. Logistic systems are an indispensable part of businesses across various industries. Despite current trends of robotization and automation in logistics, one of the main elements of logistic systems remains people—the workers. It is precisely the workers, through their constant interaction with the system, who understand it best. Over time, they develop improved processes that facilitate their interaction with the system—and thus their work. A simulation-based approach is commonly used to improve logistics processes. System modeling and subsequent simulations with cost-free parameter adjustments are effective tools for identifying improvements within the system. However, modeling logistic systems must not overlook the interaction and interventions of workers in the system. Modeling this interaction is essential if the model reflects the actual system. Modeling human behavior is complex and must consider various factors, distinguishing between effective system improvements and inefficient attempts. The process of modeling human interaction with logistic systems, based on modeling elements and heuristics, is presented through an example from the mining environment—specifically, the interaction of workers with the mining transport system. Modeling worker interactions with the transport system produced a set of rules and processes that improve the system's performance or ensure the safety of system components. In subsequent simulations and the design of effective improvements, time can be saved as some basic improvement proposals are already implemented in the model according to worker interaction with the system.

Keywords: Modeling · Logistic systems · Human interaction

1 Introduction

A logistics system manages, ensures, and executes movement within logistics chains. Describing a logistics system means defining its structure—elements, connections, functions, activities, and objectives [1]. In the case of a logistics system, elements and connections are associated with the tangible and intangible flow of material or services, finances, and information. Each of these flows somehow affect the others. The logistics system is a hierarchical system composed of several entities that must be coordinated [1, 2]. The logistics system is a specific open mixed system with a dynamic character. These systems mainly comprise social, economic, and technical elements [2].

J. Feliks et al. (Eds.): CLC2024 2024, 2026.
https://doi.org/10.1007/978-3-032-28931-5_18

Many elements in logistics systems consist of inanimate components that can easily be generalized and standardized into rules. These are technical elements such as vehicles, storage equipment, robots, machines, or computers [1]. However, an indispensable part of logistics systems includes living elements—people/workers. Despite trends toward automation and robotics, logistics systems that combine innovative solutions with human approaches and established behaviors are still more common. The presence of humans and the human factor have not disappeared from logistics systems, and system solutions must respect this. The human factor is valued for its flexibility and cognitive and motor skills. Robots and machines, in most cases, replace these human advantages at a high cost [3].

Modeling is one of the approaches to studying, analyzing, designing, and planning systems. In modeling, the term "system" can be defined as a set of elements that interact. The result of their interaction is a response or behavior observed over time. Logistics systems are complex systems, so in modeling, some non-essential elements can be omitted [4]. However, when examining logistics systems, it is necessary to identify human behavior and procedures and incorporate them into the solution. Models must integrate human factors that influence the system. It is essential to monitor improvements, such as expanding and enriching job tasks, and consider limitations associated with employee age. For managers, the challenge is to fully utilize the knowledge and experience of older employees while accounting for the gradual decline in their working abilities. Therefore, modeling production and logistics systems should also focus on modeling human-system interaction [3].

Modeling human-system interaction provides a realistic view of how a logistics system operates. People who have been in contact with the system for a long time look for methods to make their work easier and faster, thus creating their algorithms for tasks within the system. Heuristic analysis is a method that, through answers to appropriate questions, defines the cognitive activities of a person [5]. The human factor can be modeled in the context of cognitive and manual activities between humans and machines [6].

Authors Malindzak and Cacciabue developed their approaches for modeling human activities in logistics [5] and human-machine interaction [6]. This paper aims to create a procedure for modeling human-system interaction based on the principles of both approaches to utilize each other's advantages. The procedure developed in this paper is then applied to model the interaction of workers with a mining transport system.

2 Methodology

The procedure for modeling interaction is based on heuristic principles. Heuristics can create or improve solutions, but the complexity of the problem limits their capabilities. Moreover, they may end up at a local optimum, overlooking other, potentially better solutions. Metaheuristics are used to escape local optimal solutions by applying higher-level strategies [7]. The heuristic model of logistics activities is a model of human cognitive activities primarily used in management processes. A set of rules is obtained through retrospective analysis of information processes, and a set of possible states is acquired through prospective analysis. The process starts from an initial state, and subsequent states are created by analogy. If a rule has been validated by practice and was suitable for the previous state, it will also be suitable for the following states. Then, to complete the qualitative model, it is necessary to define the set of rules as accurately as possible, based on which the model can be created [5, 8]:

- Technological rules,
- Constraints,
- Expert rules,
- Optimality criteria.

The second building block of the human-system interaction modeling process is the modeling of human-machine interaction. The interaction between a human and a machine always occurs in the context of the device or machine, the working environment, and the operator. It is challenging to constrain human behavior within limits, structure the presentations of models in modeling, and then perform simulations. It is not impossible to separate human behavior from the system's dynamics. In this case, integration is necessary. The human factor can be modeled in the context of cognitive and manual activities between humans and machines. Qualitative simulations of human activities describe their structure, dependencies, logical connections, and dynamic development. They bring understanding into workers' procedures and serve as the foundation for transforming the interaction into models. Retrospective analyses assess the impact and processes of human activities, which are then completed down to the roots. On the other hand, perspective analyses can predict and evaluate the outcomes of human activity interactions within systems. Combining analyses makes it possible to create a model of predicted human-machine interactions based on past occurrences in the process [6, 9].

The proposed approach is based on the idea that rules and constraints will be created after identifying inanimate elements and connections. The set will draw from the knowledge of employees and define their interaction with the system. The creation of the set is a retrospective understanding of the system from the workers' perspective and activities. Within the set of rules and constraints, it is essential to distinguish which rules have the potential to be used in further modeling. The rules must have the potential for future application within the system. A schematic representation of the human-system interaction modeling process is shown in Fig. 1.

The human-system interaction model is formed from organized data and information about human activities within the system that are not captured in company documents. Its representation can be a table or a set in the form of

$$\{\text{Constraint 1 to n; Rule 1 to n}\} \quad (1)$$

where *n* is the number of constraints and rules.

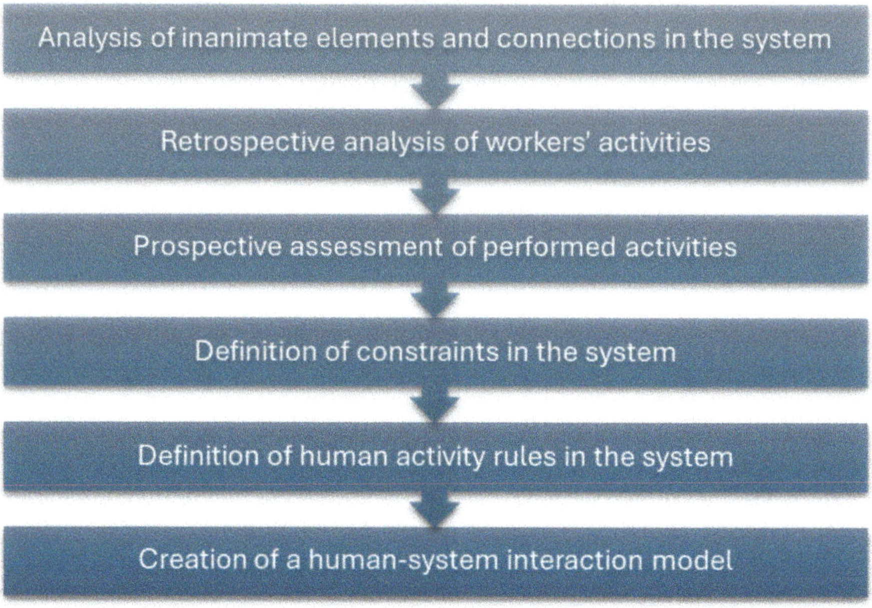

Fig. 1. Procedure for modeling human-system interaction.

3 Case Study

The procedure for modeling human-system interaction was applied to the transport system of a mining operation. The main goal was to perform a simulation analysis of the system. However, the data obtained from the system analysis left many questions needing to be answered before creating the simulation model. Therefore, the preparation of data and information before creating the simulation model was extended to include modeling the interaction between living elements—workers—and the transport system.

The transport system is based on rail transport, with the entrance and exit from the mine occurring on a single track. Along the length of this track, there are four passing points for train sets. The rail track is used for the delivery of materials needed for mining, the supply of materials for the construction of excavated chambers, the arrival and departure of personnel in the mine, and the transport of raw materials. This results in a significant transportation load on the track in both directions. The system analysis defined the directions on the railway track, the wagon stock, the requirements for the transport system, and the activities carried out within the transport system. From the available documents, it was easy to determine how the train sets are composed for different activities, where the loading and unloading take place, the distances within the transport system, how many workers come into direct contact with the transport system, and the speeds of the train sets on various segments.

However, the first simulation model attempts failed the validation process. These models did not reflect the actual transport system. It was necessary to identify the system in greater depth. A better understanding of the single-track railway's loading, unloading, passing, and transportation processes was required. Since the processes in the transport system are heavily influenced by the actions of the drivers and shunters, the focus shifted to modeling the interaction between the workers and the transport system. The foundation of the modeling process involved frequent visits to the company, interviews, and observations of the work of individual employees. The definitions of the individual constraints and activities of the workers are outlined in the following sections.

Workers use only two of the four available passing points for train sets. The passing points closer to the middle of the track are reserved as parking areas for the large wagon stock in the mine. The passing points closer to the ends of the track are used for locomotive position changes. This is due to an identified rule regarding pulling wagons on the track, which was established during the system analysis. However, the wagons are pushed at the loading or unloading sites, and there is no space available to change the position of the locomotive.

The loading process is limited to one batch of raw material being loaded at a time from one loading point to prevent mixing the quality of different batches. Only one raw material delivery can simultaneously be made to one of the two loading points. Loading is further controlled visually and with light signals between the driver and the machine operator at the loading site. If loading is done with a loader, lower mining carts with a capacity of 8 m^3 must be used instead of the standard 10 m^3 carts. Unloading occurs on a tipping bridge, where two raw material boxes exist. However, the entire train set can only unload smoothly into the first box. Unloading into the second box is interrupted by splitting the train set in half due to the overall length of the tipping bridge. This division significantly extends the unloading process because the mechanisms are in suboptimal condition, and workers must use various tools to improve their functionality.

The last process that caused inaccuracies in the initial simulation models was loading and unloading materials for constructing mined chambers. These loading and unloading activities are not treated as separate processes but as one cohesive process. The process involves one driver, two locomotives, and three wagons of the appropriate type. One wagon is loaded from the discharge of the concrete plant on the surface. At the same time, the second wagon is unloaded underground, and the third wagon is on its way to the concrete plant. The driver transports the empty wagon to the concrete plant, disconnects it from the locomotive, and connects the full one. The driver then moves to a sidetrack, positions the full wagon in front of the empty one, and pushes it under the concrete plant. Once the wagon is positioned under the concrete plant, it is detached, and the full wagon enters the mine. The full wagon is then pushed to the unloading site near a mobile pump in the mine. Meanwhile, the pump operator underground empties the third wagon and, with the help of the second locomotive, moves it to the sidetrack. The driver then moves to the second locomotive with the empty wagon and exits the mine. This system could be more efficient with the involvement of an additional driver, but this is often impossible due to personnel constraints.

A comparison with the original simulation models that did not pass validation can be made regarding the prospective evaluation of the rules. These models used all passing points whenever needed. The alternation between inbound and outbound transport from the mine was not influenced. With passing point adjustment according to the workers' needs, greater inertia is introduced into the inbound and outbound processes to avoid collisions on the track. Additionally, the locomotive position change for pushing the train set caused delays not present in the original simulation models, as the system analysis indicated no pushing of train sets, only pulling. In the case of loading, the original models assumed a continuous supply of material to the loading sites and the use of maximum capacities. However, workers' procedures revealed limitations not directly evident in the system analysis. Initially, both loading sites were considered for use simultaneously, which does not align with the system. A difference arose in the unloading of raw materials into the second box. The most significant discrepancy in the original simulation models was the approach to loading and unloading construction materials, as these were considered as two separate processes. In the actual system, however, these processes are interconnected and are carried out simultaneously.

In Table 1, the human interaction model with the transport system of the mining operation is shown. The primary model consists of only the first two columns, while the third column is auxiliary, helping to clarify the context of the rule in the model.

Applying the rules identified in the human-transport system interaction model contributed to validating subsequent simulation models of the transport system. Implementing human-system interaction modeling led to subsequent improvements in the system, such as acquiring an additional loader or revising maintenance plans for the mechanisms.

4 Conclusion

Logistics systems are an integral part of businesses in every industry sector. Often, describing a business means describing its logistics systems. These systems contain many elements and connections, and many methods and tools analyze them. Some of these methods and tools fail to define the connections between human elements—workers—and the system.

The paper aimed to develop a process for modeling human-system interaction based on ideas from the logistical heuristic model of the company system and human-machine interaction. These approaches intersect in the idea of retrospective analysis of what people do and the prospective idea that if a process has proven prosperous several times, it will succeed in future situations. The procedure developed in Chap. 2 integrates these ideas but does not separate them from established system analysis. Using prospective analysis reduces the number of generated rules, and the resulting model is conceptualized from sets of constraints and rules of the model.

The developed human-system interaction modeling procedure is applied in a case study to a mining transport system. Its application was crucial for validating the simulation model of the mining transport system. The human-system interaction model created rules commonly used by workers in the system. These rules were not captured in documents; therefore, they could not be identified in any way other than through interviews and observations of workers in the system. The context of the rules further supports the human-system interaction model to clarify the thought processes of the creator, as shown in Table 1. The work on the modeling procedure is in the beginning stages, and the procedure should be applied several times soon.

Table 1 Human-transport system interaction model.

Constrain	Model rule	Context
Passing points in the middle of the track are parking areas	Not passing train sets in the middle of the track	Creating space for an extensive wagon stock
Pushing wagons to loading and unloading sites	Regular repositioning of the locomotive at passing points at the ends of the track	It accelerates subsequent loading/unloading and transportation in the opposite direction
Not mixing different batches during loading	Delivering the next batch to the loading site only after the previous one is entirely removed	It ensures easier sorting of raw materials by quality
One loader for loading	Loading from only one loading site	The distance between loading sites is too great for simultaneous servicing with one loader
Only one loading site uses a chute for loading	Using smaller wagons at other loading sites	Other loading sites are serviced by a loader, which limits the shovel's lifting capacity
Length of the tipping bridge	Interrupting the unloading of raw material into the second box by splitting the train set	The length of the bridge allows smooth unloading only into the first box
The poor condition of the mechanisms	The locomotive driver must have tools available to fix the mechanism issues	Reducing the time required to resolve mechanical issues
Cohesive process of loading and unloading construction materials	During chamber construction, the transport on the track is limited to the system of one driver, two locomotives, and three wagons	During construction, loading and unloading are performed simultaneously, and the transport of wagons fully utilizes the track

Acknowledgement. This work is supported by the Scientific Grant Agency of the Ministry of Education, Science, Research, and Sport of the Slovak Republic and the Slovak Academy of Sciences as part of the research project VEGA 1/0430/22 and VEGA 1/0588/21.

References

1. Malindzak, D.: Teória logistiky/definície, paradigmy, princípy, štruktúry, 1st edn. KARNAT, Kosice (2007)
2. Kavka, L.: Systémová analýza logistických procesů. Acta logistica moravica **2**(1), 70–80 (2012)
3. Winkelhaus, S., Grosse, E.H.: Logistics 4.0: a systematic review towards a new logistics system. Int. J. Prod. Res. **58**(1), 18–43 (2020)

4. Birta, L.G., Arbez, G.: Modelling and Simulation, 3rd edn. Springer, London (2019)
5. Malindzak, D., Takala, J.: Projektovanie logistických systémov, 1st edn. Expres Publicit s.r.o, Kosice (2005)
6. Cacciabue, P.C.: Modelling and Simulation Of Human Behaviour in System Control, 3rd edn. Springer Science & Business Media, London (1998)
7. Griffis, S.E., Bell, J.E., Closs, D.J.: Metaheuristics in logistics and supply chain management. J. Bus. Logist. **33**(2), 90–106 (2012)
8. Rosova, A., Ondov, M., Mudarri, T.: Podniková logistika II, 2nd edn. TU, Kosice (2023)
9. Oliveira, M.L.M., Montevechi, J.A.B., Pinho, A.F., Miranda, R.C.: Using hybrid simulation to represent the human factor in production systems. Int. J. Simul. Modell. **16**(2), 263–274 (2017)

Excellence in Logistics and ERP Project Management—A Contemporary Framework for Digital Transformation and Operational Excellence

Marian Krupa(✉) and Marek Karkula

AGH University of Krakow, Krakow, Poland
{makrupa,mkarkula}@agh.edu.pl

Abstract. This paper examines the evolving concept of excellence in logistics and Enterprise Resource Planning (ERP) project management within the context of digital transformation, Industry 4.0, and sustainable supply chain practices. The study integrates traditional excellence frameworks with contemporary technological advances, artificial intelligence integration, and circular economy principles to provide a comprehensive understanding of modern logistics excellence. Through analytical-synthetic methodology and extensive literature review, this paper establishes connections between classical management excellence theories and current digital transformation trends, offering practical frameworks for achieving operational excellence in complex supply chain environments. The research demonstrates that contemporary logistics excellence requires mastery of both traditional management principles and emerging digital capabilities, including AI-powered analytics, IoT integration, blockchain transparency, and sustainable business practices.

Keywords: Logistics Excellence · ERP Project Management · Digital Transformation · Industry 4.0 · Artificial Intelligence · Sustainable Supply Chain · Business Process Management · Operational Excellence · Circular Economy

1 Introduction

Excellence is an ideal that people strive for in their ideas and actions. For Aristotle it meant "complete, containing all due parts; so good that in its kind it cannot be better; what has achieved its goal". Saint Thomas Aquinas, Barbaro, Galen, Wolff, Baumgarten presented a similar way of thinking about excellence: "what is completed, to which nothing can be added, and at the same time from which nothing should be taken away". On the other hand, "excellence" means a way of life, development-oriented sensitivity. The permanent attitude of striving for perfection is the bottom line for those who are willing to "go the extra mile", to be at least "a little better" [14].

J. Feliks et al. (Eds.): CLC2024 2024, 2026.
https://doi.org/10.1007/978-3-032-28931-5_19

In the contemporary business environment, the pursuit of excellence has evolved significantly with the advent of digital technologies, Industry 4.0 innovations, and sustainability imperatives. The traditional concepts of excellence now intersect with artificial intelligence, blockchain technology, Internet of Things (IoT), and circular economy principles, creating new paradigms for organizational performance and competitive advantage [25]. Modern logistics operations face unprecedented complexity, requiring integration of advanced technologies with human expertise to achieve optimal performance levels.

The attitude aimed at achieving the best possible results is not only conditioned by axiological premises but also constitutes a kind of bottom line for many business activities, logistics and IT ventures. Undoubtedly, these include ERP software implementation projects, which have become increasingly complex as organizations adopt cloud-based solutions, AI-enhanced analytics, and real-time integration capabilities [7].

The aim of this study "Excellence in Logistics and ERP Project Management" is to answer the following questions: 1) What is the idea of "excellence" in contemporary management within the digital transformation context?; 2) How can we define "Excellence" from a modern logistics perspective incorporating Industry 4.0 technologies?; 3) How can we evaluate the level of "Logistics Excellence" using current performance indicators and digital metrics?; 4) To what extent can the idea of Logistics Excellence impact ERP Project Management in the era of artificial intelligence and sustainable supply chains?; 5) How do emerging technologies and sustainability principles reshape traditional excellence frameworks?

Research was conducted based on the analytical-synthetic method, complemented by systematic literature review of contemporary sources spanning 2020–2024, focusing on digital transformation, sustainable logistics practices, and modern ERP implementation methodologies.

2 The Idea of "Excellence" in Contemporary Management

2.1 Historical Development of Excellence in Management

Management is a desire for excellence in human action. The idea of excellence in management has been an inspiration for many, both scientists and entrepreneurs [19, 30]. "Excellence", in relation to various aspects of economic and social activity, has been presented among theoreticians and researchers of Scientific Organization and Management, at least, since the beginning of the 20th century in the form of: "Optimal Efficiency" (Babbage, Taylor, Gilbreth, Châtelier, Emerson); lack of idleness, delays and waste (Gantt, Clark, Kent, Rytel, Hauswald, Drzewiecki); "Perfect Efficiency" (Adamiecki); "Creative-Defensive" organization, "Standards and Norms" principle (Emerson); coordination, harmony, balance and right proportion between various elements (Thompson, Plenge, Bogdanow); model of Mass Production based on the division of labor, attractive wages and lack of complex organization hierarchy (Ford); perfect matching of jobs profile to the real skills of employees (Fayol) and the concept of "Nadler's Ideal System.

Within the humanization trend (1930s), in turn, we distinguish patterns based on: employee participation in the decision-making process, integrating conflict, the existence of authority (Follett); ergonomic adjustment of workstations (Mayo); management through a philosophy based on the subjectivity of the employee entity (Sheldon); decentralization of the decision-making process with full incorporation of social responsibility (Bata).

The 1950s opened a new chapter in the search for the idea of excellence in management theory and practice. As a part of TQM (Deming, Jurand), it was, among others, the key Japanese "Kaizen" concept that was applied - as a basis for continuous improvement, constant pursuit of perfection. In 1982, the publication "In Search of Excellence" by Peters and Waterman [22], pointed to the so-called "soft", attributes determining success in business, which at the same time, constituted organizational excellence in organizational practice.

2.2 Digital Transformation and Management Excellence

Today's management excellence paradigms have been fundamentally transformed by digital technologies and data-driven decision-making processes. Contemporary excellence frameworks integrate artificial intelligence, machine learning, and predictive analytics to achieve unprecedented levels of operational efficiency and customer satisfaction [18]. Digital transformation has redefined traditional excellence models by introducing real-time monitoring, automated process optimization, and intelligent resource allocation mechanisms.

The emergence of Industry 4.0 technologies has created new dimensions of excellence, where cyber-physical systems, IoT sensors, and cloud computing platforms enable organizations to achieve levels of precision and responsiveness previously unattainable [32]. Modern excellence frameworks must therefore incorporate both traditional management wisdom and cutting-edge technological capabilities to remain relevant in today's competitive landscape.

3 "Excellence" from the MODERN Logistics Perspective

The idea of excellence in logistics process management is best presented by the well-known "8R Principle", now enhanced with digital capabilities and sustainability considerations. It means, in general, that logisticians should deliver the right product, in the right quantity, at the right quality, in the right place, at the right time, for the right customer, at the right cost and in the right way (to cause as little damage to the environment as possible). This model presents a mix of management indicators that should be met simultaneously to deliver excellent logistics service to customers while maintaining environmental responsibility.

Contemporary logistics excellence extends beyond the traditional 8R framework to include digital transparency, real-time visibility, predictive capabilities, and circular economy principles. The integration of IoT sensors, AI-powered analytics, and blockchain technology enables logistics providers to achieve new levels of service excellence while optimizing environmental impact and resource utilization [3].

We can describe the complexity of modern logistics excellence by the Enhanced Total Cost Optimization model, which now incorporates sustainability metrics, digital transformation costs, and stakeholder value creation alongside traditional cost and service level considerations (Fig. 1).

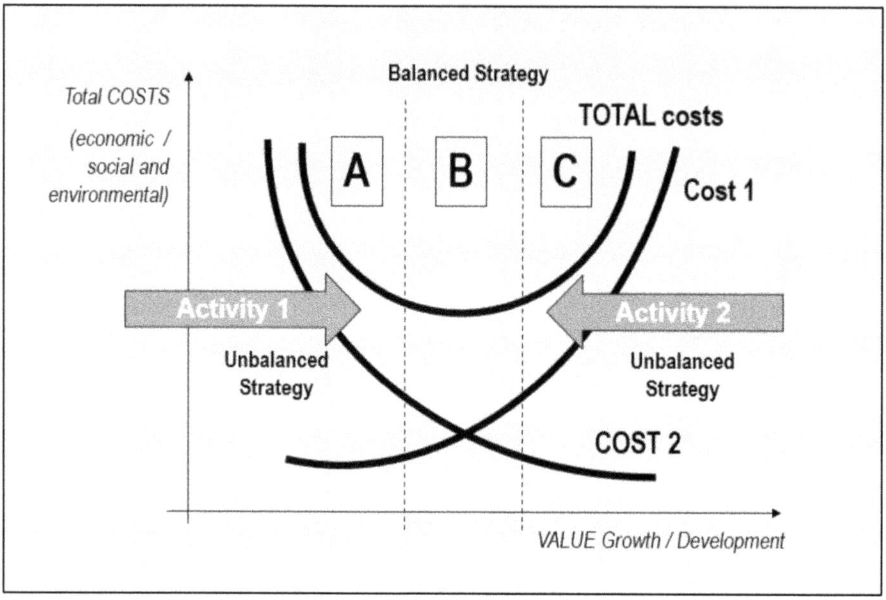

Fig. 1. Total cost optimization model in logistics excellent management

In this enhanced model, striving for logistics perfection means achieving optimal balance among multiple competing objectives including cost efficiency, service quality, environmental impact, and digital maturity. Assuming the existence of numerous optimization factors (expanded 8R Principle plus sustainability and digital metrics), the formula requires advanced mathematical equations supported by artificial intelligence and machine learning algorithms.

From the global perspective, Logistics Excellence has evolved beyond the original McKinsey and U.S. Chamber of Commerce framework [24] to incorporate digital transformation indicators, sustainability metrics, and circular economy performance measures. The modern model consists of 15 Global Logistics Excellence Keys within seven sections: Traditional sections enhanced with digital capabilities and three new sections focusing on sustainability, digital maturity, and stakeholder value creation.

3.1 Digital Technologies in Logistics Excellence

Contemporary logistics excellence leverages advanced optimization methods and techniques supported by Business Intelligence (BI), Artificial Intelligence (AI), and IoT technologies [20, 26, 28]. Examples include: Enhanced Traveling Salesman Problem (TSP) solutions using quantum computing algorithms, Advanced Value Analysis (VA) incorporating real-time Key Logistics Performance Indicators (KLPI), Dynamic K-Center Problem solutions using machine learning and edge computing, Production process optimization through digital twins and simulation (advanced Flexsim and cloud-based platforms), Cost optimization using predictive analytics and Break-Even Point analysis, and route optimization powered by AI and real-time traffic data [13].

To sum up, logistics excellence refers to optimization methods and techniques supporting decision-making dilemmas, which are usually complex and multidimensional. By implementing right method with reference to the appropriate decision dilemma, it is possible to deliver the right answer in terms of Logistics Excellence.

3.2 Artificial Intelligence and Machine Learning in Logistics

Artificial intelligence has emerged as a transformative force in logistics and supply chain management, offering unprecedented opportunities to enhance operational efficiency, improve decision-making, and create sustainable competitive advantages. AI technologies enable logistics providers to process vast amounts of data, identify patterns, predict future trends, and optimize operations in real-time [25]. Contemporary AI applications in logistics include demand forecasting using machine learning algorithms, predictive maintenance for transportation fleets and warehouse equipment, intelligent routing and load optimization, automated quality control using computer vision, chatbots and virtual assistants for customer service, and risk assessment and supply chain resilience planning. These applications demonstrate how AI can enhance both operational efficiency and customer satisfaction while reducing costs and environmental impact. Machine learning algorithms analyze historical data to predict demand fluctuations, optimize inventory levels, and identify potential supply chain disruptions before they occur [3]. The integration of AI with traditional logistics excellence frameworks [8] creates intelligent systems that can adapt to changing conditions, learn from past experiences, and continuously improve performance without human intervention. This predictive capability enables proactive rather than reactive management, embodying the essence of modern logistics excellence frameworks.

3.3 Blockchain Technology and Supply Chain Transparency

Blockchain technology has emerged as a powerful enabler of supply chain transparency, traceability, and trust. By providing immutable records of transactions and product movements, blockchain enables organizations to verify authenticity, ensure compliance, and build stakeholder confidence in their supply chain operations [9]. The implementation of blockchain in logistics creates transparent, auditable supply chains where every transaction is recorded and verified. This transparency enables better risk management, improved compliance with regulatory requirements, and enhanced customer trust through verified product provenance and ethical sourcing claims.

Key blockchain applications in logistics include product traceability and provenance verification, supply chain finance and payment automation, regulatory compliance and audit trails, quality assurance and recall management, and sustainable sourcing verification. These applications demonstrate blockchain's potential to enhance both operational efficiency and stakeholder trust [10].

Smart contracts on blockchain platforms can automate many logistics processes, reducing manual intervention, eliminating errors, and improving transaction speed. This automation contributes to logistics excellence by reducing costs, improving accuracy, and enabling real-time settlement of transactions across complex supply chain networks.

3.4 Industry 4.0 and Smart Manufacturing Integration

Industry 4.0 technologies represent a fundamental shift toward intelligent, connected, and autonomous manufacturing and logistics systems. The integration of cyber-physical systems, IoT devices, cloud computing, and artificial intelligence creates smart factories and logistics networks that can optimize performance in real-time while adapting to changing market conditions and customer requirements.

Smart manufacturing concepts extend beyond the factory floor to encompass entire supply chain networks, creating interconnected ecosystems where equipment, systems, and stakeholders communicate and collaborate seamlessly. This connectivity enables predictive maintenance, quality optimization, energy efficiency, and waste reduction throughout the manufacturing and logistics value chain [29]. Internet of Things (IoT) sensors and devices provide real-time visibility into all aspects of manufacturing and logistics operations. These technologies enable continuous monitoring of equipment performance, environmental conditions, product quality, and operational efficiency. IoT data streams feed into advanced analytics platforms that identify optimization opportunities and predict potential issues before they impact operations. Smart sensors can monitor temperature, humidity, vibration, pressure, and other critical parameters throughout the supply chain, ensuring product quality and compliance with regulatory requirements. This real-time monitoring capability enables proactive management and continuous optimization of logistics operations while reducing costs and improving customer satisfaction.

3.5 Sustainable Logistics, Circular Economy Integration and Green Supply Chain Metrics

Modern logistics excellence increasingly incorporates sustainability principles and circular economy concepts, transforming traditional linear supply chain models into regenerative systems that minimize waste and maximize resource efficiency. The circular economy approach challenges organizations to rethink their logistics operations, focusing on product lifecycle management, reverse logistics, and sustainable sourcing practices [2].

Sustainable logistics excellence requires integration of environmental, social, and economic considerations into all operational decisions. Organizations must balance efficiency objectives with sustainability goals, creating value for all stakeholders while minimizing environmental impact. This approach necessitates new performance metrics, stakeholder engagement strategies, and technological solutions that support sustainable business practices [12, 35].

The implementation of sustainable logistics practices requires comprehensive measurement systems that track environmental impact, resource efficiency, and social responsibility indicators alongside traditional performance metrics. Green supply chain metrics encompass carbon footprint measurement, waste reduction tracking, energy efficiency monitoring, and circular economy effectiveness assessment [21].

Contemporary green logistics frameworks integrate advanced analytics and IoT sensors to provide real-time monitoring of environmental parameters throughout the supply chain. Organizations can track emissions, energy consumption, water usage, and waste generation at granular levels, enabling data-driven decisions that optimize both operational efficiency and environmental performance. These systems enable automated sustainability reporting and regulatory compliance while building stakeholder trust and competitive advantage [1, 16].

4 The IT Project Management Excellence

In general, managing project requires reference to: 1) the so-called "good practices" within the selected methodology and 2) effectiveness in achieving business goals within "The Project Management Triangle" [5, 11]. Contemporary project management excellence has evolved beyond traditional scope, budget, and time constraints to include quality, risk, sustainability, stakeholder satisfaction, digital maturity, and agile delivery capabilities.

In the first case, Project Management good practices assume compliance with key principles and general standards such as: Prince 2, PMBoK. Based on the comparative research conducted by Vartiac regarding: 1) Success Factors for Managing Project Quality (SFMPQ) and 2) the world's best known business excellence model – the Deming Prize [DP], the following factors, activities supporting project excellence were indicated in Fig. 2.

SFMPQ	linkage	DP
1. A good plan	to create a good plan it is necessary to innovate the work process	2. New product development and/or work process innovation
2. Appropriate Communication	clear management policies only exist in a parallel with an appropriate communication	1. Management policies and their deployment regarding quality management
3. Manage Stakeholders	it is not possible to manage Stakeholders without knowing them, providing them with information and giving them opportunity of development	6. Human resources development
4. Good Measurement	in order to measure progress, it is necessary to implement system that measures the quantity and quality of the processes	4. Establishment of systems for managing quality, quantity, delivery, costs, safety, environment, etc.
5. Constant Review	regular reviews of the progress against the schedule are based on the collection and analysis of important information	5. Collection and analysis of quality information and utilization of information technology
6. Act early	to act early means fixing the problems in the daily work, i.e. maintenance and improvement at the operational level	3. Maintenance and improvement of product and operational qualities

Fig. 2. Linking SFMPQ and the DP criteria and excellence project management

From functional perspective, both referenced models (SFMPQ and DP) support similar criteria in achieving Project Excellence. They are both good standards for managers.

On the other hand,"The Project Management Triangle" points out on three success factor, such as: scope, budget and time. This simple approach indicates the necessity of delivering products: accordingly to the scope which was negotiated before signing the agreement, withing the budget and within the time frame.

Contemporary project management excellence increasingly relies on agile and hybrid methodologies that combine traditional project management rigor with adaptive, iterative approaches. These methodologies are particularly suited for ERP implementation projects that must accommodate rapidly changing business requirements, emerging technologies, and evolving user expectations [6]. Modern project management frameworks emphasize continuous delivery, stakeholder collaboration, and adaptive planning. Digital project management tools powered by AI provide real-time project analytics, risk assessment, and resource optimization, enabling project managers to make data-driven decisions and maintain project excellence throughout complex implementation cycles. This is the practical and complete approach towards Project Management Excellence.

5 The ERP Architecture Excellence in the Digital Era

Today, any optimization activity in logistics areas must be supported by properly designed and customized software solutions. The most critical of these is the modern ERP platform, now enhanced with cloud computing, artificial intelligence, and IoT integration capabilities. Contemporary ERP systems have evolved from traditional on-premise solutions to sophisticated cloud-native platforms that provide scalability, flexibility, and real-time analytics.

ERP (Enterprise Resource Planning) business software now refers to "a comprehensive ecosystem of interconnected cloud-based solutions supporting the entire enterprise and its extended value network" [4]. For organizations operating with modern ERP systems, this means total digital transformation of the enterprise, incorporating all key business processes, external partner integration, and Industry 4.0 technologies including IoT sensors, blockchain networks, AI platforms, and predictive analytics engines.

5.1 Cloud-Based ERP and AI Integration

Contemporary ERP systems leverage cloud computing platforms to provide unprecedented scalability, flexibility, and cost-effectiveness [31]. Cloud-native ERP solutions enable real-time collaboration, mobile access, automatic updates, and seamless integration with emerging technologies. AI integration provides intelligent automation, predictive analytics, adaptive user interfaces, and continuous learning capabilities [15].

AI-powered ERP systems can automate routine tasks, identify patterns in business data, predict future trends, and recommend optimal actions. Machine learning algorithms continuously improve system performance by learning from user behavior and business outcomes, creating increasingly intelligent and responsive enterprise platforms. Predictive analytics capabilities enable proactive decision-making and risk management across all business functions [27].

5.2 Integrated Business Flows in Digital Ecosystems

ERP project is an extremely complex business venture, also, from a logistical point of view. It leads to business flows analysis in terms of materials, information and capital in one system's architecture (Fig. 3).

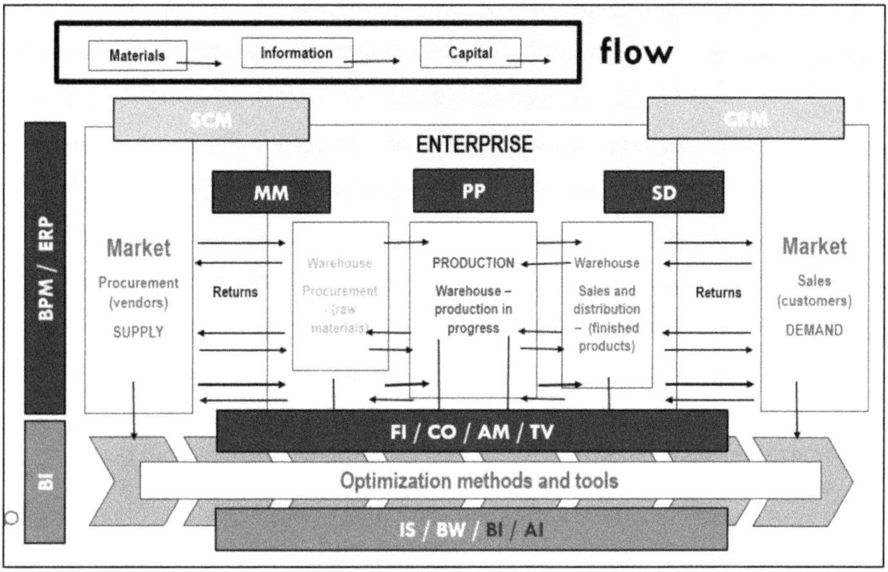

Fig. 3. The integrated flows of material, information and capital and business software solutions

As a matter of fact, those flows are represented in the ERP as business processes / internal system's workflows which need to be customized by appropriate transactions / functions with reference to the given organization's requirements and standards. We have to remember that the value, the quality of the ERP system is proportional to the actual efficiency of the business processes implemented in the company that is using such system for its daily operations. In other words, ERP software solution is just a technical tool that allows the business to work more efficiently, more perfectly through properly, first defined and optimized processes. So, ERP does not replace the optimization methods and techniques developed by logisticians, instead, it delivers sophisticated platform to make them work efficiently.

5.3 Business Process Management and Analytics Integration

From that enhanced perspective, modern ERP systems have evolved beyond their traditional limitations. While earlier ERP implementations focused primarily on orderly registration of economic activity, contemporary platforms provide sophisticated tools for optimization, innovation, and strategic decision-making [36]. To achieve state-of-the-art business ERP ecosystems, organizations now integrate multiple complementary technologies: Advanced Business Process Management Systems (BPMS) with AI-powered process optimization, Next-generation Business Intelligence (BI) platforms with real-time analytics and predictive capabilities, Machine Learning and AI engines for continuous improvement and adaptive automation.

Business Process Management (BPM) has evolved significantly with the integration of digital technologies and real-time analytics capabilities. Contemporary BPM platforms leverage AI-powered process mining, automated workflow optimization, and predictive analytics to continuously improve logistics operations and supply chain performance [23]. These advanced BPM capabilities enable organizations to identify bottlenecks, optimize resource allocation, and adapt quickly to changing market conditions.

Modern BPM methodologies emphasize process transparency, stakeholder collaboration, and continuous improvement. Digital process mapping tools provide real-time visibility into process performance, enabling managers to make data-driven decisions and implement improvements quickly and effectively [17].

6 Conclusions

ERP Project Management in the contemporary business environment is an extremely complex and multifaceted venture that must be analyzed from management, logistics, project management, architecture, sustainability, and digital transformation perspectives. This comprehensive analysis leads to the following evolved definition: ERP Logistics Project Management Excellence means the integrated managerial activity of delivering the right technologies (cloud-native ERP, AI-powered BI, intelligent BPMS, IoT integration, blockchain security), with the right functional scope (including sustainability metrics and circular economy support), at the right quality (user-centered design and continuous improvement), in the right architecture (cloud-native, mobile-first, API-driven), at the right time (agile delivery with continuous value realization), for the right end-users (including all stakeholders in the digital ecosystem), within budget (considering total cost of ownership and value realization), and in the right way (minimizing environmental impact and maximizing stakeholder value) with reference to the integrated flow of material, information, capital, data, and knowledge in Industry 4.0 ecosystems.

Contemporary logistics excellence requires organizations to master both traditional management principles and emerging digital capabilities. The integration of AI, blockchain, IoT, and sustainability principles creates new opportunities for operational excellence while also introducing new complexities and challenges. Success in this environment requires adaptive leadership, continuous learning, and commitment to both technological innovation and environmental stewardship.

The future of logistics excellence lies in the successful integration of human expertise with artificial intelligence, traditional efficiency metrics with sustainability indicators, and local optimization with global ecosystem thinking. Organizations that can master this integration will achieve sustainable competitive advantage in an increasingly complex and interconnected world.

Furthermore, the evolution toward sustainable and digitally enabled supply chains requires new competencies, governance models, and performance measurement frameworks. Excellence in this context means not only achieving operational efficiency but also contributing to broader societal goals including environmental protection, social responsibility, and economic sustainability. The organizations that embrace this holistic approach to excellence will be best positioned to thrive in the digital economy while creating positive impact for all stakeholders.

References

1. Alesiuniene, K.S., Simanaviciene, Z., Stankevicius, A.: Increasing the effectiveness of food supply chain logistics through digital transformation. Int. J. Manag. Econ. **57**(2), 85–98 (2021). https://doi.org/10.14807/ijmp.v12i6.1748
2. ASCM: The role of green logistics in supply chain management. https://www.ascm.org/topics/green-logistics/. Last accessed 17 Oct 2024
3. Attah, R.U., Garba, B.M.P., Gil-Ozoudeh, I., Nwosu, C.K.: Enhancing supply chain resilience through artificial intelligence: analyzing problem-solving approaches in logistics management. Int. J. Manag. Econ. **18**(4), 245–267 (2024). https://doi.org/10.51594/ijmer.v6i12.1745
4. Auksztol, J., Balwierz, P., Chomuszko, M.: SAP. Zrozumieć system ERP. WN PWN, Warszawa (2012)
5. Beheshti, H.M., Blaylock, B.K., Henderson, D.A., Lollar, J.G.: Selection and critical success factors in successful ERP implementation. Compet. Rev. **24**(4), 357–375 (2014). https://doi.org/10.1108/CR-10-2013-0082
6. Carton, F., Adam, F., Sammon, D.: Project management: a case study of a successful ERP implementation. Int. J. Manag. Proj. Bus. **1**(1), 106–124 (2008). https://doi.org/10.1108/17538370810846441
7. Cichosz, M., Wallenburg, C.M., Knemeyer, A.M.: Digital transformation at logistics service providers: barriers, success factors and leading practices. Int. J. Logist. Manag. **31**(2), 226–249 (2020). https://doi.org/10.1108/IJLM-08-2019-0229
8. Dash, R., McMurtrey, M., Rebman, C., Kar, U.K.: Application of artificial intelligence in automation of supply chain management. J. Strateg. Innov. Sustain. **14**(3), 43–53 (2019). https://doi.org/10.33423/jsis.v14i3.2105
9. Deloitte: Using blockchain to drive supply chain transparency and innovation. https://www.deloitte.com/us/en/services/consulting/articles/blockchain-supply-chain-innovation.html. Last accessed 17 Oct 2024
10. DHL: Leveraging blockchain technology to transform logistics. https://www.dhl.com/discover/en-hk/logistics-advice/logistics-insights/blockchain-technology-in-logistics. Last accessed 27 Oct 2024
11. Françoise, O., Bourgault, M., Pellerin, R.: ERP implementation through critical success factors' management. Bus. Process. Manag. J. **15**(3), 371–394 (2009). https://doi.org/10.1108/14637150910960620
12. Gordon, B.: The role of circular economy in sustainable logistics practices. https://benjamingordon.me/the-role-of-circular-economy-in-sustainable-logistics-practices/. Last accessed 17 Jan 2025
13. Goswami, S.S., Mondal, S., Sarkar, S., Gupta, K.K., Sahoo, S.K., Halder, R.: Artificial intelligence-enabled supply chain management: Unlocking new opportunities and challenges. Artif. Intell. Appl. **3**(1), 124–145 (2025). https://doi.org/10.47852/bonviewAIA42021814
14. Hill, N.: Think & Grow Rich. Fawcett Books, New York (1990)
15. John, I.: Cloud-based ERP and AI integration for cost optimization. https://www.researchgate.net/publication/393253599_Cloud-Based_ERP_and_AI_Integration_for_Cost_Optimization. Last accessed 18 Sep 2025
16. Junge, A.L.: Digital transformation technologies as an enabler for sustainable logistics and supply chain processes–an exploratory framework. Braz. J. Oper. Prod. Manag. **16**(3), 462–472 (2019). https://doi.org/10.14488/BJOPM.2019.v16.n3.a9
17. Kernytska, A.: Beyond optimization: The evolving role of business process management in industry transformation. Zeszyty Naukowe Wyższej Szkoły Bankowej w Poznaniu **103**(4), 111–122 (2023). https://doi.org/10.58683/dnswsb.635

18. Khan, S.A.R., Ahmad, Z., Sheikh, A.A., Yu, Z.: Digital transformation, smart technologies, and eco-innovation are paving the way toward sustainable supply chain performance. Sci. Prog. **105**(4), 1–26 (2022). https://doi.org/10.1177/00368504221145648
19. Krupa, M.: W poszukiwaniu idei doskonałości organizacyjnej. In: Materiały Konferencyjne Szkoły Głównej Handlowej, Przedsiębiorstwo partnerskie, Kazimierz Dolny (2002)
20. Mangan, J., Lalwani, C., Butcher, T.: Global Logistics and Supply Chain Management. Wiley & Sons (2008)
21. Nwokocha, G.C.: Achieving operational excellence in North American supply chains: assessing the impact of digital transformation and sustainable practices. Int. J. Membr. Sci. Technol. **31**(4), 1820–1835 (2024). https://doi.org/10.15379/ijmst.v11i1.3826
22. Peters, T., Waterman, R.H.: In Search of Excellence. Lessons from America's Best-Run Companies. Harper-Collins Publishers, USA (1982)
23. Prokopenko, O., Kichuk, Y., Ptashchenko, O., Reznik, N.: Logistics concepts to optimise business processes. Estudios de Economia Aplicada **39**(5), 1–15 (2021). https://doi.org/10.25115/eea.v39i3.4712
24. Red Prairie: The 10 keys to global logistics excellence. https://www.scdigest.com/assets/reps/scdigest_global_logistics_excellence.pdf. Last accessed 27 Nov 2024
25. Richey, R.G., Jr., Chowdhury, S., Davis-Sramek, B., Giannakis, M., Dwivedi, Y.K.: Artificial intelligence in logistics and supply chain management: a primer and roadmap for research. J. Bus. Logist. **44**(4), 532–549 (2023). https://doi.org/10.1111/jbl.12364
26. Russell, R.S., Taylor III, B.W.: Operations Management: Along the Supply Chain, 6th edn. John Wiley & Sons (2009)
27. Samara, T.: Intelligent integrated information systems as an innovation capability. Available at SSRN 5494387 (2024). https://doi.org/10.2139/ssrn.5494387
28. Szczepankiewicz, W.: Logistyka marketingowa. AE w Krakowie, Kraków (1996)
29. Tarigan, M.K., Simatupang, T.M., Basuki, R.: Building resilience through digital transformation: a systematic literature review and comprehensive framework for large enterprises. Int. J. Bus. Innov. Res. **36**(5), 1–28 (2025). https://doi.org/10.1504/IJBIR.2025.145665
30. Tatarkiewicz, W.: O doskonałości. Państwowe Wydawnictwo Naukowe, Warszawa (1976)
31. Tyagi, P., Singh, D.S.P.: Ensuring seamless data flow in SAP TM with XML and other interface solutions. Iconic Res. Eng. J. **8**(4), 156–171 (2024). https://www.irejournals.com/formatedpaper/1706532.pdf. Last accessed 17 Jan 2025
32. Vaka, D.K.: Achieving digital excellence in supply chain through advanced technologies. Educ. Adm.: Theory Pract. **29**(4), 891–906 (2023). https://doi.org/10.53555/kuey.v29i4.5638
33. Vartiak, L.: Achieving excellence in project management. In: 4th World Conference on Business, Economics and Management, Procedia Economics and Finance, vol. 26, pp. 292–299 (2015). https://doi.org/10.1016/S2212-5671(15)00855-2
34. Westerveld, E.: The project excellence model®: linking success criteria and critical success factors. Int. J. Project Manage. **21**(6), 411–418 (2003). https://doi.org/10.1016/S0263-7863(02)00112-6
35. World Economic Forum: how the circular economy secures manufacturing supply chains. https://www.weforum.org/stories/2024/02/how-manufacturers-could-lead-the-way-in-building-the-circular-economy/. Last accessed 17 Jan 2025
36. Žabjek, D., Kovačič, A., Štemberger, M.I.: The influence of business process management and some other CSFs on successful ERP implementation. Bus. Process. Manag. J. **15**(4), 588–608 (2009). https://doi.org/10.1108/14637150910975552
37. We are ProjectSmart. https://www.projectsmart.co.uk. Last accessed 25 Sep 2024

AI-based Logistics Applications

Internal Auditing in Logistics: AI-driven and Advanced Analytics Solutions

Hasan Güney(✉), Ahmet Ustaoğlu, Ahmet Yesevi Türker, Tuğçe Elçi, and Deniz Kantar

Borusan Logistics, İstanbul, Türkiye
{hasan.guney,ahmet.ustaoglu,ahmetyesevi.turker,tugce.elci, deniz.kantar}@borusan.com

Abstract. Internal auditing is important in ensuring compliance, increasing operational efficiency and protecting assets in logistics industry. However, the increasing complexity and data density of logistics operations make traditional control methods insufficient. This requires the integration of artificial intelligence (AI) and advanced analytics into internal auditing. The large volume of data generated in logistics exceeds the processing capabilities of manual controls. Artificial intelligence solves this issue by quickly and accurately analyzing large data sets. Moreover, AI enhances the detection of anomalies and risks through complex algorithms, enabling proactive and predictive audit approaches that detect problems before they escalate. In this paper, artificial intelligence is used to scale down large volume of data. Thus, instead of handling large volume of data manually, a more efficient process can be carried out in problematic, erroneous or suspicious transactions. The paper gives solutions to situations such as fraud in additional service and fuel payments, identify anomalies in trip durations, and pinpoint route discrepancies. To achieve these goals, unsupervised learning tools such as Isolation Forest, DBSCAN, and Local Outlier Factor, along with statistical methods like Z-Score, have been employed. The use of artificial intelligence in internal audit helps increase transparency in shipping operations, maintain cost control and ensure operational efficiency. In this way, it enables faster, more accurate and effective internal audit processes in the logistics industry and has the potential to set a new standard in the industry by enhancing safety and accuracy in transportation activities.

Keywords: Advanced Analytics · Isolation Forest · Supply Chain · Unsupervised Learning

1 Introduction

In today's business environment, operations are closely monitored for efficiency, cost control, and security. Ensuring the safety and accuracy of financial transactions is particularly crucial. Illegal or improper transactions such as anomalies, fraud, and smuggling can negatively impact both the financial health and reputation of companies. In logistics processes, the scope and complexity of operations increase the risk of such issues.

J. Feliks et al. (Eds.): CLC2024 2024, 2026.
https://doi.org/10.1007/978-3-032-28931-5_20

Therefore, detecting anomalies in the logistics sector to prevent fraud not only reduces financial losses but also strengthens the sustainability and reliability of the industry.

Among the commonly used methods for anomaly detection are rule-based approaches, statistical modeling, and unsupervised learning techniques, which are applied across different sectors for fraud and error detection. Within this scope, our company has conducted various anomaly detection studies specific to the logistics sector [1, 2].

This paper focuses on two main areas: fuel card payments to truck drivers and additional service payments provided to suppliers. A portion of the drivers' payments is made via fuel cards issued by partner fuel companies. In the first study, fuel card payments to drivers were analyzed based on the dates of payments and the names of fuel cardholders, and potential anomalies in this area were examined in detail using rule-based anomaly detection steps. The second section addresses payments made to suppliers for additional services provided beyond their agreed entitlements. Potential fraud, misuse, and "ghost payments" in this area were evaluated and anomalies were detected using unsupervised learning methods.

The statistical and artificial intelligence models used in both areas are described in detail in the following sections.

2 Methodology

Two different datasets were used to detect anomalies in fuel payment and additional services, with the aim of identifying fraud, misuse, or erroneous entries. The first dataset, the fuel payment dataset, contains information related to fuel payment transactions. The second dataset, the additional service dataset, includes data regarding additional service payments. Due to the different nature and operations of these two datasets, separate analyses and systems were developed to handle each. Therefore, fuel and additional services were analyzed using distinct methods. In fuel payments, analyses focused on identifying discrepancies in fuel loading dates and examining the relationship between suppliers and cardholders. For additional services, unsupervised learning tools were used to identify potential fraudulent transactions.

2.1 Fuel Payments

The primary objective for fuel payments was to detect any erroneous or potentially fraudulent data. To achieve this, initial checks were conducted to verify the consistency of fuel payment dates with trip timelines. Following this, several analyses were performed to detect inconsistencies in card ownership. The dataset used in these analyses contained a total of 357,141 rows.

Fuel Payment Dates

Under normal conditions, fuel payments are expected to occur after the trip has begun. Payments made between the start and end of a trip are considered normal, while those made before the trip starts are deemed anomalous. This is because fuel payments should be tied to a specific trip, and making payments before the trip begins could indicate suspicious behavior. In this analysis, records where the fuel payment date was earlier than the trip start date were classified as anomalies. The analysis revealed that 6,798 out of 357,141 records met this criterion and were identified as anomalies.

Card Ownership

This analysis aimed to detect anomalies related to fuel cards assigned to drivers. The first step was to examine the number of active cards held by each cardholder. Ideally, each cardholder should have no more than one active card. Having multiple active cards for a single cardholder was flagged as an anomaly. This analysis resulted in 2,012 anomalies out of a total of 23,859 records related to card ownership.

An additional analysis was conducted to investigate any relationship between cardholders and suppliers who loaded funds onto these cards. This analysis examined similarities in names to identify potential connections between the supplier and the cardholder. For this purpose, two custom functions were developed: SimilarityFinder and DifferenceFinder. The SimilarityFinder function identifies common words by comparing the names of the cardholder and supplier. The DifferenceFinder function focuses on detecting differences between two strings. Using these functions, the data was divided into three categories: similar_df, same_df, and different_df. The same_df dataset, created using adjusted parameters in the SimilarityFinder function, lists records where the card user and supplier have identical names. The similar_df dataset, also created with the SimilarityFinder function but with different parameters, contains records with only surname similarities. The different_df dataset, created by the DifferenceFinder function, includes records where there is no name or company-related string similarity between the card user and supplier. The main focus of this analysis is the different_df dataset, as a lack of connection between the individual loading funds onto the card and the cardholder could indicate suspicious activity.

2.2 Additional Service Payments

The main goal in additional service payments was to determine whether a transaction could be classified as fraudulent. Therefore, an unsupervised learning approach was employed to detect anomalies in additional service payments. The additional service dataset consists of 59,020 rows and contains information on additional service type, the amount paid, the type of vehicle used, departure and arrival cities, and other relevant details. A small number of missing records that would not impact the model were removed from the dataset. Date information was also separated into day, month, and year to suit the model.

During the data preprocessing phase, separating additional services based on customer and route specifics was an important step. Certain additional services are customer-related (e.g., handling fees, supplier commission, and cancellation fees), while others are route-related (e.g., extra kilometers, stopover fees, and ferry crossing fees). To ensure optimal model performance, these services were separated based on customer and route.

In the data preprocessing phase, categorical variables were transformed into numerical form using One-Hot Encoding. Examples of these categorical variables include customer name, additional service type, and the cities for departure and arrival [3].

During Feature Engineering, the statistical method Z-score was employed. This approach was applied to both customer and route data to identify services with above-average spending. Z-score values were also added to the model to enhance its performance [4].

Four models with varying structures were used in unsupervised learning, depending on data density: Isolation Forest, K-Nearest Neighbors (KNN), Density-Based Spatial Clustering of Applications with Noise (DBSCAN), and Local Outlier Factor (LOF) [5–7]. Isolation Forest is a model based on decision trees; KNN identifies anomalies based on the nearest neighbor distance; DBSCAN and LOF are density-based models that detect anomaly points. All models were applied to the data, and each model separately identified anomalies. The diversity in detected anomalies was expected due to the different working principles of each model [8]. This approach was aimed at ensuring that the results included the data points with the least error risk.

3 Experimental Results

After the necessary filtering and data cleaning processes, a total of 13,699 rows remained in the customer dataset, and 45,321 rows remained in the route dataset. For evaluating the results, the Silhouette Score metric, which measures how well-separated clusters are and the similarity of points within clusters, was utilized [9, 10]. However, this metric was not directly applicable for Local Outlier Factor (LOF), as it is not a clustering algorithm but instead calculates an "outlier score" for each data point. Similarly, Isolation Forest was not suitable for the Silhouette Score because it is designed specifically for anomaly detection rather than clustering. Thus, the Silhouette Score was applied only to KNN and DBSCAN.

For the Isolation Forest and LOF models, the ROC AUC (Receiver Operating Characteristic—Area Under Curve) metric was used [10]. This metric compares the true positive and false positive rates to assess the model's ability to distinguish anomalies effectively. The tables below present the average results of these metrics for the route and customer datasets.

Table 1. Silhouette scores for KNN and DBSCAN

Element	Silhouette score
KNN	0.797
DBSCAN	0.817

The scores shown in Table 1 and Table 2 indicate that the anomaly class has been correctly differentiated. The data points identified as anomalies in these models were appropriately labeled. Finally, data points marked as anomalies by all models were labeled as isSuspicious. Tables 3 and 4 show the number of anomalies detected in each model within the customer and route datasets, along with the total number of records labeled as isSuspicious.

Table 2. ROC AUC for isolation forest and LOF

Element	ROC AUC score
Isolation forest	0.894
LOF	0.843

Table 3. Anomaly counts in the customer dataset by model

Model	Normal	Anomaly
Isolation forest	11,399	2,300
KNN	12,004	1,695
DBSCAN	11,707	1,992
LOF	11,324	2,375
All Data	13,425	272

Table 4. Anomaly counts in the route dataset by model

Model	Normal	Anomaly
Isolation forest	41,741	3,580
KNN	40,276	5,045
DBSCAN	44,765	556
LOF	38,680	6,641
All data	45,282	39

As shown in Tables 3 and 4, a total of 272 suspicious records were identified in the customer dataset and 39 in the route dataset. These values are notably lower than the anomaly counts detected by each model individually, indicating that each model flagged different records as anomalies. This diversity can be attributed to the unique functioning principles of each model, emphasizing the need to carefully review records consistently marked as anomalies across multiple models. Through this study, the original dataset, which contained a total of 59,018 records, was reduced to 311 critical records that require close monitoring. Under normal circumstances, it would be necessary to manually review all 59,018 records; however, by employing artificial intelligence, this number was reduced to 311, making the review process significantly more manageable and efficient.

4 Discussion

This study's analysis of anomaly detection in fuel and additional service payments within the logistics sector can be seen as an important step toward strengthening financial security in the industry. The findings provide elements that enable businesses to conduct cost control and process management more effectively. Combining rule-based and unsupervised learning methods has allowed for a broader perspective in evaluating different types of fraud risks.

In alignment with prior studies in the field, this research demonstrates that anomaly detection not only prevents financial loss but also enhances operational efficiency. However, some findings in this study are limited by the complexity of logistics processes. For instance, relying solely on data-based anomaly detection for fuel card payments may lead to overlooked errors or false positives. Additionally, in cases where rule-based analyses are insufficient, a more extensive data cleaning and preprocessing process may be required to improve the performance of unsupervised learning models.

Apart from the fuel card payments and additional service payments mentioned above, hybrid models that combine unsupervised learning and statistical modeling can also be applied in other areas. For instance, with GPS tracking, GPS-Time Analysis can be used to analyze idle time, time spent throughout the entire transportation period, and location data, allowing for anomaly detection based on time spent at each location and throughout the trip. Similarly, GPS-Location Analysis can monitor the direction and route followed by the vehicle using GPS data, making it possible to detect deviations from the expected route and identify potential anomalies.

In conclusion, this study serves as an important example of how to build a reliable and comprehensive anomaly detection system for the logistics sector. The applicability of these findings provides a strong foundation for future studies, contributing to the sector's sustainability and security.

5 Future Works

The statistical model and artificial intelligence algorithms used in this study can serve as a foundation for future research. In particular, developing hybrid models through combinations of different algorithms or optimization adjustments tailored to sector dynamics could enhance anomaly detection. The intended future work involves communicating potential anomalies identified by unsupervised models or statistical methods to business units and the internal audit team for review. These instances would then be labeled as "anomaly" or "normal," allowing us later for supervised learning with labeled data. This approach would improve data quality, strengthen the connection between features and anomaly status, and enable the anomaly detection model to operate with a higher success rate.

6 Conclusion

In light of the topics discussed above, we can conclude that operational efficiency, cost control, and security are paramount in today's business world; hence, detecting anomalies, fraud, and improper transactions is crucial for financial integrity and reputation.

This study, focusing on the logistics sector, has provided important findings aimed at preventing fraud by identifying potential irregularities in fuel and additional service payments.

Rule-based analyses for fuel payments revealed financial risks by detecting fuel payments made before trips began and mismatches in cardholder names. For additional service payments, the use of four unsupervised learning models identified transactions with potential fraud risks. The preprocessing and modeling processes applied within this scope have enabled the identification of high-risk transactions in logistics payments.

In summary, this hybrid analysis approach, specified for various scenarios in the logistics sector, offers an effective method for enhancing security by enabling early detection of anomalies such as fraud, ghost transactions, and smuggling.

Acknowledgements. Hasan Güney, Ahmet Ustaoğlu, Ahmet Yesevi Türker, Tuğçe Elçi and Deniz Kantar are members of Data and Artificial Intelligence Services in Borusan Logistics.

References

1. Ester, M., Kriegel, H.-P., Sander, J., Xu, X.: A density-based algorithm for discovering clusters in large spatial databases with noise. In: Proceedings of the Second International Conference on Knowledge Discovery and Data Mining. AAAI Press, Portland (1996)
2. Jin, W., Tung, A.K.H., Han, J.: Mining top-n local outliers in large databases. In: Proceedings of the Seventh ACM SIGKDD International Conference on Knowledge Discovery and Data Mining. ACM Press, San Francisco (2001)
3. Brownlee, J.: One-hot encoding for categorical data. In: Machine Learning Mastery (2020)
4. Field, A.: Discovering Statistics Using IBM SPSS Statistics. SAGE Publications, London (2013)
5. Cover, T.M., Hart, P.E.: Nearest neighbor pattern classification. IEEE Trans. Inf. Theory. **13**(1), 21–27 (1967)
6. Liu, F.T., Ting, K.M., Zhou, Z.-H.: Isolation Forest. IEEE Int. Conf. Data Min. (2008)
7. Liu, F.T., Ting, K.M., Zhou, Z.-H.: Isolation-based anomaly detection. ACM Trans. Knowl. Discov. Data. (2012)
8. Quinlan, J.R.: C4.5: Programs for Machine Learning. Morgan Kaufmann, San Mateo (1993)
9. Rousseeuw, P.J.: Silhouettes: a graphical aid to the interpretation and validation of cluster analysis. J. Comput. Appl. Math. **20**, 53–65 (1987)
10. Kaufman, L., Rousseeuw, P.J.: Finding Groups in Data: an Introduction to Cluster Analysis. John Wiley & Sons, Hoboken (1990)
11. Fawcett, T.: An introduction to ROC analysis. Pattern Recogn. Lett. **27**(8), 861–874 (2006)

Evaluation of Artificial Intelligence Methods for Lead Time Prediction in Non-Cycled Areas of Automotive Production

Cornelius Hake[1,2](✉), Jonas Weigele[1,3], Frederik Reichert[3], and Christian Friedrich[2]

[1] Dr. Ing. h.c. F. Porsche AG, Stuttgart, Germany
cornelius.hake1@porsche.de
[2] Hochschule Karlsruhe—University of Applied Sciences, Karlsruhe, Germany
[3] Hochschule Esslingen—University of Applied Sciences, Esslingen, Germany

Abstract. The present study examines the effectiveness of applying Artificial Intelligence methods in an automotive production environment to predict unknown lead times in a non-cycle-controlled production area. Data structures are analyzed to identify contextual features and then preprocessed using one-hot encoding. Methods selection focuses on supervised machine learning techniques. In supervised learning methods, regression and classification methods are evaluated. Continuous regression based on target size distribution is not feasible. Classification methods analysis shows that Ensemble Learning and Support Vector Machines are the most suitable. Preliminary study results indicate that gradient boosting algorithms LightGBM, XGBoost, and CatBoost yield the best results. After further testing and extensive hyperparameter optimization, the final method choice is the LightGBM algorithm. Depending on feature availability and prediction interval granularity, relative prediction accuracies of up to 90% can be achieved. Further tests highlight the importance of periodic retraining of AI models to accurately represent complex production processes using the database. The research demonstrates that AI methods can be effectively applied to highly variable production data, adding business value by providing an additional metric for various control tasks while outperforming current non AI-based systems.

Keywords: Lead Time Prediction · Automotive Production · Machine Learning · Gradient Boosting · Artificial Intelligence

1 Introduction and Literature Review

Production planning and control (PPC) is an essential component for controlling and operating a production system. In contrast to vehicle assembly, the vehicles in the Test and Finish Centre (TFC) are no longer cycled due to the individual scopes of testing and work. As a result, the sequence of vehicles can be arbitrary, which leads to a need to optimize the sequence in PPC. The control of a TFC is an extremely complex task that depends on many influencing factors, especially in the production of vehicles with

J. Feliks et al. (Eds.): CLC2024 2024, 2026.
https://doi.org/10.1007/978-3-032-28931-5_21

a high number of variants and a low number of units. For this reason, the exact lead time of a vehicle through the TFC cannot be determined with certainty at the present time [1].

In this context, the term lead time is defined as the interval between the sixth and eighth reporting points, which are assigned when a defined change of location of the vehicle occurs within the production chain. The sixth reporting point marks the conclusion of assembly and the transfer of the vehicle to the TFC, while the eighth reporting point signifies the completion of the vehicle and the transfer of ownership from the production to the sales department [1]. The ability to predict the period between the sixth and eighth reporting points with a high degree of accuracy presents several potential advantages. From the perspective of the TFC, a further parameter is established, which serves to optimize the decision-making process for the scheduling and retrieval of individual vehicles. Furthermore, customers benefit from the enhanced precision of delivery dates. From the standpoint of sales, there is the potential for optimization in the planning of transportation. The explainable decision-making process of the AI model enables a root cause analysis, which provides information on the characteristics that are currently decisive in determining whether a vehicle has a longer or shorter lead time through the TFC. This offers the potential to optimize the TFC processes in a subsequent step.

The results of the literature research show that there are several works that deal with the prediction of lead times in production based on AI. It is important to note, that only the author's company own past research and that of Krzoska et al. specifically consider lead times in automotive production [2]. Nevertheless, all studies pursue the goal on the precise prediction of lead times in industrial production using AI methods.

Previous research by the author's company investigated two methods for predicting lead times in a TFC with AI: regression and classification using LightGBM (LGBM). The contextual features analyzed included four configuration variants (order type, color selection, steering wheel position, and special refinement) and all inspection codes recorded before the sixth reporting point. However, the limited selection of configuration variants reflects subjective feature relevance, potentially missing complex relationships in production. While LGBM can independently select relevant features, the previous approach did not fully exploit this potential. The sufficiency of the four configuration variants and the inspection codes as features is questionable, as shown by the model training results in Table 1. The LGBM regression achieved a prediction accuracy of 22%, while the LGBM classification accuracy ranged between 65 and 71%, depending on the class distribution for the time intervals. No relevant optimization of the LGBM hyperparameters was performed. The results indicate that AI can predict vehicle lead times in a TFC, but the current results are not sufficient for series application. Further optimization and research on feature selection, method selection, and model-specific hyperparameters are needed.

The study from Krzoska et al. focuses on the prediction of rework times in automotive production. The aim is not to predict the total lead time of vehicles in a TFC, but rather the individual rework times in the TFC to rectify specific vehicle faults. To this end, process-related quality data from a MES is analyzed and the prediction of rework times is tested using vehicle-specific regression models. The regression tree methods Reduced Error Pruning Tree (REPtree) and M5 Model Trees (M5P) were analyzed. Results show that simply including MES data is insufficient, with prediction accuracy ranging from 48.3% to 58.3%. Manually removing outliers improves the accuracy from 64.5% to 69.6%. Introducing a severity rating factor can further improve accuracy to 84.9%. REP-tree consistently outperforms M5P. However, the use of a weighting factor involves subjective human judgment and risks system manipulation. The study only considers individual rework times and shows relative errors up to 23.7%. For cumulative errors, all processing times must be summed to predict total lead time. Further research is needed for AI methods to accurately predict total lead time in a TFC [2].

Table 1 Previous research results regarding class distribution and prediction accuracy

Algorithm	Method	Time intervals [d]	Acc [%]
LGBM	Regression	$< 1; 1 - 3; > 3$	22
LGBM	Classification	$< 1; 1 - 3; > 3$	69
LGBM	Classification	$< 1; 1 - 2; > 2$	71
LGBM	Classification	$< 1; 1 - 3; 3 - 7; > 7$	65

Further work in this field relates to other industries to predict lead times in production. Matzka [3] and von der Hude [4] discuss two different approaches to realize this. On the one hand, it is possible to use a trained regression model to form exact and continuous value predictions for the lead time. On the other hand, discrete predictions can be made in the form of time intervals using classification methods.

The studies by Krzoska et al. [2], Lingitz et al. [5], Murphy et al. [6], Barzizza et al. [7], Aslan et al. [8] and Gyulai et al. [9] exclusively examine regression methods for the formation of lead time predictions. The studies by Lim et al. [10] and Welsing et al. [11] are limited to research in the field of classification methods. The study by Bender [12] examines both regression and classification methods. Regardless of regression or classification, ensemble learning methods, such as random forest or gradient boosting, achieve the best results in six out of nine papers. Only the studies conducted by Krzoska et al. [2] and Welsing et al. [11] have demonstrated that regression trees and cross-validated decision trees are the most effective in terms of prediction accuracy. The study by Lim et al. [10] reaches the conclusion that Support Vector Machines (SVM) result in superior prediction outcomes in comparison to the other AI models. The work by Barzizza et al. [7] demonstrates the potential of combining clustering with machine learning models, particularly Random Forest, to improve the accuracy of lead time predictions. Krätschmer [13] and Bentéjac et al. [14] highlight the relevance of optimizing model-specific hyperparameters to improve the performance and generalization capability of AI models. However, only the studies by Krzoska et al. [2] and Lingitz et al. [5] take this into account in their investigations. In general, the studies by Barzizza et al. [7], Aslan et al. [8], Murphy et al. [6] and Gyulai et al. [9] show that AI models are superior to traditional systems for predicting production lead times.

In this study, two major contributions are made to the existing literature on the subject:

1. Method development for AI-based prediction of lead times for vehicles produced in low unit numbers and individual test scopes.
2. Comparison of AI-based and current rule-based prediction of lead times.

2 Methods and Materials

2.1 Experimental Setup

Python is used as the programming language for the algorithms, which are computed on a workstation with an Intel Core I9 12900KF processor (16 cores, max. 5.20 GHz), 64 GB RAM, and an Nvidia RTX 3060 GPU with 12 GB GDDR6 graphics memory [15, 16]. To perform the analysis, it is necessary to collect, store and access the production data for the AI application, as illustrated in Fig. 1.

It is important to use a unique identification key with a vehicle reference to link data from various sources and assign it to a specific vehicle. The data can be available in metric, ordinal, nominal and unstructured form in each source, which is why standardized data preparation must be carried out for further processing.

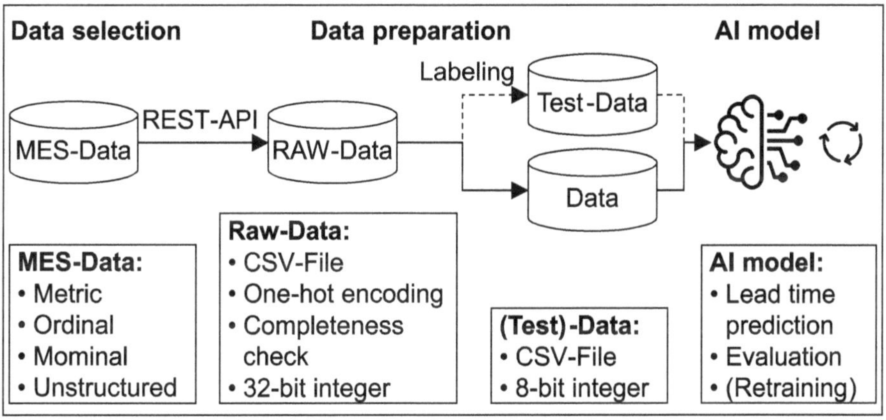

Fig. 1 Illustration of the data pipeline; dashed line: test, continuous line: series application

2.2 Obtaining and Preprocessing the Training Data

The first step is to select contextual data. A data model is searched for data influencing the vehicle lead time in the TFC. A distinction is made between characteristics known at the time of TFC entry and those recorded during the process. Known characteristics at entry include configuration variants, specific product key (vehicle derivative), and entry time (weekday and hour) as illustrated in Fig. 2. These define product complexity and shift time, considering shift changes and downtime. Additional characteristics captured during progression include priority, parking locations, and inspection type/location combination. Specific inspection codes for each vehicle influence the processing time in the TFC. Table 2 provides an overview of relevant characteristics, data types and processing types. Non-series vehicles are excluded from the analysis. The product key distinguishes between internal combustion engine vehicle (ICE) and battery electric vehicle (BEV) data.

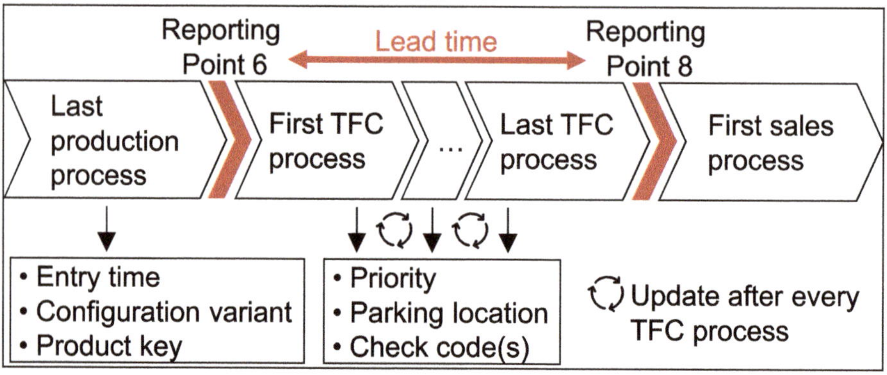

Fig. 2 TFC process with explanation of data availability and lead time calculation

Table 2 Overview of data characteristics, their distribution per vehicle after TFC and preparation techniques

Characteristic	Data type	Preparation method	Example	n per vehicle
Priority	Structured, ordinal	Not necessary	0 or 1	1
Inspection code	Structured, nominal	One-Hot-Encoding	1234-9999F	9422
Parking location	Unstructured, text	One-Hot-Encoding	Parking01-Level01	68
Product key	Structured, nominal	One-Hot-Encoding	A12BC	1
Entry time	Structured, nominal	One-Hot-Encoding	Friday-19	1
Configuration variant	Structured, nominal	One-Hot-Encoding	AA1	2673

Raw data includes both structured and unstructured forms. Since ML algorithms cannot directly process unstructured data, data preparation is essential. Features must remain unbiased by additional information. One-hot encoding is used to resolve this problem. One-hot encoding creates a unique, binary data set that allows all ML algorithms to process the same data set, making the computational results comparable. One-hot encoding increases the dimensionality of the data set. During dimension reduction, columns filled with null values are removed as they are irrelevant, thus reducing the dimensionality. An upstream check ensures the completeness of data points in the exported CSV files, and vehicles with incomplete characteristics are removed [17].

The data preparation described above is used to prepare several data sets for the training and testing process. The AI models to be investigated are trained with all data sets presented in Table 3. The cases of feature availability are the known features when entering the TFC (data set = Limited) and all available features when passing through the TFC (data set = Unlimited). For the different production lines, a distinction is made between ICE and BEV vehicles. For development reasons, a common data set is also created for both production lines (Derivative = All).

Table 3 Overview of data records by derivatives with the number of characteristics and data points

Data set	Derivative	Abbreviation	n Features	n Vehicles
Limited	All	LA	1,326	80,495
Limited	ICE	LI	1,326	45,706
Limited	BEV	LB	1,326	34,789
Unlimited	All	UA	12,166	80,495
Unlimited	ICE	UI	12,166	45,706
Unlimited	BEV	UB	12,166	34,789

2.3 Selection Process and Methodology for AI Model Evaluation

Two prediction approaches can be considered [18]: exact continuous value prediction using regression models and discrete prediction using classification methods [3, 4]. Regression models with supervised learning require a linear data distribution, but since TFC lead times follow an exponential distribution, linear regression is inappropriate. Non-linear regression models are excluded due to their high resource requirements. Classification methods considered include decision trees, ensemble learning, naive Bayes, k-nearest neighbor, SVMs, and neural networks. Neural networks are resource intensive and lack explanatory power, while naive Bayes assumes feature independence, leading to poor generalization. K-nearest neighbor requires all data points, resulting in high resource requirements and longer classification times. Ensemble learning methods and SVMs are suitable due to good generalization and fast classification times. However, ensemble learning with decision trees loses explainability with more trees. Nevertheless, ensemble learning performs feature selection during training, which improves model performance and allows feature relevance analysis [3, 17, 19]. A preliminary experiment is necessary to make the final selection decision. For this purpose, the specific methods of ensemble learning, and SVMs are first selected below [20].

Ensemble learning methods such as random forest (bagging) and gradient boosting machines (boosting) are particularly suitable. Bagging reduces model variance by aggregating multiple independent models but can introduce bias and reduce prediction accuracy [21]. Boosting algorithms, such as LGBM, XGBoost (XGB), and CatBoost (CB), address this by weighting observations. Bentéjac et al. [14] highlight these algorithms for their generalizability, training performance, and predictive precision. Therefore, the analysis focuses on these specific ML algorithms regarding their suitability for the prediction of lead times [22–24]. For configuring SVMs, choosing the kernel and strategy for multi-class classifications is crucial. The RBF kernel is ideal for complex data requiring non-linear separation. One-vs-All or One-vs-One methods are used, depending on the SVM library. ThunderSVM (TSVM), which supports OVO for multi-class classification, is suitable for GPU-based computations with the given setup.

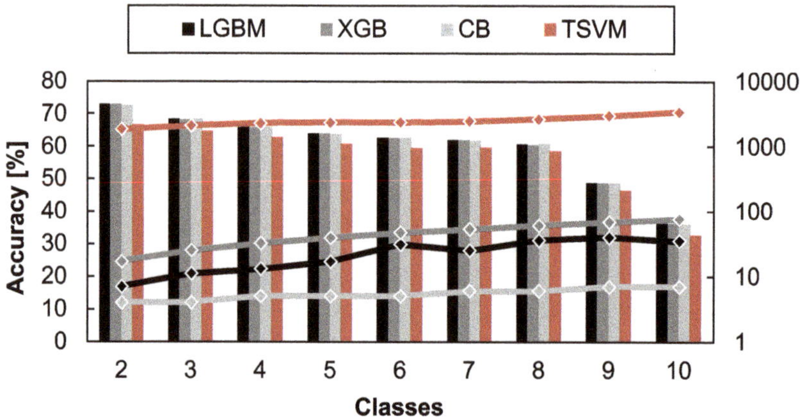

Fig. 3 Results of the prediction accuracy (bins) and training times (lines) in the preliminary experiment

In a preliminary experiment on the LA data set, LGBM, XGB, and CB achieved up to 73% accuracy, while TSVM reached only 66% as illustrated in Fig. 3. Gradient boosting machines trained faster and performed better, with TSVM taking 3348 s to train, 100 times slower than GBDT models. Due to suboptimal results TSVM is excluded from further analysis.

2.4 Labeling the Data

In the supervised learning methodology, each data point requires an associated target variable or label that represents the expected output based on its attributes. For discrete labeling of the data points, groups are formed based on the actual lead times of the vehicles. Using domain-specific knowledge, the groups are defined based on time intervals by analyzing frequency distributions and the general relevance for the company to assign the data points. To predict the lead times between the sixth and eighth reporting points, both the operational benefit and sufficient data distribution within the classes must be considered. The exact time intervals of the classes are shown in Table 4. Initially, nine test series are formed with an increasing number of classes. These start with two different classes and extend up to ten different classes. The classification for the lead time prediction starts with two classes: Vehicles that reach the eighth reporting point within 24 h from the sixth reporting point (55.7%), and those that do so after 24 h (44.3%).

With increasing class division, the class with the most data points is subdivided more finely. With eight classes, except for the direct runners, no class contains more than 10% of the data. The labels must be assigned to each data set as a target value for the training process.

Table 4 Distribution of classes and the corresponding time intervals

Number of classes	Intervals [d]									
	1	2	3	4	5	6	7	8	9	10
2	[≤ 1]	[> 1]								
3	[≤ 1]	[1–3]	[> 3]							
4	[≤ 1]	[1–3]	[3–7]	[> 7]						
5	[≤ 1]	[1–3]	[3–7]	[7–28]	[> 28]					
6	[≤ 1]	]1–3]	[3–7]	[7–28]	[28–84]	[> 84]				
7	[≤ 1]	]1–2]	[2–3]	[3–7]	[7–28]	[28–84]	[> 84]			
8	[≤ 1]	[1–2]	[2–3]	[3–7]	[7–14]	[14–28]	[28–84]	[> 84]		
9	[≤ 0.5]	[0.5–1]	[1–2]	[2–3]	[3–7]	[7–14]	[14–28]	[28–84]	[> 84]	
10	[≤ 0.25]	[0.25–0.5]	[0.5–1]	[1–2]	[2–3]	[3–7]	[7–14]	[14–28]	[28–84]	[> 84]

2.5 Training Procedure for AI Methods

This section describes the preparation, training, and validation of gradient boosting machines. First, a preliminary experiment with standard hyperparameters determines achievable target classes for which complex hyperparameter optimization is performed later. The data set is partitioned into 80% training, 10% validation, and 10% test data, ensuring that each class is equally represented by stratified splitting and using the random state value 42 for consistency. After data preparation, model-specific default hyperparameters are defined, including multi-class classification via loss functions. Models are initialized with developer-defined default hyperparameters. The training process records training time, feature relevance, number of decision trees, and the AI model. To improve the performance, robustness, and generalization of GBDT models, it is recommended to optimize model-specific hyperparameters such as learning rate from 0.01 to 0.3, number of leaves from 32 to 1024, max depth from 2 to 200, leaf est. iterations from 1 to 10, and leaf reg. from 1 to 10 [4]. This involves testing different hyperparameter combinations using grid search and k-fold cross-validation, and then evaluating their impact on training results. Due to the high resource requirements, it's performed on the LA and UA data sets, with the results applied to the ICE and BEV data sets. A fivefold cross-validation grid search is used, with 80% of the data for training and 20% for evaluation. Accuracy is chosen as the evaluation metric for the cross-validation results.

3 Results

3.1 Training Results

Finally, LGBM, XGB and CB are trained and tested for prediction accuracy. The test data labels are predicted by each model, and the actual labels are used to calculate the percentage of correct predictions, resulting in a relative prediction accuracy. Models that achieve at least 80% prediction accuracy on the test data are further evaluated. The evaluation includes parameters such as prediction accuracy, training time, and number of decision trees generated.

The analysis shows that within the UA data set, the experimental design goals are only met for class numbers between two and four with prediction accuracies up to 90%. For finer classifications from nine to ten classes, the prediction accuracy decreases significantly to 32%. The LGBM model slightly outperforms the XGB and CB models in quality by two to four classes, using 460 instead of 999 (XGB) or 997 (CB) decision trees, highlighting its efficiency. In terms of training times, XGB takes 141 s, while CB is the fastest at 9 s, with LGBM also performing well at 16 s. To efficiently meet the test objectives, target classes must be limited to two to four, highlighting the predictability of wider intervals. Overall, the LGBM model leads in prediction accuracy and efficiency. The analysis of the training results of the gradient boosting machines using the LA data set indicates an adjustment of the target value of the relative prediction accuracy to 70% due to the reduced number of features. The results reveal that the target value is only achieved with a classification into two classes. Here, all three models examined show almost equivalent performance in terms of relative prediction accuracy up to 73%. Here too, the LGBM model requires 308 instead of 838 (XGB) or 711 (CB) decision trees for comparable prediction performance. The evaluation of the training times is negligible since all models were trained in less than 20 s. In all cases, the F1 score reaches values above 0.92. From the synergy of reduced number of decision trees, solid prediction accuracy and short training time, the LGBM model emerges as the best in this comparative analysis. The results show that the desired accuracy is only achieved for class numbers of two to four using the UA, UI and UB data sets. For the data sets LA, LI, and LB, the prediction target is only achieved for two classes. Based on these findings, the resource-intensive hyperparameter optimization is only performed for those data sets for which the determined number of classes is available. Hyperparameter optimization of the LGBM, XGB, and CB models yields different optimal values. For LGBM and XGB, the default learning rates are generally optimal, while CB requires different combinations for different data sets. Tightening the tree evolution constraint helps prevent overfitting. K-fold cross-validation results show comparable performance for all models, with LGBM and XGB slightly outperforming CB. Training times were not compared because the optimization was performed only once. The k-fold cross-validation results for the LA data set in Table 5 show that all models perform similarly with minor deviations. LGBM has a deviation of ± 1.2%, XGB ± 1.4%, and CB ± 1.3%. LGBM and XGB achieve an average result of 71.3% accuracy, slightly better than CB's 71.0% accuracy. In conclusion, all three models show good generalization capabilities, with LGBM performing best overall.

Table 5 Results in form of the accuracy in % of the k-fold cross-validation for each gbdt model

Model	k = 1	k = 2	k = 3	k = 4	k = 5	Average
LGBM	71.7	72.5	70.0	70.5	72.1	71.3
XGB	71.8	72.7	69.9	70.2	72.2	71.3
CB	71.8	72.3	69.9	69.9	71.1	71.0

3.2 Evaluation of the Prediction Quality

A comparison is made between the actual, rule-based and target, AI-based systems for 465 vehicles. The planned date of the actual system is created once before it is entered into the TFC and is not continuously updated as in the target system. As a result, LGBM achieves up to 65% higher prediction accuracy than the actual system for data records without characteristic limits as it is illustrated in Fig. 4. The systems are best compared using the feature-limited data sets LA, LI, and LB, as in these cases the systems work with the same data basis. The results show that the LGBM model of the ICE production line achieves 32.8% better prediction accuracy than the actual system with a data preparation and classification time of 146 s. In the case of the BEV production line, a prediction accuracy 11.5% better than the actual system can be achieved with a data preparation and classification time of 156 s. The lower accuracy value is due to process-related changes implemented during this work. The test results of this comparison illustrate how powerful AI models can be compared to conventional systems. In all cases, the use of AI can achieve a significantly higher prediction accuracy of vehicle lead times in the TFC.

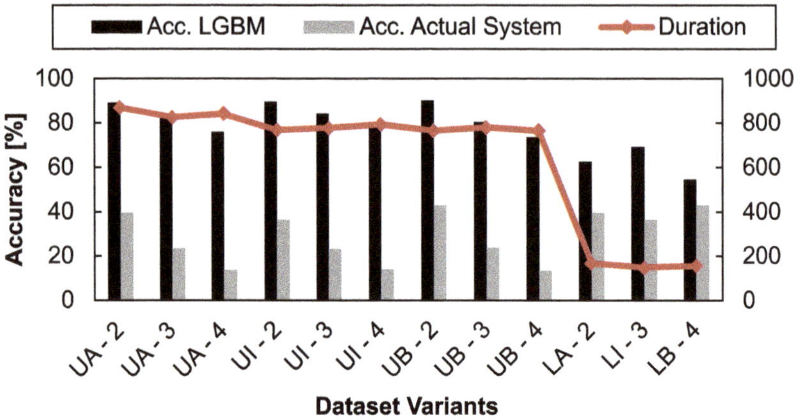

Fig. 4 Results from the comparison with the actual, rule-based system

4 Conclusion and Future Work

The research results show that AI-based prediction of lead times with GBDT models is possible for vehicles produced in low unit numbers and individual test scopes. The work also highlights the limitations of the system. Due to a process change in the BEV production line, the training data no longer accurately reflects the real production processes. As a result, the test results of the GBDT model are worse than the development results, which shows the high requirements for a good training database. The result illustrates the relevance of periodic retraining of the AI models when they are used in a real production environment. LGBM, as the GBTD model with the best test results, offers for the first time the ability to predict the lead time of a vehicle in the TFC with up to 90% accuracy. This provides an additional decision-making parameter for production planning and control in the TFC. The test results show the improvement of the AI-based system compared to the actual, rule-based system. Using hyperparameter optimization, the average improvement in prediction accuracy was only 2%.

A subsequent analysis of the relevance of the characteristics identified procedural inefficiencies. Two further research tasks arise from the research results. The first research task relates to the development of improved algorithms for explainable AI. Explainable decision-making enables a detailed analysis of the causes, creates transparency, and thus ensures that AI applications are widely accepted. The second research task is to develop an algorithm for the automated optimization of the time periods between the retraining of the AI models. The retraining ensures constant, high prediction quality. At the same time, training AI models is resource intensive and time consuming. Optimizing the number of retraining ensures that the prediction quality remains high while operating as economically as possible. This shows the relevance and motivation for further research in the field of AI applications in production.

References

1. Markus, K: *Produktionsleitsysteme für die Automobilindustrie.* Berlin, Heidelberg, Springer Vieweg (2021)
2. Sven, K., Michel, E., Jacqueline, S., Jochen, D.: Data Mining zur Nacharbeitsdauerprognose/data mining-based forecasting of rework duration—predictive rework control and work process optimization in the automotive assembly. Werkstattstechnik. Bd. 107, Nr. 10, S. 773–778 (2017). https://doi.org/10.37544/1436-4980-2017-10-95
3. Stephan, M.: *Künstliche Intelligenz in den Ingenieurwissenschaften.* Wiesbaden, Springer Vieweg (2021)
4. Marlis, V.D.H.: Predictive Analytics und Data Mining. Wiesbaden, Springer Vieweg (2020)
5. Lukas, L., Viola, G., Fazel, A., Dávid, G., András, P., Wilfried, S.: Lead time prediction using machine learning algorithms: a case study by a semiconductor manufacturer. Proc. CIRP. Bd. 72, S. 1051–1056 (2018). https://doi.org/10.1016/j.procir.2018.03.148
6. Richard, M., Anthony, N., Vincent, H., Nikolaos, P.: Machine learning technologies for order flowtime estimation in manufacturing systems. Proc. CIRP. Bd. 81, S. 701–706 (2019). https://doi.org/10.1016/j.procir.2019.03.179
7. Elena, B., Nicolò, B., Marta, D., Alberto, M., Luigi, S.: An ensemble machine learning algorithm for lead time prediction. J. Mach. Intell. Data Sci. Bd. 5 (2024)
8. Ahmet, A. et al.: Hierarchical ensemble deep learning for data-driven lead time prediction. Int. J. Adv. Manuf. Technol. Bd. 128, Nr. 9–10, S. 4169–4188 (2023)
9. Dávid, G., András, P., Gábor, N., Viola, G., Wilfried, S., László, M.: Lead time prediction in a flow-shop environment with analytical and machine learning approaches. IFAC-PapersOnLine Bd. 51, Nr. 11, S. 1029–1034 (2018). https://doi.org/10.1016/j.ifacol.2018.08.472
10. Zhi Hui, L., Umi Kalsom, Y., Haslina, S.: Manufacturing lead time classification using support vector machine. In: Lecture notes in computer science. S. 268–278. (2019). https://doi.org/10.1007/978-3-030-34032-2_25
11. Michael, W., Jan, M., Klaus, T., Andreas, G., Günther, S., Stefan, M.: Combining process mining and machine learning for lead time prediction in high variance processes. In: Lecture notes in production engineering, S. 528–537. (2020). https://doi.org/10.1007/978-3-662-62138-7_53
12. Johannes, B., Jivka, O.: Prototyping machine-learning-supported lead time prediction using AutoML. Proc. Comput. Sci. Bd. 180, S. 649–655 (2021). https://doi.org/10.1016/j.procs.2021.01.287
13. Andreas, K: *Retrospektive Diagnose von Fehlerursachen an Antriebsstrangprüfständen mithilfe künstlicher Intelligenz.* Wiesbaden, Springer Vieweg (2024)
14. Charles, B., András, C., Guillermo, M.-M.: A comparative analysis of gradient boosting algorithms. Artif. Intell. Rev. Bd. 54, Nr. 3, S. 1937–1967 (2020). https://doi.org/10.1007/s10462-020-09896-5
15. Intel Corporation.: Intel Core i9–12900KF Processor (30M Cache, up to 5.20 GHz) Specifications. [online] (2024) [viewed: 2024–10–31]. Available from: https://www.intel.de/content/www/de/de/products/sku/134600/intel-core-i912900kf-processor-30m-cache-up-to-5-20-ghz/specifications.html
16. NVIDIA Corporation.: GeForce RTX 3060 Family. NVIDIA. [online] (2024). [viewed: 2024-10-31]. Available from: https://www.nvidia.com/de-de/geforce/graphics-cards/30-series/rtx-3060-3060ti/
17. Martin, K: *Big Data.* Skript zur Vorlesung Big Data im Masterstudiengang Smart Factory an der Hochschule Esslingen, Göppingen (2023)

18. Bernhard, P. et al.: *Menschzentrierte KI-Anwendungen in der Produktion: Praxiserfahrungen und Leitfaden zu betrieblichen Einführungsstrategien.* Fraunhofer-Institut für Arbeitswirtschaft und Organisation IAO, Stuttgart (2021). [online]. Available from: https://www.researchgate.net/profile/Martin-Braun-6/publication/349590365MenschzentrierteKI-AnwendungeninderProduktion-PraxiserfahrungenundLeitfadenzubetrieblichenEinfuhrungsstrategien/links/6047c1314585154e8c8ab95e/Menschzentrierte-KI-Anwendungen-in-der-Produktion-Praxiserfahrungen-und-Leitfaden-zu-betrieblichen-Einfuehrungsstrategien.pdf. (Accessed 30 Mar 2024)
19. Thomas, K.: *Lernende Entscheidungsbäume.* Informatik-Spektrum. Bd. 44, Nr. 5, S. 364–369 (2021). https://doi.org/10.1007/s00287-021-01398-0
20. Christian, K.: Erklärbarkeit maschineller Lernverfahren, Feature Engineering mit Black-Box-Modellen. Masterarbeit, Hochschule Darmstadt (2019)
21. Bernhard, B.: *Maschinelles Lernen—Grundlagen und Anwendungen.* Wiesbaden, Springer Vieweg (2023)
22. Ke, G., Meng, Q., Finley, T., Wang, T., Chen, W., Ma, W., Ye, Q., Liu, T.Y.: LightGBM: a highly efficient gradient boosting decision tree. In: Proceedings of the 31st International Conference on Neural Information Processing Systems (NIPS'17). pp. 3149–3157. Red Hook, NY, USA: Curran Associates Inc. (2017)
23. Tianqi, C., Carlos, G.: XGBoost. KDD '16: The 22nd ACM SIGKDD International Conference on Knowledge Discovery and Data Mining (2016). https://doi.org/10.1145/2939672.2939785
24. Prokhorenkova, L., Gusev, G., Vorobev, A., Dorogush, A.V., Gulin, A.: CatBoost: unbiased boosting with categorical features. arXiv (Cornell University) (2017). https://doi.org/10.48550/arxiv.1706.09516

Artificial Intelligence in Logistics: New Opportunities for Optimization and Innovation with Chatgpt

Kristína Kleinová(✉), Martin Straka, and Peter Kačmáry

Institute of Logistics and Transport, Technical University of Kosice, Kosice 042 00, Slovakia
{kristina.kleinova,martin.straka}@tuke.sk

Abstract. This article explores the potential of utilizing artificial intelligence to streamline processes in the logistics sector, which currently faces numerous challenges, including the need to enhance logistics procedures. With the rising demand for more efficient and flexible supply chains, artificial intelligence is crucial for improving demand forecasting, route optimization, inventory planning, and warehouse operations management. The goal is to initiate a discussion on the application possibilities of the ChatGPT model in logistics, emphasizing the benefits of this advanced technology – from faster calculations and flexible responses to changing market conditions to the ability to process and analyze large volumes of data in real time. The article's primary aim is to highlight that integrating ChatGPT into logistics systems could significantly boost efficiency and sustainability in logistics operations. However, it is essential to remember that the ChatGPT model may contain errors and inaccuracies, underscoring the importance of subsequent verification of its outputs. Implementing ChatGPT could bring comprehensive solutions to complex logistics operations and create new opportunities for innovation and transformation in this sector, thus supporting performance improvement and economic efficiency.

Keywords: artificial intelligence · ChatGPT · streamline processes · logistics

1 Introduction

Technologies let instant global communication, easy access to information, and new ways to work, study, and enjoy ourselves. The Internet has transformed business by broadening audiences, enhancing customer service, and enabling online shopping. It also drives social change, helping people advocate for important causes. Today, technologies connect us more than ever, forming the foundation of our interconnected world and opening new opportunities.

The number of people connected to the Internet worldwide dropped to 5.4 billion in 2023 [1]. In today's world, the Internet functions as both a means of connecting with others and a portal to an extensive range of knowledge, career paths, and learning materials. Individuals who can fully utilize these online resources are better positioned to engage in societal activities and take advantage of digital innovations [1–3]. Demis

J. Feliks et al. (Eds.): CLC2024 2024, 2026.
https://doi.org/10.1007/978-3-032-28931-5_22

Hassabis, co-founder and CEO of Google DeepMind, believes that AI has the potential to become one of the most essential and beneficial technologies ever created [4].

The European Union's approach to artificial intelligence focuses on fostering excellence and trust, aiming to develop research and industrial capacities while ensuring safety and the protection of fundamental rights [5]. The societal impact of artificial intelligence is further examined in the publication "Societal Impacts of Artificial Intelligence: Ethical, Legal, and Governance Issues" by Yuzhou QIAN and others [6, 7]. Given growing influence of AI and its models, monitoring and analysing their development is essential to effectively maximizing their benefits and minimizing risks.

2 Review of the Subject Literature

ChatGPT is and chatbot using AI, developed by OpenAI that uses NLP to hold conversations and made text like human. The GPT in it's name stands for creative pre trained transformer.

ChatGPT can simulate conversations that resemble human communication, making it a useful tool for everyday tasks. In this context, this model can support decision-making and optimize supply chain operations, as discussed in Smart Supply Chain Management in Industry 4.0.The logistics sector has been transformed by innovations such as AI, autonomous vehicles, automated warehouses, and predictive analytics. Digitalization enables companies to streamline operations, cut costs, and plan routes efficiently, reducing empty runs. AI-driven systems also coordinate tasks in real time, lowering the need for manual labor. Emerging technologies like drones and autonomous vehicles hold potential to improve last-mile delivery and customer satisfaction [8, 9]. Most industries, including logistics, have undergone significant changes thanks to the development of modern technologies, such as the AI mentioned above. Technologies like vehicles with autonomous systems, automation in warehouse, analytics smart transport are gradually becoming a standard part of this sector. "Digitalization is essential for logistics companies. The days when warehouses operated on a 'paper and pencil' basis are long gone. Through digitalization, companies can significantly get better their processes, reduce finances, and manage overall logistics operations more efficiently," explains author [10].

Numerous examples show that automation and artificial intelligence are increasingly being applied in both research and industry. For instance, Google DM uses AI to predict the structures of millions of potential new materials. According to a study published in Nature, approximately thousants of these materials could soon be created in laboratory settings.

DeepMind researcher Ekin Dogus Cubuk notes that the goal is to compress this traditional 10-to 20-year development cycle through significant progress in experimental automation, autonomous synthesis, and machine learning models. The AI system was trained using a dataset of about 50,000 known materials, an initiative launched at laboratory in 2011. DeepMind is now making its results publicly accessible to the scientific community to accelerate further innovation in materials science. As project director Christine Pershon explains, industries are often hesitant to invest in new materials due to high costs and associated risks, which prolong the path to cost-effective production. Reducing this timeline, even modestly, could represent a major breakthrough in the field [4, 11].

2.1 Supply Chain Management and Artificial Intelligence

The emergence of AI is poised to reshape logistics and SCM at their core. Rapid growth in e-commerce, persistent labour shortages and frequent disruptions have created new pressures on supply networks. To address these challenges, many companies are increasingly adopting digitalization and advanced digital technologies to enhance operational efficiency, cut expenses, and improve adaptability [12].

Within supply chains, AI has the capacity to transform product development by strengthening market analysis, accelerating the design phase and enhancing control. These advancements can shorten product launch timelines. It is important to note, however, that AI is intended to complement rather than replace human ingenuity and professional expertise [13].

Modern supply chains have become more intricate, and disruptions related to resource shortages, environmental events, or political conflict demand the exploration of alternative solutions [14]. AI empowers companies to stay ahead by speeding up the journey from idea to insight [15]. COVID-19 pandemic underscored the critical need for flexibility in supply chain management. The crisis triggered production shutdowns, operational delays, and rising costs, reinforcing the importance of digital transformation and the deployment of intelligent software. Such systems enable swift adaptation to demand fluctuations, optimize inventory levels, increase supply chain agility, and promote better coordination among stakeholders [16].

This research explores how AI can drive efficiency gains in examined sector. These technologies, including modern AI models, are examined as potential solutions to existing logistical challenges. With ChatGPT's debut, AI chat tools have surged in use, thanks to their versatility in handling tasks useful for work and everyday life—from coding and email composition to report generation, image creation, and complex spreadsheet calculations [17].

OpenAI launched the ChatGPT chatbot on December 1, 2022. Based on the GPT, this system generates contextually relevant text from user prompts, enabling people to engage with AI on complex topics regardless of their educational background [18]. In March 2023, OpenAI released GPT-4, which surpasses average human benchmarks on many standardized appraisals and scores within the top five percent on numerous metrics [19, 20]. GPT-4 can write, interpret, and execute code, allowing for sophisticated data analysis and visualization [19].

It can produce highly advanced content, leading some to speculate about its potential to substitute certain forms of human research [21]. While GPT-4 currently represents the most powerful model, other firms, including Google, are developing similarly advanced systems, suggesting that even more capable AI tools will soon become widely available to professionals [18]. Future generations of AI could trigger profound technological shifts, influencing employment patterns, strategies—changes that will inevitably impact the structure and operation of SCM as well [22]. Overall, artificial intelligence systems can be applied across numerous stages of the supply chain. By improving decision-making and operational efficiency, AI offers the ability to streamline workflows and enable organizations to respond more effectively to real-time challenges.

Direct Applications of AI in Supply Chain Management (SCM)

Artificial intelligence is redefining how companies approach consumer analysis and extract consumer insights, offering data-driven foundations for product innovation. Through the analysis of extensive datasets—ranging from social media activity and online reviews to purchasing patterns—AI systems can detect shifting consumer choices and identify trends as they emerge or decline. This level of insight allows businesses to design products more precisely aligned with market expectations, reducing the risk of costly experimentation approaches and increasing the probability of market success [23].

By processing customer feedback, sensor readings, and production data, AI can detect patterns that indicate potential defects or inconsistencies. Automating these procedures not only minimizes human error but also enhances operational efficiency. Furthermore, AI can simulate diverse testing conditions, allowing companies to optimize product performance before physical prototypes are produced [13, 23].

AI-driven analytics can markedly improve the accuracy of demand forecasts in supply chains. By integrating discounts records with current indicators, AI models can project future demand with high precision. This enables firms to better plan production schedules, manage inventory levels, and coordinate supply chain activities, resulting in lower storage costs, reduced overproduction, and higher customer satisfaction [10, 23].

AI can contribute to the design of advanced strategies by accounting for cost structures, route constraints, weather conditions, and environmental impacts. By evaluating large volumes of operational informations, AI can recommend optimal transportation routes that minimize travel time and fuel consumption, thereby increasing delivery efficiency [16, 24]. Another promising area for AI is the improvement of coordination and communication across the supply chain. AI can support seamless information exchange between engineers, designers, suppliers, and manufacturers, which accelerates decision-making and limits delays or miscommunications. This improved collaboration shortens development cycles and promotes more synchronized operations across different stages of production [23, 25].

ChatGPT represents a useful tool for managing operational costs and enhancing efficiency, particularly by handling repetitive and time-intensive tasks. While it can automate many functions, it is important to recognize that it cannot replace human cognitive skills—such as contextual reasoning, nuanced interpretation, and complex decision-making—which remain essential to supply chain management. Nevertheless, ChatGPT marks a significant advancement in AI development and holds considerable potential to support logistics and related fields. Its conversational abilities and capacity to deliver relevant information can contribute to streamlining various processes. Still, ethical and technical considerations have to be cautiously addressed to ensure responsible integration. As the technology continues to evolve, its benefits are likely to expand while its limitations are reduced, creating new opportunities for application in supply chains [26].

3 Coordinating Distribution, Transportation and Logistics

The application of AI models such as ChatGPT can significantly accelerate computational tasks in logistics and transportation by enhancing decision-making processes. Specifically, these models can:

- Rapidly process and evaluate large datasets to generate better strategies.
- Recommend the most efficient routes by analyzing warehouse capacity, transportation costs, and geographic distances.
- Support calculations for determining the number of vehicles.
- Simulate delivery scenarios and anticipate the effects of fluctuations in demand.
- Optimize orders and route planning to lower operational expenses and shorten delivery times [12, 16, 23, 24].

By improving route optimization, ChatGPT contributes to more sustainable and cost-effective logistics operations, including:

- Reducing costs through shorter travel distances, which decrease fuel usage and vehicle wear.
- Saving time by enabling faster deliveries along optimized routes.
- Lowering environmental impact by cutting CO2 emissions associated with unnecessary mileage.
- Increasing vehicle utilization by minimizing the number of vehicles required to meet delivery demands.
- Mitigating delays by factoring in real-time traffic data and route restrictions during planning [9, 15, 23, 24].

A study [27] explored the use of ChatGPT to optimize distribution routes. In this research, ChatGPT was assigned the task of choosing the right optimization technique and processing the relevant data to complete the route planning process. The objective was to assess whether the model could deliver results more rapidly or efficiently compared to a previous approach based on the ratio index method, which had achieved cost reductions of nearly 50%.

Within the study, ChatGPT was provided with the necessary instructions and datasets—including address lists, distance matrices, and operational constraints—and tasked with performing route optimization. By incorporating elements of the traveling salesman problem (TSP) and clustering algorithms, ChatGPT was able to reduce total travel distances and associated costs, thereby contributing to more streamlined logistics operations. However, minor computational inaccuracies were observed during the optimization process, emphasizing the need for subsequent verification and adjustment of the generated results. Although ChatGPT did not outperform the earlier optimization method in terms of cost savings, it displayed the capability to process instructions quickly and handle complex questions efficiently. These findings suggest that, despite certain limitations, ChatGPT can support logistics-related tasks such as transportation route optimization within supply chains. The study highlights that while artificial intelligence solutions may not always surpass traditional methods in achieving cost reductions, they can greatly enhance responsiveness and adaptability in logistics planning. Overall, ChatGPT provides effective tools for managing routine logistics operations, contributing to process optimization and higher supply chain efficiency. The integration of artificial intelligence into logistics therefore represents an important step toward modernization, enabling improved performance and supporting the future development of the logistics industry.

4 Discussion

ChatGPT is capable of performing easy calculations and rational thinking; however, its accuracy tends to decrease when dealing with complex mathematical problems or detailed operational tasks. Numerous studies have emphasized the necessity of verifying the outputs generated by the model, as its responses can be unreliable when addressing topics on which it has not been adequately trained. It is important to acknowledge that ChatGPT remains an evolving technology and end to undergo refinement. These inherent limitations and the risk of potential errors highlight the need for users to approach its outputs with caution, validating results through additional sources and fact-checking [28, 29]. Several drawbacks associated with the use of ChatGPT must also be considered. The quality and accuracy of input data are critical factors—high-quality data increases the likelihood of reliable outputs, whereas inaccurate or incomplete data can lead to flawed results. Furthermore, the model may exhibit knowledge gaps in highly specialized fields, which can result in incorrect conclusions or suboptimal decisions. Concerns about data security and the transparency of decision-making algorithms are equally important. Increasing transparency could strengthen user trust and improve understanding of how the system functions [30].AI has considerable potential to enhance logistics operations Automation can speed up deliveries and improve the overall customer experience. Nevertheless, the shift toward automation may influence labor markets as machines assume certain repetitive tasks. At the same time, the growth of AI technologies is generating new employment opportunities in fields related to AI development, implementation, and oversight.

Similar to the transformative effects of automation in the automotive industry—where autonomous vehicles and advanced navigation systems have reduced transportation costs and improved safety—AI could bring about comparable changes in logistics. This article indicates that robotization in the examined sector has led to lower operational costs and enhanced safety performance, outcomes that could potentially be replicated in logistics through the strategic use of AI [31, 32].

5 Conclusion

AI-driven language models leverage learning techniques to address routine logistics questions, including the optimization of transportation routes, making them crucial for enhancing sustainable efficiency in logistics. These systems improve the precision, speed, and overall effectiveness of logistics operations while simultaneously lowering operational costs. The application of artificial intelligence in micro-logistics frameworks can result in substantial reductions in transportation expenses and elevate the performance of logistics processes. In this context, digital transformation and the integration of cutting-edge technologies have become indispensable tools for driving organizational change and maintaining competitiveness.

Acknowledgments. "Funded by the eu nextgenerationeu through the recovery and resilience plan for Slovakia under the project no. 09i03-03-v02-00010."

References

1. International Telecommunication Union: Universal and meaningful connectivity by. https://www.itu.int/en/mediacentre/Pages/PR-2023-09-12-universal-and-meaningful-connectivity-by-2030.aspx (2030). Last Accessed 24 Oct 2024
2. Zhang, G., Yang, Y., Yang, G.: Smart supply chain management in industry 4.0: the review, research agenda and strategies in North America. Ann. Oper. Res. **322**, 1075–1117 (2023). https://doi.org/10.1007/s10479-022-04689-1
3. Liu, X., Zhao, M., Li, S., Zhang, F., Trappe, W.: A security framework for the internet of things in the future internet architecture. Futur. Internet. **9**, 27 (2017). https://doi.org/10.3390/fi9030027
4. DeepMind: About. https://deepmind.google/about/. Last Accessed 25 Oct 2024
5. European Commission: European approach to artificial intelligence. https://digital-strategy.ec.europa.eu/en/policies/european-approach-artificial-intelligence. Last Accessed 25 Oct 2024
6. Qian, Y., Siau, K.L., Nah, F.F.: Societal impacts of artificial intelligence: ethical, legal, and governance issues. Soc. Impact. **3**, 100040 (2024). https://doi.org/10.1016/j.socimp.2024.100040
7. EDIZone informačný portál: Umelá inteligencia v retaile a jej prínosy a nástrahy. https://www.edizone.sk/spravy/umela-inteligencia-v-retaile-a-jej-prinosy-a-nastrahy/. Last Accessed 26 Oct 2024
8. Sukhpal, S.G., Rupinder, K.: ChatGPT: vision and challenges. Internet Things Cyber-Phys. Syst. **3**, 262–271 (2023). https://doi.org/10.1016/j.iotcps.2023.05.004
9. Arsenijevic, U., Jovic, M.: Artificial intelligence marketing: Chatbots. In: International Conference on Artificial Intelligence: Applications and Innovations (IC-AIAI 2019), pp. 19–193. IEEE, Belgrade, Belgrade, Serbia (2019). https://doi.org/10.1109/IC-AIAI48757.2019.00010
10. IPRICE RECARE: Umelá inteligencia a budúcnosť logistiky. https://www.kancelarie.sk/novinky/umela-inteligencia-a-buducnost-logistiky. Last Accessed 27 Oct 2024
11. Živě aktuality: Zrýchlenie vývoja či testovania nových materiálov – pomôcť by mala umelá inteligencia. https://zive.aktuality.sk/clanok/nhyFjNn/zrychlenie-vyvoja-ci-testovania-novych-materialov-pomoct-by-mala-umela-inteligencia/. Last Accessed 28 Oct 2024
12. Supply Chain Management Review: The future of supply chains – next generation technology and beyond. https://www.scmr.com/article/the_future_of_supply_chains_next_generation_technology_and_beyond/software. Last Accessed 28 Oct 2024
13. Supply Chain Management Review: Harmony innovation – collaborative AI product creation. https://www.scmr.com/article/harmony-innovation-collaborative-ai-product-creation1/software. Last Accessed 28 Oct 2024
14. Supply Chain Management Review: AI simulation supply chain forecasting tools. https://www.scmr.com/article/ai-simulation-supply-chain-forecasting-tools/software. Last Accessed 29 Oct 2024
15. International Business Machines Corporation: Generative AI in supply chain. https://www.ibm.com/thought-leadership/institute-business-value/en-us/report/Generative-AI-supply chain. Last Accessed 29 Oct 2024
16. Supply Chain Management Review: Transforming supply chain management – intelligent software. https://www.scmr.com/article/transforming-supply-chain-management-intelligent-software/software. Last Accessed 30 Oct 2024
17. ZDNet: Best AI Chatbot. https://www.zdnet.com/article/best-ai-chatbot/#google_vignette. Last Accessed 30 Oct 2024
18. Henriksen, C.H.: Artificial intelligence for supply chain management: disruptive innovation or innovative disruption? J. Supply Chain Manag. **59**, 65–76 (2023). https://doi.org/10.1111/jscm.12304

19. Bubeck, S., Chandrasekaran, V., Eldan, R., et al.: Sparks of Artificial General Intelligence: Early experiments with GPT-4. arXiv preprint https://arxiv.org/abs/2303.12712 (2023). https://doi.org/10.48550/arXiv.2303.12712.
20. One Useful Thing: It is starting to get strange. https://www.oneusefulthing.org/p/it-is-starting-to-get-strange. Last Accessed 30 Oct 2024
21. Hitch, D.: Artificial intelligence (AI) augmented qualitative analysis: the way of the future? Qual. Health Res. (2023). https://doi.org/10.2139/ssrn.4451740
22. Budhwar, P., Chowdhury, S., Wood, G., et al.: Human resource management in the age of generative artificial intelligence: perspectives and research directions on ChatGPT. Hum. Resour. Manag. J. **33**, 1097–1097 (2023). https://doi.org/10.1111/1748-8583.12524
23. Supply Chain Management Review: Harmony innovation – collaborative AI product creation. https://www.scmr.com/article/harmony-innovation-collaborative-ai-product-creation1/software. Last Accessed 05 Feb 2024
24. Richey, G.R., Chowdhury, S., Davis-Sramek, B., Giannakis, M., Dwivedi, Y.K.: Artificial intelligence in logistics and supply chain management: a primer and roadmap for research. J. Bus. Logist. **44**, 532–549 (2023). https://doi.org/10.1111/jbl.12364
25. International Business Machines Corporation: Generative AI in supply chain. https://www.ibm.com/thought-leadership/institute-business-value/en-us/report/Generative-AI-supply chain. Last Accessed 05 May 2024
26. Fresh Space: ChatGPT – nový spoločník v sfére umelej inteligencie. https://zaft.sk/chatgpt-novy-spolocnik-v-sfere-umelnej-inteligencie/. Last Accessed 30 Oct 2024
27. Kleinová, K., Straka, M.: Streamlining distribution routes using the language model of artificial intelligence. Sustainability. **16**, 6890 (2024). https://doi.org/10.3390/su16166890
28. Hua, S., Jin, S., Jiang, S.: The limitations and ethical considerations of ChatGPT. Data Intell. **6**, 201–239 (2024). https://doi.org/10.1162/dint_a_00243
29. Nasr, M., Shokri, R., Houmansadr, A.: Comprehensive privacy analysis of deep learning: passive and active white-box inference attacks against centralized and federated learning. In: IEEE Symposium on Security and Privacy, vol. 1, pp. 739–753 (2019). https://doi.org/10.1016/j.iotcps.2023.04.003
30. Inglada Galiana, L., Corral Gudino, L., Miramontes González, P.: Ethics and artificial intelligence. Revista Clínica Española (English Edition). **224**(3), 178–186 (2024). https://doi.org/10.1016/j.rce.2024.01.007
31. Voß, S.: Successfully using ChatGPT in logistics: are we there yet? In: Daduna, J.R., Liedtke, G., Shi, X., Voß, S. (eds.) Computational Logistics. Proceedings of the ICCL 2023, Berlin, Germany, September 6–8, 2023. Lecture Notes in Computer Science, vol. 14239, pp. 1–15. Springer, Cham (2023). https://doi.org/10.1007/978-3-031-43612-3_1
32. Kobrinskii, B.A.: Artificial intelligence: problems, solutions, and prospects. Pattern Recognit. Image Anal. **33**, 217–220 (2023). https://doi.org/10.1134/S1054661823030203

Transport Optimization and Planning

Development of Public Transport Use in Selected Cities within Slovakia

Peter Kačmáry(✉), Simona Špirková, Kristína Kleinová, and Maroš Mesároš

Faculty BERG, Institute of Logistics and Transport, Technical University of Košice, Park Komenského 14, 04200 Košice, Slovakia
peter.kacmary@tuke.sk

Abstract. The structure of a transport system in larger cities is crucial for the fast and efficient movement of goods and people. In cities, public transport tends to be problematic and is overwhelmingly represented by the individual car transport. It is the endeavour of every city to operate public transport in a way that encourages people not to use their own cars. This paper is focused on the analysis and prediction of public transport use in selected cities in Slovakia. The analysis enters data of annual values for the last 10 years. The prediction involves a combined forecasting approach, which is the result of merging the four methods described in the paper.

Keywords: Public transport analysis · forecasting · data research

1 Introduction

Development of public transport in cities is a key element for developing a sustainable urban environment, improving the quality of life of residents and promoting economic efficiency. Public transport provides people with access to work, education, healthcare and recreational activities, while reducing congestion, emissions and dependence on private cars use [1, 2].

These are the key aspects of targeting public transport in cities:

(a) Efficiency and accessibility:

- Frequency and reliability: public transport must be frequent and reliable to be attractive to passengers. This includes frequent services and minimising delays.
- Connectivity: It is important to ensure good connectivity between different modes of transport, such as buses, trams, metro, trains and bicycles, to facilitate transfers and increase the fluidity of travel.
- Accessibility: Public transport should be accessible to all residents, including people with disabilities, the elderly and people with reduced mobility.

(b) Sustainability and environmental impact:

J. Feliks et al. (Eds.): CLC2024 2024, 2026.
https://doi.org/10.1007/978-3-032-28931-5_23

- Reducing emissions: cities are focusing on reducing greenhouse gas emissions and air pollution through the use of clean vehicles such as electric buses, trams and metros, or biofuel or hydrogen vehicles.
- Promoting non-motorised transport: Developing infrastructure for pedestrians and cyclists is important to reduce dependence on motor vehicles and promote healthier lifestyles.

(c) Cost-effectiveness:

- Operation and maintenance costs: Optimising public transport operations is a key to reducing costs, which includes efficient route planning and scheduling, as well as maintenance and renewal of vehicles and infrastructure.
- Affordability: Ticket prices should be accessible to all segments of the population to avoid social exclusion and promote inclusiveness.

(d) Technological innovation:

- Intelligent Transport Systems (ITS): the introduction of technologies such as GPS, passenger information systems and automated driving systems can improve efficiency, safety and passenger information.
- Digitisation and mobile apps: The use of apps to plan journeys, purchase tickets and provide up-to-date transport information improves passenger convenience and satisfaction.

(e) Safety and comfort of passengers:

- Passenger safety: it is important to provide a safe environment on public transport, including lighting, CCTV monitoring and the presence of staff.
- Passenger comfort: Comfortable and clean vehicles, ample space and modern facilities contribute to a positive passenger experience.
- Adaptability to changing needs: Public transport should be flexible and responsive to the changing needs of the population, including population growth, changes in working patterns and urban sprawl.

Transport is an important sector in the economy and one of the most important segments of the tertiary sector. In addition to its basic function of moving goods, goods and people, the transport sector generates a number of employment opportunities not only in the transport activity itself, but also in downstream activities. With the development of transport, on the one hand, industry, services, the standard of living of the population, etc., develops, but on the other hand, transport has a negative impact on the environment and is one of the main consumers of energy [3, 4].

The task of transport development forecasting as a whole is the detection and evaluation of traffic volume, traffic intensity, traffic ratios, etc. on existing transport communication facilities in order to examine not only the sufficiency of transport infrastructure capacity in the future, but also to forecast other activities that are affected by transport [3, 4].

2 Methods and Methodology

Forecasting in transport, like forecasting in any other economic field, is much more difficult than forecasting in, for example, the natural sciences and engineering. While in natural sciences constants predominate and forecasting is about detecting deviations from a desired state, in economic and social sciences there are very few constants and human behavior, which is highly variable, has to be taken into account. Although some events are sometimes repeated (e.g. economic or seasonal cycles), in reality this is only an appearance, because the repetition takes place under constantly changing conditions. However, no expert studies and conclusions on the evolution of public transport were available at the time of writing this article, so the forecast itself is based on a number of quantitative methods to increase its objectivity.

After the analysis of the time series, it was decided to use the following three selected quantitative methods:

- Method of regression analysis—linear regression; this method is generally known for creating an overall linear trend of values in a time series, which follows from the Gaussian least squares method.
- ARIMA method; this method is also generally known described by Box and Jenkins.
- The harmonic weights method; this method is described in more detail in the work [5].

These methods are expected to be able to correctly evaluate and interpret the relevant forecast. The next step will be to merge the results of the particular methods to obtain a combined i.e. overall forecast (CLC 2012 + book) of the evolution of public transport use in each chosen city over the next two years. Public transport service companies from Bratislava, Košice, Prešov and Žilina were selected for the analysis. These are the four largest cities in Slovakia in the order as listed.

3 Results

The analysis included data from the period of 11 years (i.e. 2013–2023), which are published in the annual reports of individual transport undertakings. In these reports, it is clear that the drop in passenger ridership was significant in the pandemic years 2020–2021. However, this may negatively affect the forecast of passenger ridership in the coming years, as we already observe that in many aspects of life, people's behaviour has returned to the pre-COVID-19 period. For this reason, the calculated forecasts are divided in two forms.

(a) The forecast includes data from the COVID-19 crisis period;
(b) The forecast does not include data from the COVID-19 crisis period. In this case, data from the 2020–2021 period are idealized i.e. linearly linked from 2019 to 2022 period.

3.1 Forecast with COVID-19 Data

In this case of forecasting, the data obtained reflects the aforementioned pandemic crisis COVID-19, which limited passenger transport around the world on the basis of restrictions announced by national governments. Furthermore, it is evident that over the last

10 years there has been a tendency towards a decline in the use of public transport, with the exception of the city of Žilina (Table 1. and Fig. 1.).

Table 1. Number of passengers transported on the urban public transport [6–9]

Year	Public transport company/dopravný podnik (DP)			
	DP Bratislava (DPB)	DP mesta Košice (DPMK)	DP mesta Prešov (DPMP)	DP mesta Žilina (DPMŽ)
	NPT × 10^3 [a]	NPT × 10^3	NPT × 10^3	NPT × 10^3
2013	230680	88838	34730	10630
2014	248557	83868	33233	10809
2015	251078	80346	32796	11015
2016	246506	83144	32082	11411
2017	243044	82540	31618	11650
2018	236265	83003	31422	12053
2019	250672	82656	30823	12241
2020	192254	59756	21423	7786
2021	166408	52894	14632	16806
2022	223664	68638	25790	21022
2023	247585 [b]	74963	29599	23562

[a]NPT × 103—Number of passengers transported (in thousands)
[b]Estimated value (the volume has not been published yet)

A number of factors may be influencing the decline in public transport ridership:

- Restrictions on public transport services, resulting in increased service times, delays, changes or cancellations of routes due to saving costs or driver shortages, etc.;
- Increasing ticket prices;
- Uncomfortable vehicles in the fleet (e.g. vehicles without working air conditioning, vintage vehicles, etc.);
- Shared transport (introduction of shared means of transport such as scooters, bicycles, motorbikes or small cars) or low-cost taxi applications.

The following table (see Table 2.) shows the values of the partial and total passenger forecasts for each city.

The diagram of passengers numbers transported on the urban public transport is in Fig. 1..

The following table (see Table 2.) shows the values of the partial and total passenger forecasts for each city.

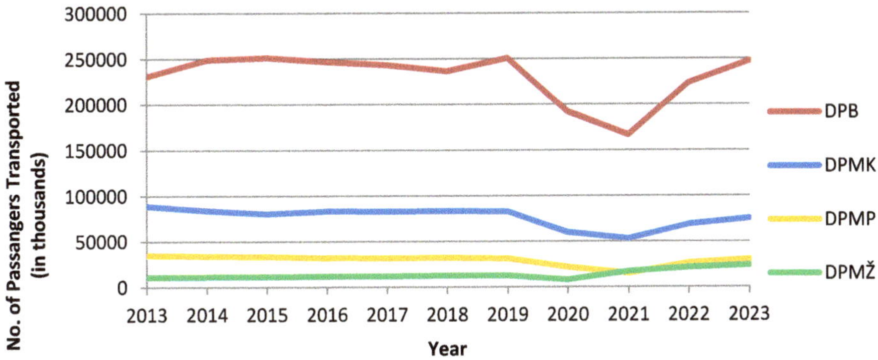

Fig. 1. Number of passengers transported by the urban public transport

Table 2. Forecasts of passengers transported by the urban public transport

	DP Bratislava (DPB)	Forecasts		
Year	NPT $\times 10^3$	LR	ARIMA	HW
2013	230680			
2014	248557			
2015	251078			
2016	246506			
2017	243044			
2018	236265			
2019	250672			
2020	192254			
2021	166408			
2022	223664			
2023	247585			
2024	**236388**	210432	241058	257675
2025	**234813**	207069	230636	266734
	DP mesta Prešov (DPMP)			
	NPT $\times 10^3$			
2013	34730			
2014	33233			
2015	32796			
2016	32082			
2017	31618			

(continued)

Table 2. (*continued*)

	DP mesta Prešov (DPMP)			
	NPT × 10^3			
2018	31422			
2019	30823			
2020	21423			
2021	14632			
2022	25790			
2023	29599			
2024	**26867**	21721	27728	31152
2025	**27463**	20521	29375	32494
Year	DP mesta Košice (DPMK)	Forecasts		
	NPT × 10^3	LR	ARIMA	HW
2013	88838			
2014	83868			
2015	80346			
2016	83144			
2017	82540			
2018	83003			
2019	82656			
2020	59756			
2021	52894			
2022	68638			
2023	74963			
2024	**71255**	62278	74824	76663
2025	**71092**	59921	75375	77979
	DP mesta Žilina (DPMŽ)			
	NPT × 10^3 *			
2013	10630			
2014	10809			
2015	11015			
2016	11411			
2017	11650			
2018	12053			

(*continued*)

Table 2. (*continued*)

	DP mesta Žilina (DPMŽ)			
	NPT × 10^3 *			
2019	12241			
2020	7786			
2021	16806			
2022	21022			
2023	23562			
2024	**21851**	19884	18480	27190
2025	**22032**	20940	13859	31298

3.2 Forecast without COVID-x19 Data

The following forecast of the status of transported passengers is based on the assumption that we would not have been affected by the global COVID 19 crisis. Data, during the pandemic period, was replaced by linear interpolation.

Table 3. Number of passengers transported by the urban public transport [6–9]

Year	Public transport company/dopravný podnik (DP)			
	DP Bratislava (DPB)	DP mesta Košice (DPMK)	DP mesta Prešov (DPMP)	DP mesta Žilina (DPMŽ)
	NPT × 10^3 [a]	NPT × 10^3	NPT × 10^3	NPT × 10^3
2013	230680	88838	34730	10630
2014	248557	83868	33233	10809
2015	251078	80346	32796	11015
2016	246506	83144	32082	11411
2017	243044	82540	31618	11650
2018	236265	83003	31422	12053
2019	250672	82656	30823	12241
2020	192254	59756	21423	7786
2021	166408	52894	14632	16806
2022	223664	68638	25790	21022
2023	247585 [b]	74963	29599	23562

[a]NPT × 103–number of passengers transported (in Thousands)
[b]Estimated value (the volume has not been published yet)

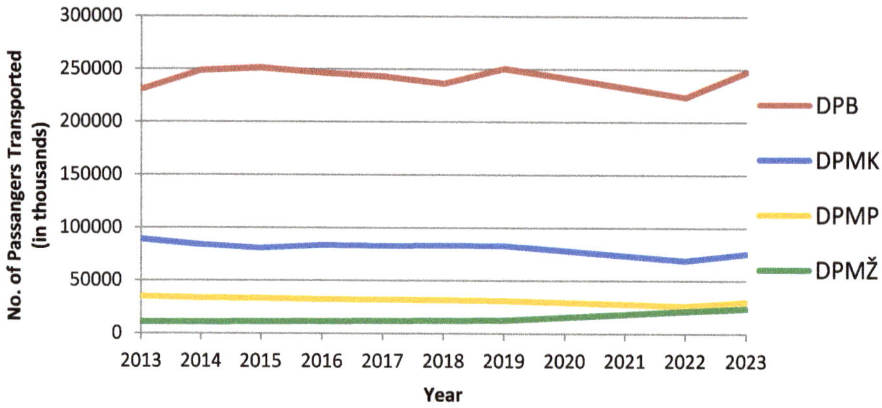

Fig. 2. Number of passengers transported by the urban public transport

The following table (Table 4.) shows the values of the partial and total passenger forecasts for each city (Table 3.), (Fig. 2.).

Table 4. Forecasts of passengers transported by the urban public transport

	DP Bratislava (DPB)	Forecasts		
Year	NPT × 10^3	LR	ARIMA	HW
2013	230680			
2014	248557			
2015	251078			
2016	246506			
2017	243044			
2018	236265			
2019	250672			
2020	241669			
2021	232666			
2022	223664			
2023	247585			
2024	**243680**	237181	242391	251468
2025	**244083**	236523	241094	254632

(*continued*)

Table 4. (*continued*)

	DP mesta Prešov (DPMP)			
	NPT × 10^3			
2013	34730			
2014	33233			
2015	32796			
2016	32082			
2017	31618			
2018	31422			
2019	30823			
2020	29145			
2021	27467			
2022	25790			
2023	29599			
2024	**29550**	26532	32239	29879
2025	**28975**	25822	31086	30017
Year	DP mesta Košice (DPMK)	Forecasts		
	NPT × 10^3	LR	ARIMA	HW
2013	88838			
2014	83868			
2015	80346			
2016	83144			
2017	82540			
2018	83003			
2019	82656			
2020	77983			
2021	73310			
2022	68638			
2023	74963			
2024	**71120**	71120	81059	74756

(*continued*)

Table 4. (*continued*)

Year	DP mesta Košice (DPMK)	Forecasts		
	NPT × 10^3	LR	ARIMA	HW
2025	**74754**	69651	80292	74320
	DP mesta Žilina (DPMŽ)			
	NPT × 10^3 *			
2013	10630			
2014	10809			
2015	11015			
2016	11411			
2017	11650			
2018	12053			
2019	12241			
2020	*15168*			
2021	*18092*			
2022	21022			
2023	23562			
2024	**24566**	21687	25518	26493
2025	**23230**	22913	17018	29759

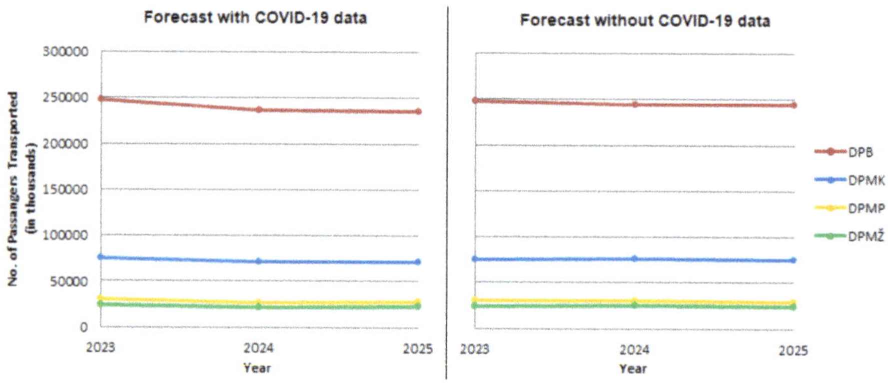

Fig. 3. Comparison of forecasts from both approaches

4 Conclusion

As it can be seen from the last figure (Figure 3.), the results of the forecasts are very close, whether or not the COVID-19 period data were taken into account. In order to justification of well-chosen methodology, a test was carried out that forecasted the number of passengers carried (i.e. it is the same object) for 2018 and 2019 i.e. for the pre-COVID years. The accuracy of the forecasts was found to be as follows (see Table 5.).

Table 5. Testing forecast accuracy for years 2018–2019

	DP Bratislava (DPB)	DP mesta Košice (DPMK)	DP mesta Prešov (DPMP)	DP mesta Žilina (DPMŽ)
MAPE	3,06%	1,12%	2,29%	7,57%
RMSE	7621	971	749	926
Theil's U	0,0156	0,0059	0,0119	0,0396

The accuracy of the forecast was very good in the pre-COVID years and is expected to be satisfactory in the present. Some public transport companies suffer from underfunding by the municipal authorities, which is ultimately reflected in reduced passenger comfort and is contrary to what was stated in the introduction section. That is why the number of transported passengers will continue to decrease in these cities, but not such intensive as in past. Therefore, any such cost-saving steps should be strongly re-evaluated, as it is important to create a sustainable, efficient and accessible transport system in cities that promotes social inclusion, reduces environmental impact and improves economic efficiency [10]. Cities that focus on the development of modern and sustainable public transport are better prepared to meet the challenges of the future and improve the quality of life of their citizens.

Acknowledgement. The research is part of the project grant number KEGA 010TUKE-4/2023 Application of educational robots in the process of teaching the study program industrial logistics.

References

1. Trembošová, M., Dubcová, A., Nagyová, Ľ., Cagáňová, D., Kóňová, T.: Suburbanization process of regional cities in Slovakia. Transport & Logistics: the International Journal. **22**(53), 21–31 (2022)
2. Dolinayová, A., Nedeliaková, E.: Prognózovanie pre manažérov dopravy. Lura, Bratislava (2010)
3. Dolinayová, A.: Možnosti prognózovania vývoja osobnej prepravy v meniacich sa podmienkach. Perner's Contacts. **4**(1), 42–49 (2009)
4. Kačmáry, P., Malindžák, D.: The forecast methods of sale and production in dynamically changing market economy. VŠB, Technical University Ostrava (2013)
5. Košice City Transport Company—Annual reports. https://www.dpmk.sk/dpmk/vyrocne-spravy. Last accessed 23 Sep 2024

6. Bratislava Transport Company—Annual reports. https://dpb.sk/sk/verejne-informacie. Last accessed 23 Sep 2024
7. Prešov City Transport Company—Annual reports. https://www.dpmp.sk/vyrocne-spravy. Last accessed 24 Sep 2024
8. Žilina City Transport Company—Annual reports. https://www.dpmz.sk/vyrocne-spravy. Last accessed 24 Sep 2024
9. Straka, M., Balog, M.: Allocation of microwave's check sites inintelligent transport systems. Acta Montan. Slovaca. **12**(2), 114–120 (2007)

Preliminary Analysis of the Hydrogen Refuelling Station Location Problem in Rail Transport

Katarzyna Gdowska(✉), Roger Książek, Radosław Kapłan, Magdalena Dudek, and Andrzej Raźniak

AGH University of Krakow, Krakow, Poland
{kgdowska,roger,rkaplan,potoczek,razniak}@agh.ed.pl

Abstract. Locating infrastructure facilities is a strategic, critical, and long-term decision across various sectors, including manufacturing, logistics, and defense. This challenge becomes particularly important when introducing emerging technologies with the potential to transform industries and economies. In the context of rail transport, the adoption of hydrogen propulsion underscores the strategic importance of hydrogen refueling stations (HRS), as the fixed-track network offers limited flexibility for fueling. This paper introduces the Hydrogen Refuelling Station Location Problem in Rail Transport (HRSLPR), inspired by the Economic Order Quantity (EOQ) model, and proposes a preliminary analysis. The approach integrates fuel demand, hydrogen supply conditions, and infrastructure costs into an optimization problem for which a dedicated Mixed-Integer Linear Programming (MILP) model was formulated. The model enables scenario-based analysis, considering both single-period and multi-period perspectives to capture factors such as demand evolution and variability in hydrogen production costs. In this preliminary study, a simplified single-period formulation is presented together with the corresponding MILP model. The framework aims to support decision-makers in assessing the influence of key parameters on hydrogen infrastructure planning, thereby facilitating the efficient integration of hydrogen propulsion into rail networks.

Keywords: Hydrogen refueling station · Infrastructure location problem · Optimization · Railways · Decision making · Strategic management

1 Introduction

The infrastructure location problem is a critical long-term decision that involves determining the optimal location(s) for one or more facilities within a planning area to achieve a specific objective. This process is widely used across various sectors, including production, logistics, military, and daily life [1]. It becomes essential when implementing new technologies that have the potential to impact the industry or economy [2] significantly. The research referenced in this paper focuses on integrating hydrogen propulsion technology into rail transportation, which is characterized by a fixed infrastructure with limited flexibility due to its defined track network. The layout of hydrogen refueling stations is crucial for advancing hydrogen energy and fuel cell technology within the railway sector [3–5].

J. Feliks et al. (Eds.): CLC2024 2024, 2026.
https://doi.org/10.1007/978-3-032-28931-5_24

For solving the hydrogen refueling station location problem in a railway network, an approach based on the Economic Order Quantity (EOQ) model was adopted. Within this optimization framework, factors such as fuel demand, hydrogen availability, infrastructure expenditures, and other critical constraints are systematically taken into account. Mathematical modeling, particularly linear and integer programming, is an essential tool in addressing this problem. Multiple Mixed-Integer Programming (MIP) models have been developed to identify the optimal locations of hydrogen refueling stations based on selected decision criteria. This approach enables decision-makers to analyze different scenarios (data instances) and evaluate the optimal solutions for each.

The formulated optimization problems and corresponding mathematical models represent a crucial step toward understanding how the placement of hydrogen refueling stations affects the broader adoption of hydrogen propulsion in rail transport. The use of mathematical models, combined with the analysis of optimal solutions (including the development of various problem scenarios), will enable decision-makers to assess the impact of individual parameters – such as demand, hydrogen purchase or production costs, and station types – on strategic decision-making. Since facility location problems represent long-term strategic decisions, two main modeling approaches are considered: (1) a single-period formulation, which offers an initial perspective on the interactions between demand, supply, candidate locations, and cost structures (both fixed and variable), and (2) a multi-period formulation, which additionally incorporates economic dynamics relevant to the adoption of new technologies in strategic transportation sectors, such as projected demand growth and variability in hydrogen production costs.

This paper introduces the Hydrogen Refuelling Station Location Problem for Railways (HRSLPR), formulated as a simplified single-period problem. A corresponding Mixed-Integer Linear Programming (MILP) model is developed, and the methodological approach, problem formulation, and model structure are outlined. No implementation examples are included at this stage.

2 Optimizing Decisions on Hydrogen Refuelling Station Location

Determining the optimal locations for hydrogen refueling stations (HRS) within the Polish railway network is a complex problem that can be broken down into several research subproblems, requiring an interdisciplinary approach. This integrates energy engineering, transportation engineering, applied mathematics, operations research, computer science, and logistics. Sensitivity analysis plays a crucial role in addressing input uncertainty by quantifying the impact of parameter value changes on the solution. However, sensitivity analysis does not incorporate uncertainty directly into models. Stochastic programming and scenario-based planning offer more robust frameworks, formalizing complexities and uncertainties in real-world decision-making by embedding them within the model structure [6, 7].

A critical aspect of this research is identifying and structuring the problem, given its hierarchical nature. This process defines activities at each stage, from data preparation to model solving and result analysis. Identifying parameters, such as hydrogen demand, forms the basis for defining system requirements. These parameters are input into the decision model, which computes optimal decision variables—such as the type and location of each HRS—to minimize total costs under the given conditions. Specific locations may be excluded if no station is viable. This identification process enables clear optimization criteria, boundary conditions, and interdependencies, ensuring the model is transparent, understandable, and adaptable to different datasets.

This process also reduces the dataset size, directly influencing computational complexity and required computing power. The resulting optimization model represents the problem mathematically, with updates required if system assumptions or relationships change significantly. The variety of hydrogen station types compounds the problem's complexity. HRS differ in size, capacity, refueling rate, hydrogen source, and operational characteristics. These factors impact both station placement costs and the unit cost of hydrogen. A tailored classification of HRS types is essential for the successful implementation and alignment with the rail network's requirements. The railway network itself introduces further complexity. It consists of various lines, stations, and vehicles with distinct needs. Lines are managed by different entities, with variations in traffic volumes and propulsion systems. Understanding these complexities is crucial for identifying optimal HRS locations that maximize benefit without disrupting network performance. The rail network's structure must ensure station locations are practical and strategically advantageous.

The primary challenge lies in developing a practical decision-making approach for HRS placement. This begins with reviewing existing methodologies for locating strategic infrastructure, particularly fueling stations, which inform railway-specific problems. Insights from road transport infrastructure are valuable but cannot be directly applied to railways due to their unique spatial and operational characteristics. The proposed model aims to minimize total costs by optimizing station placement while considering hydrogen demand, station costs, and network performance. Figure 1 schematically outlines the methodology for the problem identification process and the relationships among the key analysis stages.

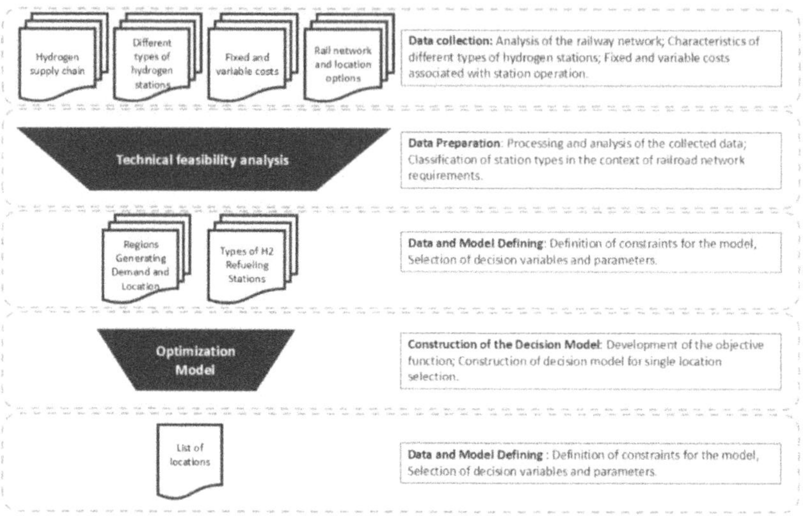

Fig. 1. HRSLPR – problem identification. *Source* own elaboration.

3 Facility Location Problem

The configuration of the hydrogen refuelling station network plays a decisive role in advancing the commercial adoption of hydrogen energy and fuel cell vehicles. The facility location problem [1], which involves determining the optimal placement of one or more facilities within a defined area to achieve a specific objective, has been widely applied across various domains, including manufacturing, logistics, public services, and defense. This problem involves the locations of factories, warehouses, emergency centers, fire stations, garbage disposal centers, logistics centers, and missile warehouses, among others. Site selection is one of the most important long-term decisions. In the substantial literature (e.g. [1, 8]), three main formulations of facility location problems lay the foundations for any subsequent research in this area:

- *p*-median problem, which uses the weighted distance between demand nodes and facilities by the associated demand quantity and calculates the total weighted travel distance between demands and facilities, and in a nutshell, is stated as follows: find the location of p facilities to minimize the total demand-weighted travel distance between demands and facilities;
- covering problem, where the demand is said to be covered if it can be served within a specified time. Two covering problems can be distinguished: the location set covering problem and the maximal covering problem. In the set covering problem, the objective is to minimize the cost of facility location while ensuring a specified level of coverage.
- *p*-centered problem, where the objective is to cover all demands using a given number of facilities in such a way that minimizes coverage distance. Rather than taking an input coverage distance s, this model determines the minimal coverage distance required to locate p facilities endogenously. The p-center problem is also known as the minimax problem, as we seek to minimize the maximum distance between any

demand and its nearest facility. If facility locations are restricted to the network's nodes, the problem is a vertex-center problem. Center problems that allow facilities to be located anywhere on the network are absolute.

In this study, we deviate from the approaches mentioned above. The newly developed model focuses on the operational aspects of railway systems, where hydrogen refueling stations (HRS) are implemented to ensure continuous system operation. Consequently, these stations must meet the total hydrogen demand while minimizing the overall costs associated with their deployment and operation.

4 Single-Period Hydrogen Refuelling Station Location Problem for Railways

The goal of the newly developed HRSLPR is to place hydrogen refueling stations at known potential locations with predetermined hydrogen demand generated in demand-generating regions within a single planning period. Thus, let L denote the set of available hydrogen refueling station locations. Let R be the set of demand-generation regions (or areas) in the network of interconnected hydrogen-fueled rail lines, each with a common starting point (central station) and a location where trainsets are prepared to depart for their respective routes. In addition, let K be the set of selected hydrogen refueling station types. Specific parameters will be defined for the sets mentioned above. Therefore, let parameter d_r denote the value of known hydrogen demand in region (area) r. In addition, let the parameter $z_{rl} = 1$ denote that the potential location (construction) of a hydrogen refueling station at location l, can allow to serve fully or partially the hydrogen demand for region (area) r. In addition, let the type of refueling station k be characterized by a fixed operating cost C_k and the maximum capacity of that station to meet the refueling demand m_k. In addition, let c_k denote the unit cost of consuming hydrogen. Let it also be known that the total investment cost of placing (construction) a hydrogen refueling station of type k. In the WRSLPR, four decision variables (y_{lk}, x_{rl}, w_{lk}, D) are defined, allowing for the derivation of the optimal solution. Let the binary variable y_{lk} be equal to 1 if, at potential location l, we decide to place (or build) a station of type k, and 0 otherwise. A refueling station of type k will be placed (built) in location l, so this refueling station of type k will satisfy share x_{rl} of the demand for hydrogen claimed by region (area) r; x_{rl} may cover up to 100% of the demand of a region. The solution will also determine the share of total demand w_{lk} for location l that is satisfied by refueling station type k. Variable D_l will reveal the total amount of hydrogen demand realized in localion l. Let the decision criterion in the WRSLPR be the minimization of the total cost of investment and operation of the hydrogen refueling system (the sum of the costs of investment, operation, and the cost associated with the consumption of hydrogen), so the solution will ensure satisfaction of the entire demand for hydrogen in the system. The notation for the MILP model formulated for this problem is presented in Table 1.

Table 1. Notation used in the HRSLPR.

Sets	
L	- Available potential locations for HRS
R	- Regions to be supplied with H2
K	- Types of HRS
Parameters	
d_r	- Demand for H2 in region r
$z_{r,l}$	$= 1$, if station located in l can serve region r, 0 otherwise
C_k	- fixed operating costs of an HRS of type k
c_k	- Unit H2 cost at the HRS of type k
I_k	- Total investment cost of HRS of type k
m_k	- Maximum capacity of HRS of type k
M	- Big constant
Decision variables	
D_l	- Demand for H2 satisfied by a station located in l
$x_{r,l}$	- The share of demand for H2 generated by region r satisfied by a station located in l
$w_{l,k}$	- Demand for H2 in location l satisfied by HRS of type k
$y_{l,k}$,	$= 1$, if in location l HRS of type k is used, 0 otherwise
$y_{l,k}$,	$= 1$, if in location l HRS of type k is used, 0 otherwise

With the HRSLPR, the following linear program (1)–(10) is associated.

$$\min : \sum_{l \in L} \sum_{k \in K} (I_k + C_k) y_{l,k} + \sum_{l \in L} \sum_{k \in K} c_k w_{l,k} \tag{1}$$

$$\sum_{l \in L} x_{r,l} = d_r, \, r \in R \tag{2}$$

$$x_{r,l} \leq M z_{r,l}, \, r \in R, \, l \in L \tag{3}$$

$$\sum_{k \in K} y_{l,k} \leq l, \, l \in L \tag{4}$$

$$\sum_{r \in R} x_{r,l} = D_l, \, l \in L \tag{5}$$

$$\sum_{k \in K} y_{l,k} m_k \geq D_l, \, l \in L \tag{6}$$

$$w_{j,k} \geq \sum_{r \in R} x_{r,l} - M(1 - y_{lk}), \, l \in L, \, k \in K \tag{7}$$

$$x_{r,l} \geq 0, r \quad \in \mathrm{R}, 1 \quad \in \mathrm{L} \tag{8}$$

$$w_{l,k} \geq 0, l \quad \in \mathrm{L}, k \quad \in \mathrm{K} \tag{9}$$

$$y_{l,k} \in \{0, 1\}, 1 \quad \in \mathrm{L}, \mathrm{k} \quad \in \mathrm{K} \tag{10}$$

Objective function (1) minimizes the total costs (i.e., investment and operating costs) of HRS used to satisfy the total demand claimed by the regions under consideration. Constraint (2) ensures that the demand for hydrogen claimed by region r is fully satisfied by HRSs located in predefined potential locations. Constraint (3) guarantees that the demand for hydrogen of region r is satisfied only by HRSs situated in locations that can serve this region. In each location, at most one type of HRS can be used, meaning that not every location must be used to satisfy the total demand (4). The total demand for hydrogen to be fulfilled by an HRS located in location l forces choosing for that location such a type of HRS which is capable of providing the required amount of hydrogen (5)–(7).

5 Conclusion

This paper presents a preliminary study of the HRSLPR to identify optimal locations for hydrogen refueling stations (HRS) within a railway network. The proposed Mixed-Integer Linear Programming (MILP) model minimizes investment, operational, and hydrogen consumption costs while ensuring complete satisfaction of hydrogen demand across regions. By considering station types, capacities, and regional demand, the model supports strategic decision-making for hydrogen infrastructure deployment. Sensitivity analysis highlights the impact of input uncertainties, and the model's flexibility accommodates varying scenarios and demand patterns. This study lays the groundwork for future research into multi-period models. It provides a valuable decision-making tool for integrating hydrogen propulsion into rail systems, supporting the adoption of sustainable rail transportation.

Acknowledgements. The research was supported by Europe's Rail Flagship Project 4 – Sustainable and green rail systems, which has received co-funding from the Horizon Europe Framework Program (HORIZON) under grant agreement no. 101101917 and the Polish Ministry of Education and Science under grant agreement no. MEiN 5337/HE/2023/2 under the program titled International Co-Financed Projects. The project is implemented at the AGH University of Science and Technology in Krakow from 2022 to 2026.

References

1. Owen, S.H., Daskin, M.S.: Strategic facility location: a review. Eur. J. Oper. Res. **111**, 423–447 (1998)
2. Moreno-Benito, M., Agnolucci, P., Papageorgiou, L.G.: Towards a sustainable hydrogen economy: optimisation-based framework for hydrogen infrastructure development. Comput. Chem. Eng. **102**, 110–127 (2017)
3. Li, L., Al Chami, Z., Manier, H., Manier, M.A., Xue, J.: Incorporating fuel delivery in network design for hydrogen fueling stations: Formulation and two metaheuristic approaches. Transp. Res. Part E: Logist. Transp. Rev. **152**, 102384 (2021)
4. Abbaas, O., Ventura, J.A.: An edge scanning method for the continuous deviation-flow refueling station location problem on a general network. Networks **79**(3), 264–291 (2022)
5. Thiel, D.: A pricing-based location model for deploying a hydrogen fueling station network. Int. J. Hydrogen Energy **45**(46), 24174–24189 (2020)
6. Averbakh, I., Berman, O.: Minimax regret p-center location on a network with demand uncertainty. Locat. Sci. **5**(4), 247–254 (1997)
7. Labbé, M., Thisse, J.-F., Wendell, R.E.: Sensitivity analysis in minisum facility location problems. Oper. Res. **39**(6), 961–969 (1991)
8. Lin, R.H., Ye, Z.Z., Wu, B.D.: A review of hydrogen station location models. Int. J. Hydrogen Energy **45**(39), 20176–20183 (2020)

Implementation of Unit Load AGV Technology in Streamlining of the In-House Logistics of an Automotive Assembly Hall

Tomáš Malčic[1(✉)] and Jiří Záruba[2]

[1] Škoda Auto University, Mladá Boleslav, Czech Republic
tomas.malcic@savs.cz
[2] VSB—Technical University of Ostrava, Ostrava, Czech Republic

Abstract. The automotive industry is currently undergoing a significant transformation due to 4th industrial revolution. The importance of effective implementation of advanced technological solutions in manufacturing and logistics processes is increasing due to a number of factors, including globalisation, competitiveness, product customisation, shortage of qualified workforce and overal SCM complexity. In order to overcome these challenges, it is essential that logistics systems are being continuously improved and innovated in the automotive industry. The implementation of automated guided vehicles (AGVs) represents a pivotal technology in the automation of production logistics processes. Similary to the automotive industry as a whole, AGV technology is subject to constant evolution and improvement as well. The aim of this paper is to present the advantages and benefits of implementing a new generation of advanced unit load AGVs within a case study of the streamlining of assembly line supply processes in the automotive industry. The case study will analyse the original supply process system, which was based on tow AGV units with a central charging station. The original technological solution will be analysed, with focus on the identification of shortcomings of the system. Subsequently, a new system for automatic assembly line supply that is based on the implementation of unit load AGVs will be presented. The contribution of new technological functionalities, including automatic loading/unloading, optimisation of supply routes, flexible charging modes and more advanced navigation, will be emphasised. Finally, a future vision of the assembly line supply system will be presented, which will fully utilize the potential of the implemented technology.

Keywords: AGV · Automation · Automotive · Industry 4.0 · Production Logistics

1 Introduction

The accelerated evolution of the automotive industry, driven by the 4th industrial revolution, has underscored the need for advanced technological solutions in manufacturing and logistics. Rising global competition, increased demand for customization, labor shortages, and complex supply chains have made continuous improvements in logistics

J. Feliks et al. (Eds.): CLC2024 2024, 2026.
https://doi.org/10.1007/978-3-032-28931-5_25

essential. Automated Guided Vehicles (AGVs) have become central to this transformation, providing automation and optimization for material handling in production. This paper investigates the implementation of an advanced generation of unit-load AGVs within an automotive assembly hall to streamline in-house logistics. The study analysis the previous AGV tow tractor system and its limitations like fixed routing, manual loading/unloading, and central charging station. It then analyzes how unit-load AGVs, with autonomous movement, decentralized charging, and automated handling, address these issues to improve operational efficiency.

2 Literature Review

Automated Guided Vehicles (AGVs) and Autonomous Mobile Robots (AMRs) are pivotal in automating logistics processes within the automotive industry, where there is increased demand for highly flexible and adaptable systems. AGVs are defined as battery-powered vehicles that transport materials autonomously along predefined paths, typically for repetitive tasks in structured environments. Controlled by onboard computers and equipped with navigation systems, AGVs rely on navigation technologies such as electromagnetic paths, laser guidance, gyroscopic sensors, and vision-based systems. AMRs, as an advanced subset of AGVs, do not require fixed paths and are equipped with sensors and intelligence algorithms that allow them to navigate dynamically and adjust to changing environments, making them suitable for handling irregular material flows. Technological advancements in AGV and AMR systems have expanded their applications from in-house transport to tasks such as loading and unloading in assembly line logistics. They have proven beneficial in managing diverse logistics requirements through a range of types, including towing vehicles, forklifts, and unit load carriers. However, implementing these technologies involves challenges, such as fleet size determination, scheduling, and logistics alignment with production needs [1].

2.1 Barries and Challenges of AGV Implementation

The utilisation of AGVs enhances the automation and productivity of logistics operations, mainly in the smart factory frameworks. This is achieved through the integration of the Internet of Things (IoT) and Cyber-Physical Systems (CPS), which enables the implementation of precise, unmanned operations. The incorporation of sophisticated functionalities, such as laser navigation and task scheduling, into AGVs has the potential to enhance operational precision and facilitate the streamlining of material handling processes [2]. In Industry 4.0 environments, AGVs also contribute to energy efficiency. However, they introduce cybersecurity vulnerabilities and collision risks, necessitating advancements in navigation, battery life, and path optimisation for seamless integration [3]. The adoption of AGVs is supported by a robust foundation in lean tools [4]. Within internal logistics, AGVs optimise material flow, reduce operational costs, and improve safety, thereby yielding a high return on investment when properly integrated [5]. Risk management frameworks, such as FMEA, identify potential supply chain disruptions and integration challenges associated with AGV implementation [6]. The implementation of AMRs in warehouse settings is influenced by a number of critical factors, including the necessity for technical expertise, feasibility studies, and the consideration of financial models such as Robotics-as-a-Service. Despite the advantages, the implementation of AMRs encounters considerable obstacles, particularly the high initial costs and integration with existing systems [7]. In comparison to conventional AGVs, AMRs provide augmented flexibility in intralogistics through decentralised control systems, allowing them to adapt dynamically to environmental changes without the necessity for centralised guidance. This transition from AGVs to AMRs is advancement in automation, enhancing adaptability and operational efficiency [8]. The efficacy of collaboration between humans and AGVs in shared workspaces is contingent upon the implementation of human factors, particularly in regard to safety protocols and ergonomic design. It is important to design workplaces with these considerations in mind in order to maintain safe and productive environments in the context of Industry 4.0 [9]. As the field progresses the objective of achieving robotisation maturity is to adopt sustainable, human-centred approaches that enhance production flexibility and operational efficiency. The role of collaborative robots and AI in supporting agile manufacturing environments is becoming increasingly significant [10].

2.2 Advanced Technologies for AGV Implementation

The incorporation of sophisticated technologies into AGV systems is a crucial step in the transition to Industry 4.0. The combination of particle swarm optimisation with a dynamic-window approach has been demonstrated to markedly enhance the efficiency, adaptability and planning speed of AGV navigation, particularly in dynamic settings [11]. In smart warehouse setups, virtual AGV models facilitate synchronised real-time control, thereby aiding in performance validation and highlighting the need for precise coordination between virtual and physical models [12]. The application of digital twins and artificial intelligence (AI) represents a significant opportunity to enhance decision-making processes within the domain of production logistics, with a particular focus on optimising cost, time, and energy efficiency [13]. A flexible simulation framework for AGV transport systems enables the efficient optimisation of fleet size and scheduling, proving beneficial in complex industrial environments [14]. The optimisation of task allocation in AGVs through a comparison of global and local optimisation methods facilitates energy-efficient operations through flexible scheduling. In particular, local optimisation facilitates the management of energy resources in AGV fleets [15]. An intelligent AGV control framework utilising machine learning enables real-time operational adjustments in dynamic environments, thereby enhancing performance through the optimal selection of algorithms based on current conditions [16]. The real-time dispatching of AGVs with consideration of battery constraints, managed through deep reinforcement learning in a Markov decision process framework, addresses issues pertaining to the capacity of charging stations and minimises costs associated with travel and tardiness [17]. The utilisation of AMRs in matrix-like assembly systems enhances flexibility by reducing dependence on traditional handling equipment and increasing logistics performance. The sensor-based autonomous navigation of AMRs enhances adaptability in complex assembly operations [18].

3 Case Study: Implementation of Unit Load AGV Technology in the In-House Logistics of an Automotive Assembly Hall

This chapter presents a case study analysing the key benefits of implementing advanced AGV technology within the in-house logistics of an automotive assembly hall. The initial analysis focuses on the original logistics system, which relied on AGV tow tractors for assembly line supply. This highlights the limitations of this technological approach. Following this, the study analyses the current logistics system after the transition to unit-load AGVs. This emphasises the improvements in operational flexibility, efficiency, and overall logistics performance. The case study concludes with a comparative evaluation of the two systems. This quantitatively assesses the benefits introduced by the new logistics system.

3.1 Analysis of the Logistic System for Assembly Line Supply with AGV Tow Tractor

The original state of the examined in-house logistics system for automotive assembly line supply used AGV tow tractors (see Fig. 1) in a train-like configuration, following fixed, unalterable circuits. Each AGV operated individually along a predefined circular route, stopping at designated points to deliver components. This system was based on the utilization of a central charging station functioning as a "train station" where AGVs recharged between each supply cycle (see Fig. 2). This setup ensured a steady, predictable flow of materials to the assembly line.

Fig. 1. Tow tractor AGV.

This system had distinct advantages. Its structured design allowed for easier deployment due to standardized routes and centralized charging. The train-like AGV style offered flexibility in load capacity, enabling the transport of various cart sizes and the ability to load multiple pallets at once, maximizing transport efficiency. However, several limitations affected the system adaptability. Each AGV runs on a fixed circuit that cannot be changed without extensive reconfiguration, and the circuit loading structure constrains thus making it difficult to adjust for variations in assembly line demand. The requirement for a dedicated central AGV charging station occupied valuable floor space, and manual intervention was required to disconnect and transfer materials, increasing handling time. The maximum length of AGV train could reach lengths of up to 11 m, obstructing narrow aisles and complicating movement in restricted spaces. Additionally, contact-based charging is less efficient than modern alternatives, potentially increasing downtime and reducing operational throughput.

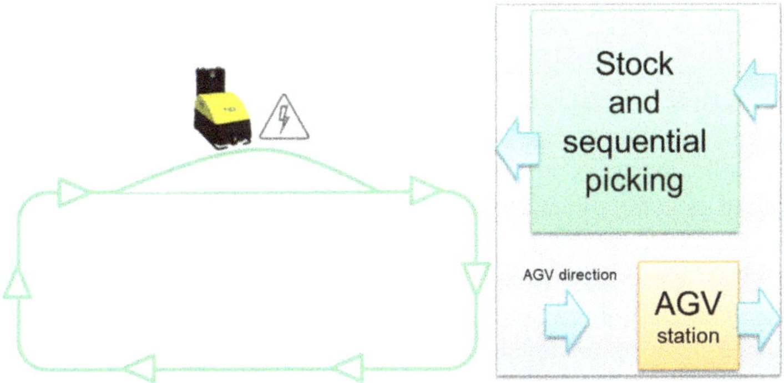

Fig. 2. Tow tractor AGV route scheme.

3.2 Analysis of the Logistics System for Assembly Line Supply with Unit Load AGV

The newly implemented in-house logistics system for automotive assembly line supply utilizes unit-load AGVs (see Fig. 3), which offers improved navigation flexibility compared to the previous system. The system also incorporates multiple dedicated charging spots throughout the warehouse and material zones, removing the dependency on a central charging station. This decentralized charging capability allows the AGVs to recharge at various locations, ensuring continuous operation without the need for centralized downtime.

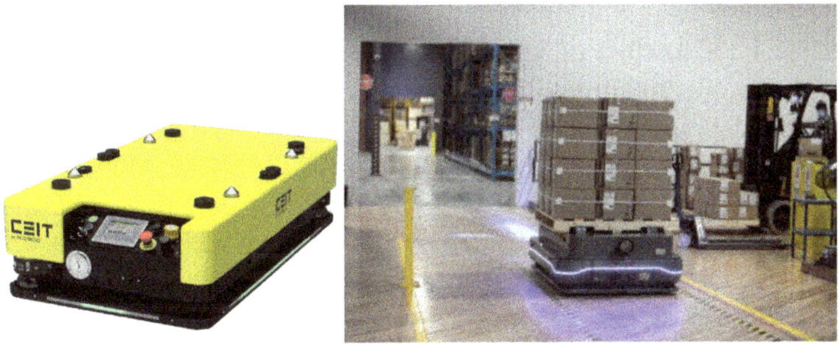

Fig. 3. Unit load AGV.

One of the main advantages of this system is the automatic pallet exchange between the warehouse and the assembly line, which streamlines the material handling process.

The ability to autonomously exchange pallets reduces the need for manual intervention, improving overall efficiency and eliminating waiting times. The compact size of the AGVs, with a maximum length of 2 m, further enhances flexibility by improving manoeuvrability in narrow aisles, making it easier for the vehicles to navigate through high traffic areas. The AGVs' omnidirectional movement capability allows them to navigate freely in all directions, enhancing operational efficiency and reducing the complexity of route planning. However, this system also presents some challenges. Despite the advanced features, the AGVs still require a joystick for manual movement in specific scenarios, which can limit the level of full automation. The use of a unified chassis for the pallets reduces flexibility, as the system is designed primarily to handle a single type of load, restricting the types of materials that can be transported. Additionally, the system's more sophisticated implementation requirements may introduce complexity in deployment, necessitating more time and resources for setup. The current configuration also allows for the transport of only one pallet at a time, limiting throughput when multiple pallets are needed for efficient material handling (Fig. 4).

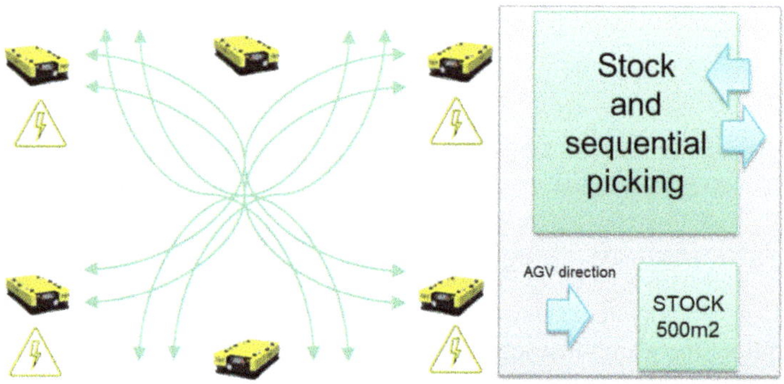

Fig. 4. Unit load AGV route scheme.

4 Results: Evaluation of the Benefits of Implementation of Unit Load AGV Technology

This chapter presents evaluation and comparison of the advantages associated with implementing unit-load AGVs, quantifying the efficiency gains achieved through the reorganization of supply processes. The comparative analysis presents the structural differences between the two systems. The initial system, which utilized AGV tow tractors with unidirectional material flow directed through a central charging station. In contrast, the implementation of unit-load AGVs, which facilitated the elimination of the central charging station allowed to repurpose this space as a storage area. Despite the advancements provided by the unit-load AGVs over the previous technology, they do not yet represent the highest level of mobile logistics automation achievable by AMRs. The current unit-load AGVs operate along predefined routes and lack the ability to alter paths or navigate around obstacles dynamically. However, the omnidirectional movement capability of these AGVs has allowed for more efficient and flexible route planning. The specific benefits observed include a 3% increase in warehouse area, achieved by converting the central charging area into storage, translating into substantial savings by reducing the need for external storage. The redesign has also reduced AGV driving time by 10%, thanks to improved routing efficiency and decentralized charging points. Further efficiency gains were observed in the reduction of AGV idle time, with an 80% decrease in waiting time due to the automation of loading and unloading tasks and the removal of fixed charging stations. This configuration has created an opportunity to apply the "taxi" principle, in which any AGV would be capable of fulfilling any supply task. This concept, if implemented, could utilize the AGVs' available capacity more effectively, as analysis shows up to 40% of their potential operational time remains unutilized. In the future, this untapped capacity could allow for fleet-wide task flexibility, where each AGV would handle a range of supply operations rather than being dedicated to specific components or workplaces. Figure 5 visually represents this scenario, comparing telemetry data across similar time intervals for four tow tractor AGVs and four unit-load AGVs. The data illustrates that unit-load AGVs are able to complete the same tasks in shorter times and, by leveraging automated loading and unloading, effectively eliminate waiting periods, potentially enabling nearly twice as many supply operations if optimized through an advanced fleet management system.

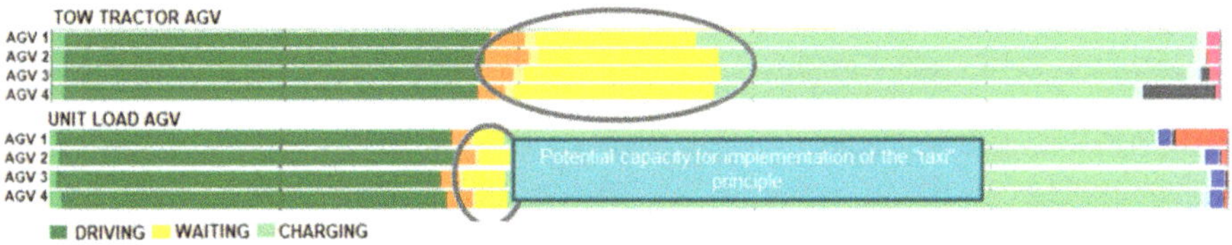

Fig. 5. Analysis of utilization of AGV capacities.

5 Conclusion

This paper has shown that implementing unit-load AGVs significantly enhances the efficiency and adaptability of in-house logistics in an automotive assembly hall. Replacing the AGV tow tractor system with unit-load AGVs yielded key benefits, including a 3% increase in warehouse area, a 10% reduction in driving time, and an 80% decrease in waiting time due to automated loading and unloading. These improvements, along with decentralized charging and omnidirectional movement, streamline logistics operations and enable flexible route planning. The study successfully quantified the advantages of the new AGV technology, but current limitations—such as the lack of dynamic rerouting and obstacle avoidance—highlight areas for future enhancement. Moving forward, integrating AMRs could elevate automation and flexibility, enabling real-time adaptability and a fully autonomous logistics system that would further improve productivity and sustainability in the automotive industry.

References

1. Čech, M. et al.: Autonomous Mobile robot technology for supplying assembly lines in the automotive industry. Acta Logist. **7**(2), 103–109 (2020). https://doi.org/10.22306/al.v7i2.164
2. Liu, X., Sun, X., Xu, J.: A case study of intelligent manufacturing for key components of vehicles. J. Phys.: Conf. Ser., 7. **1983**(1) Pro Quest Central; Publicly Available Content Database. ISSN 17426588 (2021). https://doi.org/10.1088/1742-6596/1983/1/012100
3. Tubis, A.A., Poturaj, H.: Risk related to AGV systems—open-access literature review. Energies. **15**(23), 8910. Pro Quest Central; Publicly Available Content Database (2022). https://doi.org/10.3390/en15238910
4. Palacios-Gazules, S., Giménez, G., DE CASTRO, R.: Adopting industry 4.0 technologies through lean tools: evidence from the European manufacturing survey. Int. J. Lean Six Sigma. **15**(8), 120–142 (2024). https://doi.org/10.1108/IJLSS-06-2023-0103
5. Štaffenová, K., Rakyta, M., Biňasová, V.: The use of automated guided vehicles in the internal logistics of the production company. Transp. Res. Procedia. **74**, 458–464 (2023). https://doi.org/10.1016/j.trpro.2023.11.168
6. Bekishev, Y., Pisarenko, Z., Arkadiev, V.: FMEA model in risk analysis for the implementation of AGV/AMR robotic technologies into the internal supply system of enterprises. Risks. **11**(10), 172. Pro Quest Central; Publicly Available Content Database (2023). https://doi.org/10.3390/risks11100172
7. Grover, A.K., Ashraf, M.H.: Leveraging autonomous Mobile robots for industry 4.0 warehouses: a multiple case study analysis. The. Int. J. Logist. Manag. **35**(4), 1168–1199 (2023). https://doi.org/10.1108/ijlm-09-2022-0362
8. Fragapane, G., De Koster, R., Sgarbossa, F., Strandhagen, J.O.: Planning and control of autonomous Mobile robots for intralogistics: literature review and research agenda. Eur. J. Oper. Res. **294**(2), 405–426 (2021). https://doi.org/10.1016/j.ejor.2021.01.019
9. Tubis, A.A., Poturaj, H., Smok, A.: Interaction between a human and an AGV system in a shared workspace—a literature review identifying research areas. Sustainability. **16**(3), 974. Pro Quest Central; Publicly Available Content Database (2024). https://doi.org/10.3390/su16030974
10. Hetmanczyk, M.P.: A method to evaluate the maturity level of Robotization of production processes in the context of digital transformation—polish case study. Appl. Sci. **14**(13), 5401. Pro Quest Central; Publicly Available Content Database (2024). https://doi.org/10.3390/app14135401
11. Dao, T.-K., et al.: Enhancing path planning capabilities of automated guided vehicles in dynamic environments: multi-objective PSO and dynamic-window approach. Biomimetics. **9**(1), 35. Pro Quest Central; Publicly Available Content Database (2024). https://doi.org/10.3390/biomimetics9010035
12. Nguyen, H., Nguyen, T.P., Ngo, H.Q.T.: Using ether CAT Technology to launch online automated guided vehicle manipulation with unity-based platform for smart warehouse management. IET Control Theory Appl. **18**(2), 229–243. Pro Quest Central. ISSN 17518644 (2024). https://doi.org/10.1049/cth2.12570
13. Yi, H., et al.: Production logistics in industry 3.X: bibliometric analysis, frontier case study, and future directions. Systems. **11**(7), 371. Pro Quest Central; Publicly Available Content Database (2023). https://doi.org/10.3390/systems11070371
14. López, J., Zalama, E., Gómez-García-Bermejo, J.: A simulation and control framework for AGV based transport systems. Simul. Model. Pract. Theory. **116**, 102430 (2022). https://doi.org/10.1016/j.simpat.2021.102430
15. Härdtlein, C., Stadlbauer, H., Daub, R.: Energy-oriented optimization of the task allocation for Mobile robots. Procedia CIRP. **126**, 182–187 (2024). https://doi.org/10.1016/j.procir.2024.08.321

16. Bitsch, G., Schweitzer, F.: Selection of optimal machine learning algorithm for autonomous guided vehicle's control in a smart manufacturing environment. Procedia CIRP. **107**, 1409–1414 (2022). https://doi.org/10.1016/j.procir.2022.05.166
17. Singh, N., Akcay, A., Dang, Q.-V., Martagan, T., Adan, I.: Dispatching agvs with battery constraints using deep reinforcement learning. Computers & amp. Ind. Eng. **187**, 109678 (2024). https://doi.org/10.1016/j.cie.2023.109678
18. Kathmann, T., Reh, D., Arlinghaus, J.C.: Exploiting the technological capabilities of autonomous vehicles as assembly items to improve assembly performance. SSRN Electron. J. (2022). https://doi.org/10.2139/ssrn.4211558

Operational Transport Planning Using a Heuristic Approach

Patrícia Muchová(✉) and Marek Ondov

Faculty of Mining, Ecology, Process Control and Geotechnology, Institute of Logistics and Transport, Technical University of Kosice, Kosice, Slovakia
patricia.muchova@tuke.sk

Abstract. The paper refers to the characteristics of using a heuristic approach and characterizes heuristic analysis for designing an effective method of optimizing transport planning to ensure cost-effective solutions under time-constrained conditions. The heuristic model is structured to provide acceptable solutions in cases where optimal solutions are computationally demanding or when precise data is unavailable. The criteria for creating the heuristic model outlined in the paper enable dynamic decision-making and adaptation in operational planning, ensuring flexibility in response to changing transport conditions. The proposed approach delivers rapid, practical solutions to operational challenges in transport. The heuristics implemented in the model are particularly useful when precise optimization solutions are not practically attainable and where finding the optimal solution can be computationally intensive or time-inefficient. By applying the heuristic model to operational transport management, overall transport costs can be minimized. More importantly, the model enhances the efficiency of the entire transport system, demonstrating the systemic impact of the heuristic approach. The article focuses on clarifying the creation of the heuristic model and its subsequent application to a problem within the scope of operational transport planning. The study's primary goal is to understand the heuristic approach in logistics comprehensively. Additionally, it aims to support future work by other researchers and fill research gaps by highlighting the importance of applying a heuristic approach to specific issues in the transport process.

Keywords: Operational transport planning · Heuristics · Heuristic approach · heuristic model

1 Introduction

Transportation planning is often defined as an orderly and rational process of steps that logically follow one another. In reality, planning and project development are much more complex, often involving many different activities occurring simultaneously. The planning process begins with understanding the problems the company faces and concludes with solutions to the identified issues. This is followed by recognizing issues, challenges, opportunities, and performance shortcomings of the transportation system. This initial step is crucial because a planning agency typically initiates a planning study

J. Feliks et al. (Eds.): CLC2024 2024, 2026.
https://doi.org/10.1007/978-3-032-28931-5_26

based on prior planning and analysis. Goals and objectives may also help in identify system performance measures. Data collection and analysis, the next step in the planning process, are essential for understanding the problems and potential challenges facing the transportation system and the surrounding community. This analysis process focuses primarily on understanding how the transportation system and its components operate and how changes in the system will affect its performance [1]. Transportation costs are becoming an important component for many companies, as they spend significant financial resources on moving goods from supply to demand [2]. Therefore, the transportation process is one of the activities primarily managed through operational decision-making [3]. Operational planning is the process through which operational plans are developed within an organization. It involves specific facts rather than a summary, as is the case with strategic planning. It is based on a clearly defined organizational structure, existing human resources, and assets. Conceptually, operational planning is, therefore, a simpler part of the planning process, and in many industries, it can be a relatively straightforward task [4]. Operational planning thus involves time-based and sequential planning, which depends on the level of detail implemented during the operational planning process. Operational planning is defined by a set of orders—confirmed orders for the relevant planning period. The operational plan must take into account customer requirements, as well as organizational, logistical, and technological conditions of the production or transportation process [5]. The role of the planning system is to define the company's goals, ensuring alignment for the relevant planning period between business, finance, capacities, and production, in order for the company to achieve a positive economic result [6]. Algorithms used in problem-solving are sometimes too slow or fail to find a solution within a reasonable time, while heuristics can provide a reasonable estimate that is good enough for the given problem—this led to the introduction of heuristics in problem-solving and decision-making. A heuristic solution may not be the fastest or the cheapest of all possible solutions, nor may it accurately approximate the optimal solution. Nevertheless, it provides a shortcut to achieving the goal in a relatively short time [3]. In transportation, optimization models assume centralized planning despite the decentralization of logistics systems, as it is a heuristic technique that has a significant impact on operational performance, especially in multi-route problems [7]. The goal of transportation planning is to find the optimal solution with minimal total costs [8]. In the process of solving a transportation problem, finding an initial basic feasible solution is a prerequisite for obtaining the optimal solution [1].

Heuristic approaches have been explored by authors in publications such as [9–11]. The aim of this paper is to present the application of a heuristic approach—the procedure for implementing a heuristic model for the transportation process according to the methodology of Dr.h.c. Prof. Ing. Dušan Malindžák, CSc, and its application using the main optimization criterion.

2 Methodology

The heuristic method is a practical approach that allows for achieving an approximately efficient solution, even when the optimal solution to the problem is not guaranteed. Heuristic methods are used in many real-world problems [12], such as optimization of the manufacturing process, operational transportation planning, transportation efficiency improvement, job order accumulation, and forecasting of the production process. Heuristics are iterative algorithms that do not converge to the optimal or feasible solution of a problem [13], but instead search for better solutions than those originally proposed in problems with high complexity and large search spaces [14]. The use of heuristic-based models helps better understand the behavior of systems and increases the reliability of simulations [12]. In computationally complex and ill-defined problems, a solution that is close to the optimal is preferred over the optimal solution. Heuristic is a method that "based on human experience or judgment, seems likely to provide a good solution to the problem, but it is not guaranteed to achieve the optimal solution" [15]. Although certain advances in precise optimization have been made over the years, heuristics remain highly popular in the operations research community for several reasons:

The computational complexity of many problems means that optimal solutions are unlikely to be found in a reasonable time for larger cases.

Problems may be poorly defined or data imprecise, so an optimal solution based on estimated data will almost certainly not be optimal for actual data. In such situations, a robust solution that is close to optimal in most scenarios is preferred.

The user may need several different solutions to make a final decision, especially in situations where multiple criteria need to be balanced using human judgment instead of technical measures [15].

Heuristics can help make quick decisions without the need to spend much time searching and analyzing information [16], and at the same time, heuristics are a suitable approach for the computational complexity of many problems where it is not possible to find an optimal solution in a short time. Some problems may be based on inaccurate data, so a solution based on estimated data is not optimal for actual data. In such scenarios, i.e., in computationally complex and poorly defined problems, a good solution that is nearly optimal is preferred over the optimal one. Final decision-making may also require several solutions, especially in cases where multiple criteria need to be adjusted using human judgment [2].

The heuristic approach, in comparison to analytical methods, assumes the modeling of information processing principles at various stages of human activity when solving individual tasks. Heuristic models are created based on this principle. Prof. Malindžák defined the sequence of steps [6] for creating a heuristic model as follows:

1. Definition of the problem or the initial situation,
2. Creation of possible solutions—possible variants,
3. Creation of criteria—rules for selecting the appropriate solution—heuristic analysis,
4. Synthesis of the heuristic model,
5. Verification of the model.

If we have the input information for a given decision-making process, the analysis needs to describe the algorithms, procedures, and rules by which humans process them during decision-making. The analysis for creating a heuristic model and synthesizing the decision-making system (LS) based on the heuristic model is carried out on the existing system, or we analyze an analogous benchmark system. Heuristic analysis is characterized by the decomposition of the problem based on the principle of elementary information processes. The problem or situation is decomposed according to defined rules until we reach the level of elements where we can describe and quantify it, creating a tree structure or an incomplete maze of system elements. This utilizes heuristics, expert rules, technological rules, and criteria that reflect the experience, practice, intuition, and knowledge of experts and workers in the field [6]. For the application of a heuristic approach, one of the challenges in transportation management is the operational problem of ensuring the availability of vehicles for transportation.

Heuristic rules for transportation are defined by the repetitive execution of individual processes. They are based on analogy and subsequent induction. This applies intuition, experience, practice, trial-and-error theory, and similar elements into the model, which are utilized by the decision-maker. Each company establishes these rules based on the specific problems arising in particular situations. They rely solely on the workers' experience and are not supported by any regulations—workers therefore rely on their own knowledge gained through practice. For the transportation process in this paper, a rule is defined: transportation of goods is more efficient when waste is loaded at multiple locations during a single trip.

Technological rules are determined by technological principles outlined in standards, laws, and regulations [6].

Capacity rules define the capacity requirements. These concern the capacity of transport means—their utilization, the order of operations, the number of loading and unloading points, the distribution of sales and distribution centers, and the times for starting, processing, or completing individual processes. Defining these rules involves determining the vehicle capacities, their maximum potential utilization, and maximizing the cargo space of the vehicles [17].

Expert rules determine that certain decision-making processes will still be controlled by humans, such as logisticians, planners, dispatchers, etc. These rules are organized according to specific priorities into a sequence, which forms an algorithm—a model for operational planning that is optimized according to the selected optimization criterion. For operational transportation management, expert criteria include, for example, the choice of the correct number of vehicles for transportation—if a vehicle is utilized at 50%, there is a need to try to load it so that the vehicle's capacity is maximized to 100%... [17].

The optimization criterion (OC) implicitly and explicitly defines the goal, which is the optimization of a given process [7]. For a company, cost minimization is the most important parameter in planning and management. This means that, for cost minimization, the most suitable option is selected, which meets all the constraints placed on the company's transportation process. The main criterion for creating a heuristic model thus represents the total production or transportation costs. When we define the conditions, criteria, and constraints for a future design in the form of rules (heuristics, expert rules, restrictions) and create a heuristic model from them that mimics human creative activity in processing the design, we refer to this as heuristic design. The synthesis of the given rules and the defined structure ensures the creation of an algorithm—or model—from the input data, which results in the desired plan. The principle of defining rules is outlined in the literature [7]. Heuristic models are applied when "humans know how to solve a problem, but mathematics and operational analysis do not." In such cases, we resort to modeling their actions in solving a specific problem—creating a heuristic model [6] (Fig. 1).

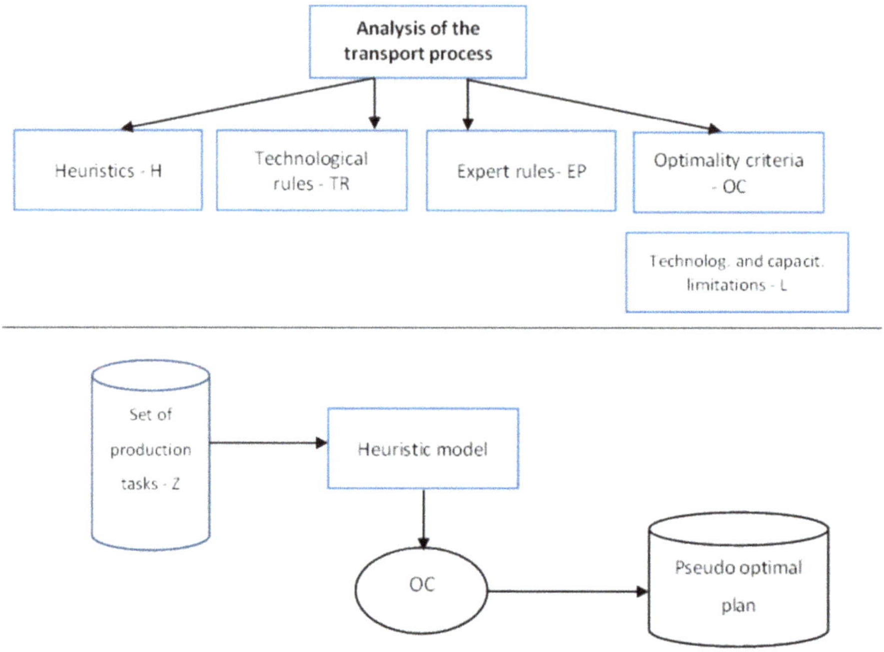

Fig. 1. Scheme of creating a heuristic model for transport process

The heuristic approach is applied to the problem within the management of the transportation process. The model is created based on data sources and information, to which various criteria correspond. In this case, the problem is characterized by routine and repetitive tasks performed by employees, during which delays occur—representing a lack of time for processing non-standard requests, such as for express product deliveries. The situation is recurring and requires a heuristic and urgent solution to the problem within the company [17]. The application of the heuristic model for operational transport planning was discussed in the work by Muchová [18].

3 Results

This section of the paper presents the results of implementing a heuristic approach in operational transport planning in a recycling company, with transportation costs as the main criterion for evaluating the efficiency of solutions. The goal of the calculations was to minimize total transportation costs in the long term, thereby increasing the efficiency of operational planning and improving resource utilization within the transport process.

Heuristic analysis was applied to the transport process of a recycling company, which arranges transportation of materials as well as the delivery of its products to consumers.

The transportation costs were calculated using the summation method [19] (Fig. 2; Table 1).

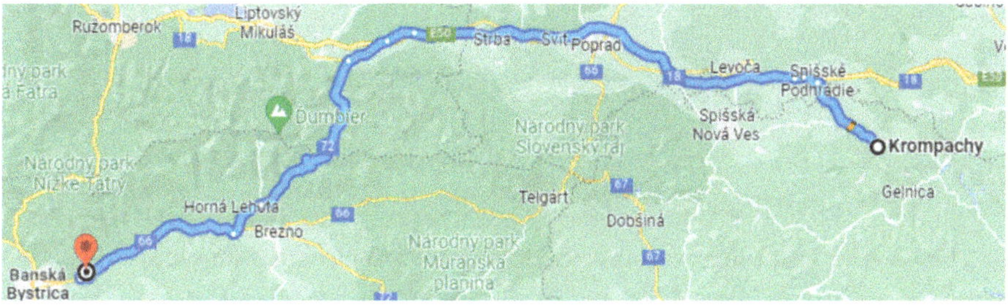

Fig. 2. Vehicle route analysis

Table 1. Heuristic analysis of transport activities

S. n.	Activities	Duration	Transportation-related costs
1.	Preparing driver documentation	1 h	Worker wage 8€/1 h
2.	Route planning	1 h	Worker wage 8€/1 h
3.	Loading	30 min	Driver's wage 10€/1 h
4.	Delivery of finished products	2 h	Driver's wage 10€/1 h
5.	Return journey with an empty vehicle	2 h	Driver's wage 10€/1 h
6.	Unloading	30 min	Driver's wage 10€/1 h
7.	Other—highway toll, fuel…	–	15,88€; 112,5€
Total—3x per week		21 h	583,14€
Total—1 year		52 weeks	30 323,28€

From the decomposition of transport activities, it is clear that the main optimization criterion is based on routine and repetitive transport processes. The current transport planning strategy is optimal for a single unloading; however, in the long term, it is inefficient in terms of overall transportation costs and truck utilization. Based on the use of the heuristic approach and from the perspective of saving long-term total costs, the results of the work are shown in Tables 2 and 3; (Fig. 3).

The table presents an overview of individual transportation activities following the application of the heuristic approach, including the duration of each activity and the associated costs. The total time and expenses are calculated for transportation conducted three times per week and extrapolated to an annual basis, providing a clear picture of the overall efficiency of transportation planning and operations.

To demonstrate the practical application of the heuristic model, the following calculation focuses on four selected loads. Table 3 presents the specific costs associated with

Fig. 3. The distribution network created by the application of the heuristic approach

Table 2. The transportation activities after applying the heuristic approach

S. n.	Activities	Duration	Transportation-related costs
1.	Preparing driver documentation	1 h	Worker wage 8€/1 h
2.	Route planning	3 h	Worker wage 8€/1 h
3.	Loading	2,5 h	Driver's wage 10€/1 h
4.	Delivery of finished products	2 h	Driver's wage 10€/1 h
5.	Return journey with an empty vehicle	4 h	Driver's wage 10€/1 h
6.	Unloading	1 h	Driver's wage 10€/1 h
7.	Other—highway toll, fuel…	–	15,88€; 135€
Total—3x per week		40,5 h	803,64€
Total—1 year		52 weeks	41 789,28€

these loads, illustrating how the model can be applied to optimize individual trips and improve overall transportation efficiency.

Table 3. Cost calculation for 4 loadings

S. n.	Activities	Duration	Transportation-related costs
1.	Preparing driver documentation	1 h	Worker wage 8€/1 h
2.	Route planning	1 h	Worker wage 8€/1 h
3.	Loading	2 h	Driver's wage 10€/1 h
4.	Unloading	0,5 h	Driver's wage 10€/1 h
5.	Other—highway toll, fuel…	–	15,88€; 112,5€
Total—3x per week		40,5 h	508,14€
Total—1 year		52 weeks	26 423,28€

4 Conclusion

Operational transportation planning involves the daily decision-making processes that ensure the smooth functioning of logistics operations within companies. These processes include vehicle allocation, route determination, scheduling, and responding to unforeseen events in transportation. In practice, these tasks are often complex and time-consuming, especially when various constraints must be considered, such as vehicle capacity, time windows, or staff availability.

Heuristic approaches are commonly used in operational planning to address these challenges effectively, as they allow the generation of feasible solutions in a short time without the need to find an optimal solution for every situation. This approach provides practical tools to support decision-making, enhances planning flexibility, and enables companies to respond more efficiently to changes and unexpected circumstances in logistics processes.

Transportation planning is currently managed operationally within companies. Routine tasks thus impose a significant time burden on employees, leading to a high staff workload. This also contributes to the continuous increase in company costs. The time required to make individual decisions presents a challenge. The conducted heuristic analysis evaluated the inefficient selection of the distribution network. The heuristic design for operational transportation management ensures smooth decision-making and reduces overall transportation costs. The proposed solution in this paper offers benefits that, in the long run, will save on transportation costs and employee time. The main goal of the work was primarily to reduce total production and transportation costs. The distribution network applied in the analysis for loading waste three times a week would cost approximately €30,323.28 annually. By utilizing a heuristic approach and selecting individual requirements, the company could, in terms of capacity, process up to four additional loadings for a single transportation. With the proposed change to the distribution network using the heuristic approach, the cost would amount to €41,789.28. If the vehicle could be fully utilized at 100% capacity on a single route and four additional loadings for a recycling company were secured, this would result in savings on total transportation costs. Although the initial costs associated with transportation planning are higher, the long-term savings in transportation costs would amount to €14,957.28 annually. By utilizing the theoretical knowledge gained from the defined heuristic analysis (defining criteria and constraints), the time for employees to process standard tasks is saved. Non-standard tasks would thus be allocated more time for processing transportation requirements, which would increase employee efficiency. The most significant contribution of this work is the regulation and elimination of overall transportation-related costs.

Acknowledgements. Research is part of the project grant number KEGA 010TUKE-4/2023 Application of educational robots in the process of teaching the study program industrial logistics.

References

1. Meyer, M.D.: Transportation Planning, 1st edn. Wiley, Institute of Transportation Engineers (2016)
2. Aickelin, U., Clark, A.: Heuristic optimisation. J. Oper. Res. Soc. **62**(2), 251–252 (2011)
3. Basalo, R., Manuel, P.: Heuristic Algorithms and Applications to Formal Methods. Complutense University of Madrid, Publications Service, Madrid (2011)
4. Etschmaier, M.M.: Operational planning and control in transportation systems. In: 9th IFAC International Proceedings on Proceedings, pp. 147–153. IFAC, Lausanne (1987)
5. Malindžák, D.: Výrobná logistika I, 1st edn. Štroffek, Košice (1997)
6. Malindžák, D., Takala, J.: Projektovanie logistických systémov. Expres Publicit, Košice (2005)
7. Malindžák, D., et al.: Logistika ako akcelerátor svetovej ekonomiky. In: Teória logistiky, p. 8. Košice (2007)
8. Juman, Z.A.M.S., Nawarathne, N.G.S.A., et al.: An efficient alternative approach to solve a transportation problem. Ceylon J. Sci. **48**, 19–29 (2019)
9. Demirag, O.C., Swann, J.L.: Capacity allocation to sales agents in a decentralized logistics network. Nav. Res. Logist. **54**, 796–810 (2007)
10. Amaliah, B., Fatichah, C., Suryani, E.: Total opportunity cost matrix – minimal total: a new approach to determine initial basic feasible solution of a transportation problem. Egypt. Inform. J. **20**, 131–141 (2019)
11. Alate, P., Isaza, G., Toro, E., Jaramillo, J.: Trends and applications of heuristic algorithms in transportation logistics. In: Trends in Sustainable Smart Cities and Territories, pp. 474–479. Springer, Heidelberg (2023)
12. Al-Shaery, A.M., Khozium, M.O., Farooqi, N.S., et al.: Problem solving in crowd management using heuristic approach. IEEE Access. **10**, 25422–25434 (2022)
13. Merbach, H.M.: Heuristics: intelligent search strategies for computer problem solving. Eur. J. Oper. Res. **21**, 278–279 (1984)
14. Lu, E.H.C., Chuang, Y.J.: A heuristic approach for bike-sharing system rebalancing problem with multiple time windows. Soft. Comput. **28**, 8337–8359 (2024)
15. Haouari, M., Aissaoui, N., Mansour, F.Z.: Network flow-based approaches for integrated aircraft fleeting and routing. Eur. J. Oper. Res. **193**, 591–599 (2009)
16. Ferreira, I.A., Oliveira, J., Antonissen, J., Carvalho, H.: A heuristic approach to support route planning for delivery and installation of furniture: a case study. Appl. Sci. **13**, 3285 (2023)
17. Muchová, P.: Heuristic analysis for the transport process. Transp. & Logist.: Int. J. **23**(54) (2023)
18. Muchová, P.: Heuristický model pre operatívne plánovanie dopravy v recyklačnej firme. Diploma thesis,. Technical University of Košice, Košice (2023)
19. Teplická, K.: Základy operačnej analýzy. Technical University of Košice, Košice, Teaching texts (2011)

Transport Packaging in the Cargo Bike Delivery Sector

Mateusz Rydlewski(✉) and Szymon Czapkowski

Department of Technical Systems Operation and Maintenance, Faculty of Mechanical Engineering, Wroclaw University of Science and Technology, Wroclaw, Poland
mateusz.rydlewski@pwr.edu.pl

Abstract. Cargo bikes are specialist vehicles structurally similar to bicycles, but equipped with a cargo space that enables the transportation of goods in densely populated urban areas. Primarily used for short-distance deliveries, and they are an environmentally friendly alternative to traditional delivery vehicles, especially in the case of delivering food, parcels and small goods in city centres. Due to the growing popularity of their use and plans to implement such systems in the country, the aspect of transport packaging used in this delivery sector was analysed. The article discusses the issue of transport packaging, analyses the design solutions of delivery bikes available on the Polish market and compares them with each other. In addition, based on the legal conditions applicable in the delivery sector for transport packaging, they were verified in terms of the possibility of their use in bicycle deliveries. The discussed topic also included the initial results of research conducted among recipients of goods in the strict centre of one of the Polish cities. The target group selected in the study consisted of business owners, mainly from the HORECA group, for whom goods can potentially be delivered using cargo bikes. The analysis of the data collected in this way allowed for the drawing of conclusions on the development of the indicated area, the determination of actions necessary to improve the logistics system in the bicycle delivery sector and the challenges posed to the packaging of shipments delivered using this type of vehicles.

Keywords: packaging transport · cargo bike · delivery sector

1 Cargo Bicycles in the delivery sector

The potential of cargo bikes and the growing importance of the aspect related to sustainable urban development pose a challenge for highly urbanized areas and create opportunities to use this means of transport to relieve or replace other vehicles. The issue of last mile deliveries in highly urbanized areas is growing in European city centers [1], and therefore the use of cargo bikes is becoming increasingly important. Studies indicate that this research area is currently very popular and is characterized by wide research interest in various areas, including, among others, the construction of business models for operators of cargo bike systems used for last mile deliveries [2], mathematical modeling to determine the location of transshipment centers for suppliers [3], modeling the delivery

J. Feliks et al. (Eds.): CLC2024 2024, 2026.
https://doi.org/10.1007/978-3-032-28931-5_27

service process taking into account aspects such as restrictions on bicycle capacity, road grade and workload balancing [4], the conditions for the financial sustainability of cargo bikes compared with electric light commercial vehicles (LCVs) [5]. Studies also show [6], that cargo bike deliveries over short distances from distribution centers and in the case of deliveries to a larger number of recipients are more profitable than conventional means, i.e. delivery vehicles. In the further part of the article, the authors focused on the aspect related to the use of transport packaging in this popular research area.

The wide range of research on cargo bikes available in the literature is reflected in the provisions of European law. Here, however, similarly to the development of the use of cargo bikes, the legal solutions of the European Union concerning cargo bikes are currently in the forming phase. Table 1 presents the types of documents adopted or planned to be adopted broadly related to cargo bikes. At the moment, the first three standards have been adopted, which in their final form are to include seven parts (DIN EN 17860-2024 Carrier Cycles) [7].

Table 1. Comprehensive series of standards of cargo bikes, source: [7].

Reference	Tytle	Status
EN 17860-1:2024	Carrier cycles—Part 1: terms and definitions	Approved
EN 17860-2:2024	Carrier cycles—Part 2: lightweight single track carrier cycles—mechanical aspects	Approved
EN 17860-3:2024	Carrier cycles—Part 3: lightweight multi track carrier cycles—mechanical aspects	Approved
FprEN 17860-5	Carrier cycles—Part 5: electrical aspects	Under approval
FprEN 17860-7	Carrier cycles—Part 7: cargo trailers	Under approval
prEN 17860-4	Carrier cycles—Part 4: heavy weight carrier cycles—mechanical and functional aspects	Under enquiry
prEN 17860-6	Carrier cycles—Part 6: passenger transport	Under approval

1.1 Classification of Cargo Bikes

Cargo bikes can be classified based on two primary criteria: the type of drive system and the construction style. In terms of the drive system, they can be divided into electric bikes and conventional (muscle-powered) bikes. The second distinguishing factor categorizes these bikes across a greater number of subcategories, such as construction, cargo space organization, weight, or dimensions of transported goods. Below are the main types of cargo bikes available on the market:

Single-track Cargo Bikes:

- Carrier cycle/luggage rack—a classic solution with a reinforced frame and a small luggage rack over the front and/or rear wheel.

- Butchers bike/cycle truck—this type has the typical features of a classic city bike but with a smaller 20 front wheel, allowing for the transport of heavier and larger cargo.
- Front loader—similar to the previous type but with a larger cargo area.
- Long John—the cargo area is located between the handlebars and the front wheel, with an extended wheelbase and high maneuverability.
- Longtail—characterized by an extended frame behind the rider's seat.

Multi-track Cargo Cycles:

- Trike—these are three-wheeled bikes with a cargo box located at the front, with the load supported by two wheels on the front axle. This type of bike offers a higher load capacity of 100 to 200 kg, but its design results in reduced maneuverability compared to the Single-track Cargo Bikes. Bikes with two wheels on the rear axle, known as "cargo trikes," are also used, with a cargo area positioned above the rear axle [8, 9].

An alternative to the types of cargo bikes described above is the option to use a trailer, providing additional cargo space when needed. Two basic types are available: a traditional two-wheeled trailer or a single-wheel version to which bags can be attached on both sides. The advantage of this solution is that it does not increase the original width of a conventional bicycle [10]. Manufacturers of various bike types adapt the cargo space to generic solutions, such as KERL 500 rails or Eurobox container systems. Figure 1 shows the classification of cargo bikes and their schematic designs.

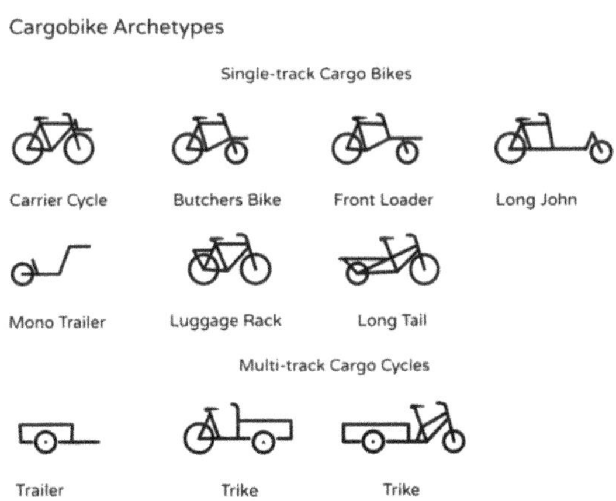

Fig. 1. Types of cargo bikes, source: own works based on [11].

2 Identification of Packaging Types Used in City Center Deliveries

Direct interviews with owners and managers of companies in the very center of Wrocław (Poland) conducted in 2024 aimed to identify challenges and opportunities for creating a system in which deliveries to selected premises could be made using cargo bikes. The study covered over 200 points located in the specified research area. More than half of the premises studied belong to the HORECA group. Currently, deliveries made using cargo bikes constitute less than 3% of all deliveries made to these points. At the same time, over 80% of all deliveries are made using combustion vehicles. As part of this study, the focus was, among other things, on the aspect related to the type of transport packaging currently used in deliveries to these premises. The results indicate that by far the largest part of the products was delivered in cartons containing groups of goods. This also fits into the most common way of packaging products by food producers, i.e. cardboard collective packaging. Another of the most common identified ways of packaging orders for these premises was a cardboard collective packaging containing individual products. This solution significantly reduces the efficiency of space in the transport of goods. The next group are foil packaging in which single goods or groups of goods were delivered. Among the methods of packaging (or transporting orders) distinguished in the study, packaging in the form of foil packaging and euro pallets were also distinguished. In each of these cases, the total number of deliveries using these solutions to points in the strict city center was negligible and had an incidental nature. Respondents taking part in the research also drew attention to other types of packaging used in the transport of products they ordered. Among the emerging possible packaging solutions, it is worth mentioning crates or plastic boxes or polystyrene containers for ready-made food used to maintain a constant low or elevated temperature during transport. The types of packaging used for the transport of goods identified in this study create considerable potential for the possibility of carrying out these deliveries using cargo bikes. This potential is also noticeable among the respondents, of whom over 40% are open to changing the form of deliveries to cargo bikes, sometimes needing more information in this area.

3 Transport Packaging in the Bicycle Delivery Sector

Based on the growing popularity of cargo bike delivery solutions and the high social acceptability of this type of solution, it was decided to determine the state of knowledge and research on the use of transport packaging in this sector. The authors focused on the conditions imposed on transport packaging and on the solutions used in the field of packaging in the cargo bike delivery sector.

3.1 Requirements for Transport Packaging

The primary role of transport packaging is to protect the product from damage, spoilage, contact with other products, and adverse external conditions (including weather). An important aspect is maximizing and optimizing cargo space to minimize transportation costs. Individual packaging, and often collective packaging as well, has aesthetic functions that distinguish the product from competitors, making it more attractive to potential buyers. In Poland and the European Union, transport packaging in the bicycle sector is subject to general packaging regulations, in particular:

- The Packaging and Packaging Waste Management Act (Packaging Management Act) [12]
- Regulations on Packaging Labeling [13] and their Safety [14]

- European Standards Concerning Transport Packaging [15]

However, there are no specific requirements for packaging in bicycle transport. In the area of food delivery using cargo bikes (excluding ready-meal delivery, which is not covered in this article), restrictive requirements and legal conditions apply exclusively to packaging that comes into contact with food [14]. However, these regulations pertain to the product and method of transport, rather than the type of vehicle used for goods transport.

3.2 Overview of Packaging Solutions for the Bicycle Delivery Sector

Several types of packaging are used for transport purposes in cargo bikes. For longtail bikes, rigid side panniers, known as cargo panniers, are available, into which products can be placed [15]. These can accommodate bulk-packed products, items in cartons (mainly for individual goods), or in plastic wraps (including foil bags). Another solution involves rigid baskets designed to create a structure for transporting goods [16]. The limitation here is the size of the transported material. Cargo bikes can also be fitted with boxes on the rear rack for ready meal and food deliveries [17] or with boxes with a rigid structure [18]. Another solution includes foldable fabric boxes with a rigid structure that secures the cargo [19].

In long john bikes, several types of solutions can be distinguished, such as open boxes made of two wooden boards on the sides of the cargo space [20]. In this type of bike, waterproof, lockable containers with a capacity of around 200–250 L can also be found. These types of containers are used for transporting logistics shipments, food, flowers, or documents. These containers are often equipped with soundproofing mats [21]. Another example in this type of bike is the use of nets or ropes to secure cargo with non-standard shapes or sizes. This enables the safe transportation of goods.

Cargo bikes with three wheels offer greater stability, load capacity, and configuration options. A large cargo with a capacity of up to 2500 L and a weight of 300 Kg can be placed on the front axle of the bike [22]. This solution allows for the use of cargo boxes to transport heavy products of various purposes. On the basis of this type of cargo bike, customized structures, such as cafes or ice cream stands, can also be implemented. Bikes with a wider rear axle, known as "cargo trikes," feature space for cargo behind the rider's seat. Large cargo boxes with a wide range of applications, from 1000 to 3000 L, or open structures for custom uses, can be placed there [23].

An important element to note is that there is no official standardization of cargo or packaging space in cargo bikes. This creates a situation that, over time, may lead to issues related to safety and standardization of solutions, potentially resulting in high costs for adapting solutions used by companies in the cargo bike industry.

4 Transport Packaging in the Bicycle Delivery Sector

After analyzing the current solutions on the packaging market in cargo bike deliveries, it can be seen that this is an area requiring further development. At this point, it is possible to use solutions proposed directly by bicycle manufacturers, such as metal baskets, high sides in the loading space or closed containers. An alternative are containers installed, for example, by food delivery companies, adapted especially to these needs. However, there is a lack of dedicated solutions, thanks to which the load on the bike could be more systematized, adjusted to the needs of users and thanks to which the use of available space could be more effective. Here, one can consider the possibility of transporting goods in cardboard packaging (or other materials), which will ensure stability during transport and protect the product during transport. Many types of cargo bikes allow for their wide use in urban transport. However, the growing development of this delivery industry, which is followed by social openness to changes in the delivery sector, create the need to systematize emerging solutions in order to ensure safety, reliability and the highest possible economic efficiency for cargo bike deliveries. Parallel actions to standardize the packaging systems used in transport together with the design solutions of cargo bikes provide an opportunity to create a highly efficient economic logistics system, which, as research shows, has an economic justification for use in deliveries in city centers. Integration of existing solutions, maximization of cargo space filling and social openness to such solutions is a great chance for the development of the cargo bike delivery sector in the future.

References

1. Naumov, V., Pawluś, M.: Identifying the optimal packing and routing to improve. Energies. **14**(4132) (2021)
2. Bošković, S., Švadlenka, L., Dobrodolac, M., Jovčić, S., Zanne, M.: An extended AROMAN method for cargo bike delivery concept selection. Decis. Mak.: Appl. Manag. Eng. **1**(1), 1–9 (2023)
3. Naumov, V., Starczewski, J.: Choosing the localisation of loading points for the cargo bicycles system in the Krakow old town. In: Reliability and Statistics in Transportation and Communications—Selected Papers from the 18th International Conference on Reliability and Statistics in Transportation and Communication, RelStat'18, pp. 353–362. IEEE, Lviv (2019)
4. Papaioannou, E., Iliopoulou, C., Kepaptsoglou, K.: Last-mile logistics network design under E-cargo bikes. Futur. Transp. **3**(2), 403–416 (2023)
5. Robichet, A., Nierat, P., Combes, F.: First and last miles by cargo bikes: ecological commitment or economically feasible? The case of a parcel service company in Paris. Transp. Res. Rec. **2676**(9), 269–278 (2022)
6. Sheth, M., Butrina, P., Goodchild, A., Mccormack, E.: Measuring delivery route cost tradeoffs between electric-assist cargo bicycles and delivery trucks in dense urban areas. Eur. Transp. Res. Rev. **11**(1) (2019)
7. ACT LAB: Cargo Bike Safety—EN 17860. https://act-lab.com/cargo-bike-safety-en-17860/. Last Accessed 08 Nov 2024
8. URVIS BIKE: Urvis—rower cargo, rower towarowy. https://omniumcargo.com/shop/product/foldable-carg. Last Accessed 08 Nov 2024
9. SALEM BREAKFAST ON BIKES: Cargo Bikes in the News!. https://urvis.bike/produkt/urvis-bike-open-box/. Last Accessed 08 Nov 2024
10. ROWERYSTYLOWE.PL.: Przyczepka do roweru transportowego. https://more-cargobike.de/en/collections/unsere-bo. Last Accessed 08 Nov 2024
11. MORGENLAB: Cargo Bike Archetypes. https://morgenlab.net/cargobikearchetypes. Last Accessed 08 Nov 2024
12. Dz.U. 2013 poz. 888. Ustawa z dnia 13 czerwca 2013 r. o gospodarce opakowaniami i odpadami opakowaniowymi. Dziennik Ustaw Rzeczypospolitej Polskiej

13. Dz.U. 2014 poz. 1298. Rozporządzenie Ministra Środowiska z dnia 3 września 2014 r. w sprawie wzorów oznakowania opakowań. Dziennik Ustaw Rzeczypospolitej Polskiej
14. EUROPEAN PARLIAMENT AND COUNCIL REGULATION (EC): No 1935/2004 of 27 October 2004 on materials and articles intended to come into contact with food
15. ISO 4180:2019: Packaging—Complete, filled transport packages—General rules for the compilation of performance test schedules. International Organization for Standardization, Geneva (2019)
16. LECTRIC E-BIKES: XL Cargo Pannier. https://lectricebikes.com/products/xl-cargo-pannie. Last Accessed 08 Nov 2024
17. LECTRIC E-BIKES: Cargo Pack. https://lectricebikes.com/products/cargo-pack. Last Accessed 08 Nov 2024
18. LECTRIC E-BIKES: Large Insulated Bag. https://lectricebikes.com/products/large-insulated. Last Accessed 08 Nov 2024
19. DECATHLON: Electric Cargo Bike. https://www.decathlon.pl/p/rower-elektryczny-trans. Last Accessed 08 Nov 2024
20. OMNIUM CARGO: Foldable Cargo Bike. https://omniumcargo.com/shop/product/foldable-carg. Last Accessed 08 Nov 2024
21. URVIS BIKE: Urvis Bike—Open Box. https://urvis.bike/produkt/urvis-bike-open-box/. Last Accessed 08 Nov 2024
22. MORE CARGOBIKE: Cargo Bikes Collection. https://more-cargobike.de/en/collections/unsere-bo. Last Accessed 08 Nov 2024
23. URBAN ARROW: Tender Cargo Bike. https://urbanarrow.com/business-bikes/tender/. Last Accessed 08 Nov 2024
24. MAXPRO MOBILITY: Maxpro Mobility Cargo Bikes. https://www.maxpromobility.com/cargo-bikes. Last Accessed 08 Nov 2024

Maturity Model for Transport System Assessment

Emilia T. Skupień(✉)

Wroclaw University of Science and Technology, Wroclaw, Poland
emilia.skupien@pwr.edu.pl

Abstract. Transport systems represent complex structures composed of interdependent subsystems encompassing organizational, technical, financial, and regulatory dimensions. Their evaluation remains a challenging task, even when limited to specific modes of transport, due to varying geographical, economic, and administrative conditions influencing system performance. Previous research indicates that the most frequently analyzed factor groups include transport efficiency, economic and social aspects, safety, political context, business environment, and ecological impact. Multi-criteria decision-making methods have gained a dominant position among tools applied to transport system evaluation. Building upon these analytical foundations, the article proposes a maturity model as an effective framework for both comparative assessment and the identification of development trajectories. The proposed model integrates the principal parameter groups affecting transport system performance, incorporating key components such as infrastructure, vehicle fleets, and transport service operations, structured across five maturity levels. Its flexible design enables application to different transport branches—road, rail, maritime, inland waterway, and air—facilitating the assessment of system status, benchmarking, and formulation of targeted recommendations for further advancement. Considering the ongoing digital transformation, particular attention is devoted to the influence of digitalization on the functioning and progression of transport systems.

Keywords: Maturity Model · Transport Systems · Transport Systems Assessment

1 Introduction

One of the key elements of logistics is transport. Transport enables shipping cargo by physical movement between different points on a map, e.g. from manufacturer to warehouse, warehouse to retailer, or directly to consumer. Transportation is a key element in optimizing logistics processes. The right choice of means of transport, routes, and modes of transport affects the efficiency and cost of the entire supply chain, and effective transportation management is inextricably linked to inventory management, production, and distribution. Transport, through its participation in the global economy, also affects the natural environment. For these reasons, it is important to be able to evaluate, compare, and develop transport systems.

J. Feliks et al. (Eds.): CLC2024 2024, 2026.
https://doi.org/10.1007/978-3-032-28931-5_28

Transportation by different modes of transport differs in terms of cost, delivery time, flexibility, accessibility, capacity, safety, and environmental impact. Although one can compare the listed parameters describing the process or system of transport, it is difficult to compare systems with each other and transport systems of different branches. It is difficult to compare systems of the same branch operating in a different geographical or economic environment. It is difficult to say whether rail transport is better than air transport, it is difficult to compare the road transport system in Germany and China in one dimension.

The aim of this article is to propose a method that enables the assessment and comparison of transport systems. The author of the article, based on the contemporary literature, determined the parameters of the transport systems that are evaluated and the methods used for this purpose. On this basis, the author proposed the assessment of transport systems through a maturity model.

2 Methodology

The methodology used in this article consists of 4 steps. 1. an examination of the factors subject to evaluation by researchers in the context of transport systems. 2. an examination of the tools/methods used for such analyses. 3. an analysis of the maturity models and the possibilities of using them to assess and compare transport systems of different modes and/or in different environments. 4. a proposal of a tool for assessing and comparing transport systems.

In order to obtain information on the parameters of transport systems that are subject to evaluation (step 1), and what tools/methods are used in this context (step 2), a literature review was performed based on the Web of Science (WoS) Core Collection database [1]. The analysis of the maturity models (step 3) was also performed according to the literature review. Step 4 was carried out through own desk research, on the basis of the results of the previous three steps.

3 Results of the Literature Review

The literature review was carried out the basis of the Web of Science (WoS) Core Collection database. The searched keywords were "transport mode", "transport system", and "transport branch", cross- referenced with "assessment", "evaluation", and "development". This means that nine search rounds were conducted. To make the results authoritative, publications older than 5 years were rejected. The author focused on English language, and reviewed papers. Also, to narrow the search, Web of Science categories were applied: Transportation Science & Technology and Transportation. Finally, 123 publications were examined.

3.1 Parameters of Transport Systems

Based on the literature identified using the WoS database, specific factors under investigation related to the assessment of transport systems were gathered. Researchers examine many factors that may be divided into transport branches, including urban transport, e.g [2, 3], or relating to transport systems as such e.g [4, 5]. Particular research papers refer to decision making and safety [6, 7] and the physical and mental health of transport users, e.g [8, 9]. Some articles refer to specific aspects, such as hazardous materials [10, 11], or technical issues of vehicles [12, 13]. There are also papers describing transport systems efficiency [4, 14]. Many publications refer to infrastructure and its accessibility [15, 16], planning [2] and organization [17]. There is also a group of papers considering the environment, natural e.g [18]. and political e.g [17].

Taking into account the examined publications in step 1, the identified factors that are assessed in the matter of transport systems, may be divided into several groups considering: (i) transport process; process parameters and elements (fleet and infrastructure), (ii) economy and legislation, (iii) human factors concerning safety and social environment, (iv) ecology and natural environment.

The elements of the transport process, i.e. infrastructure and fleet, are difficult to compare using similar criteria, so it seems reasonable to separate them. The process itself, after isolating infrastructure and fleet, can be generally reduced to management. Economic and legislative appendices can be included in the concept of management. Issues related to safety and ecology appear in every aspect, so they can be included within them.

3.2 Methods to Assess Transport Systems

The next step of the literature review was to identify the methods used for the assessment of transport systems. The same base of scientific papers was examined. Analyzing any system requires including its elements. Therefore, it would be hard to describe and assess systems using methods considering only one parameter at the time. This means that the methods used for the assessment of transport systems in quantitative manner should compare many criteria. Qualitive methods are harder to extend to other systems operating in other environment and are also more complicated to use in a general manner, because of a difficulty in circumstances repeatability, and analyzing the results requires expert knowledge.

Particularly, some of the papers describe methods in very general manner, naming them just multi-criterion decision making methods (e.g [4, 14]). or qualitative methods (e.g [17, 18]). Some of the methods were named specifically: AHP, DEMATEL ([19]), fractal analysis ([16]), modeling ([5]). Evaluation and comparison of parameters ([15]), case study, conceptual study, and survey (e.g [17, 18]).

Methods from the group of multi-criterion decision-making (MCDM) may be used to build a model enabling the comparison of transport systems and their parameters.

3.3 Methods to Assess Transport Systems

Maturity models define maturity stages that describe the analyzed objects using defined criteria. Maturity models contain three elements: defined stages (levels); evaluation criteria, and a development procedure (to achieve higher levels of maturity). Therefore, maturity models were under specific consideration of the author, as a possible tool to assess transport systems. In this manner, a literature review was conducted.

Literature review on this theme was performed using WoS database, using "maturity model" as a keyword for the topic. The time period chosen for the analysis of publications was last 20 years. For narrowing the number of results, the "transport" category was used. This gave 127 responses, which were subject to a detailed investigation.

Some of the examined papers do not show the model directly, or only mention it in abstract. But there are also papers that describe the model, its levels, and factors on those levels. For example, [20] gives an example of a complex maturity model of air traffic management. The presented model is composed of 5 levels. Level A—Informal arrangements, Level B—Defined, Level C—Managed, Level D—Assured and Level E—Optimized. A cycling maturity planning model is presented in [21]. The model consists of 6 stages, of 15 individual factors. This maturity model is proposed for the formulation of an urban cycling plan. The measurement of the safety management system is described in [22]. The model is composed of five levels of safety maturity, which are as follows: Level 1: Initiating, Level 2: Planning, Level 3: Implementation, Level 4: Managing & measuring, and Level 5: Continuous improvement. The risk maturity model of maritime transport is described in [23]. The model describes 17 attributes on 5 levels: Inadequate, reactive, compliant, proactive and optimal.

4 Maturity Model for the Assessment of Transport Systems

Maturity models were firstly based on the Capability Maturity Model (CMM). This Models were created in the 1980s for improving the predictability and quality of software development process. The main objective was to assess the ability to manage processes effectively and reduce risk [24].

Maturity models are used for assessing the current state of an examined object, identifying gaps between individual objects, and planning a development path toward higher levels of described goal. System maturity refers to its advanced state of development. It can be assumed that a mature system is characterized by the following: (i) stability—system operates reliably and predictably, (ii) functionality—system offers a full set of functions and capabilities that meet the needs of users, (iii) optimization—system is optimized for performance and efficiency, (iv) scalability—system can be easily developed or adapted to new needs, (v) compliance with standards—system is compliant with industry requirements and best practices, (vi) security—system provides appropriate mechanisms for protection against defined threats.

As stated in paragraph 3.1., the maturity model of the transport system should take into account an infrastructure, fleet, and management system. The proposal of subparametrs and their levels of maturity are shown in Table 1.

Since the model presented in Table 1 is intended to enable the evaluation of transport systems of each branch (to enable comparison of different systems with each other), the individual levels are described in relation to applicable standards or European or world averages. This way the universality of the method has been achieved, but part of the responsibility for its correct use has been transferred to the user of the method. Additionally, the levels refer to national systems. If you wish to use the method to assess local systems, you can adjust the levels to the guidelines and averages than those proposed. The concept of maturity models assumes the assessment of maturity levels

Table 1. Maturity model for transport system assessment, level of maturity and assessed factors [25]

	Maturity level				
	I	II	III	IV	V
Fleet	Fleet age exceeds the world average	Fleet age exceeds the European average	Fleet age corresponds to the European average	Fleet age is lower than the European average	Fleet age is significantly lower than the European average
	Equipment does not meet the lowest international standards	Equipment meets the lowest international standards	Equipment slightly exceeds the lowest international standards	Equipment exceeds international standards	Equipment meets the highest international standards
	Fleet number and type significantly exceeds or falls below market needs	Fleet number or type meets market needs, but the other factor significantly exceeds or falls below market needs	Fleet number and type is slightly below or slightly above market needs	Fleet number or type meets market needs, but the other factor is slightly above or below market needs	Fleet number and type meets market needs in the long term
	Number and competence of fleet and process staff do not meet the demand	Number or competence of fleet and process staff does not meet demand	Number and competence of fleet and process staff are slightly below demand	Number or competence of fleet and process staff is slightly below demand	Number and competence of fleet and process staff meet the demand
Infrastructure	Funds allocated for infrastructure maintenance do not cover current needs	Funds allocated for infrastructure maintenance cover current needs	Funds allocated for infrastructure maintenance cover current needs and minor preventive measures	Funds allocated for infrastructure maintenance cover current needs and investments	Funds allocated for infrastructure maintenance cover current needs and large-scale investments

(*continued*)

based on the criteria met. To reach a higher level, all criteria of the lower level must be met. This means that, for example, meeting all criteria at level 3 except one results in remaining at level 2.

Each of the aspects assessed in the presented model is very extensive, and using the model in this form allows for a rough assessment of the system. However, it is possible to assess systems in more detail by considering each of the aspects separately and in more detail. An approach assessing the system from a different angle is also possible. In

Table 1. (*continued*)

	Maturity level				
	I	II	III	IV	V
	Linear infrastructure does not create a network	Linear infrastructure creates a network with certain limitations	Linear infrastructure creates a coherent network and connects to the network of another region	Linear infrastructure creates a coherent network and connects with networks of other regions	Linear infrastructure creates a coherent network and connects with all neighbouring networks
	Point infrastructure is insufficient and does not allow for inter-branch transshipment	Point infrastructure is insufficient but allows for some inter-branch transshipments	Point infrastructure is sufficient and allows for some inter-branch transshipments	Point infrastructure is sufficient and allows for inter-branch transshipment	Point infrastructure fully integrates available transport modes
	Administration and security services do not operate to the appropriate extent	Administration or security services do not operate to the appropriate extent	Administration and security services operate to the extent that allows maintaining the current status quo	Administration and security services operate to the extent that allows for the improvement of the current situation	Administration and security services ensure the development of the system
Management	Local regulations do not correspond to supra-local regulations	Local regulations correspond to supra-local regulations to a minimum extent and are introduced with a delay	Local regulations are adapted to supra-local regulations with a slight delay	Local regulations are constantly being adapted to supra-local regulations	Local regulations set the directions for creating supra-local regulations

(*continued*)

the era of Industry 5.0, digital maturity plays a crucial role in the functioning of systems, i.e. facilitating an appropriate level of process support from an IT perspective. In this aspect, the following are assessed: (i) customer service, (ii) operational process support, (iii) operational fleet support, (iv) vehicle navigation, (v) system administration. The maturity of the system in terms of its digitization will be the subject of further research by the author. However, the maturity of the system can also be assessed separately in terms of its security, environmental impact, or meeting market needs.

Table 1. (*continued*)

	Maturity level				
	I	II	III	IV	V
	Regulations and procedures governing the fleet and transport process are unclear and unenforced	Regulations or procedures governing the fleet and transportation process are unclear and unenforced	Regulations and procedures governing the fleet and transportation process are complicated	Regulations and procedures governing the fleet and transportation process are clear	Regulations and governing the fleet and transportation process are clear and as simple as possible
	Information flow between system elements is very difficult	Information flow between some system elements is hindered	Information flow between system elements is efficient	Most information flows between system elements occurs online	Information flow between system elements occurs online

5 Conclusions

The transport process affects the cost, delivery time and efficiency of the supply chain. The evaluation of transport systems is important in the context of their development.

Maturity models are an effective assessment tool, and the proposed model for assessing the maturity of transport systems enables the comparison of various systems and the recognition of areas for improvement, offering a universal approach.

The proposed transport system assessment maturity model focuses on the basic components of the system, i.e. fleet, infrastructure and management. Due to the assumption of the possibility of using the model for each transport branch, adaptation of some aspects may be required. At the same time as a result of the complexity of transport systems and the numerous components they comprise, it may be necessary to extend the model and assess individual elements and/or aspects separately. In further research, the author will focus on the evaluation of the transport systems' digital maturity.

References

1. Web of science. https: //www.webofscience.com/wos/woscc/basic-search Last accessed 11 Oct 2024.
2. Teusch, J., et al.: A systematic literature review on machine learning in shared mobility. IEEE Open J. Intell. Transp. Syst. **4**, 870–899 (2023)
3. Liao, F. & Correia, G.: Electric carsharing and micromobility: a literature review on their usage pattern, demand, and potential impacts. Int. J. Sustain. Transp.**16**, 269–286 Preprint at https://doi.org/10.1080/15568318.2020.1861394 (2022)
4. Yannis, G., Kopsacheili, A., Dragomanovits, A. and Petraki, V.: State-of-the-art review on multi-criteria decision-making in the transport sector. J. Traffic Transp. Eng. (Engl. Ed.) **7**, 413–431 Preprint at https://doi.org/10.1016/j.jtte.2020.05.005 (2020)

5. Zhang, Y., Kamargianni, M.: A review on the factors influencing the adoption of new mobility technologies and services: autonomous vehicle, drone, micromobility and mobility as a service. Transp. Rev. **43**, 407–429 (2023)
6. Ferreira-Vanegas, C. M., Velez, J. I. and Garcia-Llinas, G. A. Analytical methods and determinants of frequency and severity of road accidents: a 20-year systematic literature review. J. Adv. Transp. **2022** Preprint at https://doi.org/10.1155/2022/7239464 (2022)
7. Berhanu, Y., Alemayehu, E. and Schröder, D. Examining car accident prediction techniques and road traffic congestion: a comparative analysis of road safety and prevention of world challenges in low-income and high-income countries. J. Adv. Transp. **2023** Preprint at https://doi.org/10.1155/2023/6643412 (2023)
8. Liu, J., Ettema, D., Helbich, M.: Systematic review of the association between commuting, subjective wellbeing and mental health. Travel Behav. Soc. **28**, 59–74 (2022)
9. Li, J., Xiang, T. and He, L. Modeling epidemic spread in transportation networks: a review. J. Traffic Transp. Eng. (Engl. Ed.) **8**, 139–152 Preprint at https://doi.org/10.1016/j.jtte.2020.10.003 (2021)
10. Rukavina, B.: Some issues about legal regulation regarding handling dangerous goods in maritime traffic. Pomorstvo. **34**, 302–308 (2020)
11. Guo, J. and Luo, C. Risk assessment of hazardous materials transportation: a review of research progress in the last thirty years. J. Traffic Transp. Eng. (Engl. Ed.)**9**, 571–590 Preprint at https://doi.org/10.1016/j.jtte.2022.01.004 (2022)
12. Meishner, F. & Sauer, D. U. Wayside energy recovery systems in DC urban railway grids. eTransportation **1** Preprint at https://doi.org/10.1016/j.etran.2019.04.001 (2019)
13. Liu, Z., Berg, M. & Bustad, T. Review of the existing energy labelling systems and a proposal for rail vehicles. Proc. Inst. Mech. Eng., Part F: J. Rail Rapid Transit vol. **235** 518–528 Preprint at https://doi.org/10.1177/0954409720926362 (2021)
14. Velasco Arevalo, A., Gerike, R.: Sustainability evaluation methods for public transport with a focus on Latin American cities: a literature review. Int. J. Sustain. Transp. **17**, 1236–1253 (2023)
15. Sierra Muñoz, J., Duboz, L., Pucci, P. and Ciuffo, B. Why do we rely on cars? Car dependence assessment and dimensions from a systematic literature review. Eur. Transp. Res. Rev. **16** Preprint at https://doi.org/10.1186/s12544-024-00639-z (2024)
16. Lu, M., Perez, E. J., Mason, K. & He, Y. Fractal assessment analysis of China's air-HSR network integration. J. Transp. Geogr. **114** Preprint at https://doi.org/10.1016/j.jtrangeo.2023.103786 (2024)
17. Kervall, M. & Pålsson, H. Barriers to change in urban freight systems: a systematic literature review. Eur. Transp. Res. Rev. **14** Preprint at https://doi.org/10.1186/s12544-022-00553-2 (2022).
18. Zhang, Y., Kasraian, D., van Wesemael, P.: Built environment and micro-mobility: a systematic review of international literature. J. Transp. Land Use. **16**, 293–317 (2023)
19. Karam, A., Reinau, K. H. & Østergaard, C. R. Horizontal collaboration in the freight transport sector: barrier and decision-making frameworks. Eur. Transp. Res. Rev. 13 Preprint at https://doi.org/10.1186/s12544-021-00512-3 (2021).
20. Kovacova, M., Balint, J.: Single European Sky Vision: Increase Capacity 3 Times, Reduction of ATM Cost by Half and Ensuring Improvement of Safety by 10 Times - how to Satisfy this Goal in Safety Area? In Transportation Research Procedia, vol. 28, pp. 89–98. Elsevier B.V (2017)
21. McLeod, S., Babb, C., Barlow, S.: How to 'do' a bike plan: collating best practices to synthesise a maturity model of planning for cycling. Transp Res Interdiscip Perspect. **5**, 100130 (2020)
22. Yoon, M.G., Kim, J.K.: Evaluation methodology for safety maturity in air navigation safety. J. Air Transp. Manag. **98** (2022)

23. Laine, V., Valdez-Banda, O., Goerlandt, F.: Risk maturity model for the maritime authorities: a Delphi study to design the R-Mare matrix model. WMU J. Marit. Aff. **23**, 137–163 (2024)
24. Wendler, R.: The maturity of maturity model research: a systematic mapping study. Inf. Softw. Technol. **54**, 1317–1339 (2012)
25. Skupień, E.T.: The maturity model as a tool for assessing transportation systems on the example of inland navigation. Sustainability. **17**, 1577 (2025)

The Optimization of Fuel Delivery Routes to Gas Stations—A Sensitivity Analysis of The Solution to a Change in The Level of Acceptable Risk

Grzegorz Tarczyński[1(✉)], Jacek Kaleta[2], and Marek Karkula[3]

[1] Wroclaw University of Economics and Business, Wroclaw, Poland
grzegorz.tarczynski@ue.wroc.pl
[2] Samat Polska Sp. z o.o., Ul. Opolska 18, 31-323 Krakow, Poland
jacek.kaleta@samat.pl
[3] AGH University of Krakow, Krakow, Poland
mkarkula@agh.edu.pl

Abstract. We present a discussion of selected problems related to the practical implementation of the Inventory Routing Problem (IRP) model for the needs of a company supplying LPG fuel to petrol stations. The analyzed issue concerns deliveries that take place in accordance with the Vendor Managed Inventory (VMI) concept, which consists of managing the recipient's stocks by the supplier. The research presents a decision problem, the elements of which are modules for scheduling dates and volumes of deliveries, as well as determining journey plans. The optimized long-term goal is the ratio that expresses the average length of routes traveled by tankers necessary to deliver 1 ton of goods (km/ton ratio), and the main constraint is maintaining the minimum acceptable stock levels at gas stations (above the so-called dead level of fuel tanks).

Keywords: Inventory Routing Problem · Optimization · Sensitivity Analysis · Fuel Delivery

1 Introduction

The ever-increasing competition in the freight transport market is leading transport company managers to look for cost savings. Despite a good reputation among their customers and a modern fleet, many companies are often operating on the verge of profitability in the struggle to win orders. An opportunity to improve profitability and maintain a high level of customer service may lie in the use of route planning methods, which would allow the load on the means of transport to be optimized and rationalized.

The research problem considered in this paper concerns the optimization of LPG delivery routes in a network of gas stations operated by a transport company. The evaluation of received routes and delivery schedules is multifaceted and should take place both in a short planning horizon of a few days and in a long-term horizon, e.g., monthly. This study addresses the issue of operational planning with a three-day planning horizon.

J. Feliks et al. (Eds.): CLC2024 2024, 2026.
https://doi.org/10.1007/978-3-032-28931-5_29

The optimization task carried out in this research concerns the optimal planning of the supply of various types of fuel over a longer time horizon and can be described as a Multi-Period Inventory Routing Problem. The problem is described in the literature [1, 4] but requires adaptation of the model's constraining conditions to the specifics of the Polish fuel market. The solutions proposed by the system created by the team must take into account the requirements for the proper use of equipment (balanced use of the tanker pool, and in the longer planning horizon, consideration of service breaks, resilience to emergencies resulting from equipment breakdowns) and human resources (inclusion in the model of the maximum driving time and work of drivers; inclusion in the daily schedules of driving breaks required by law; on longer routes, designation of driver shift locations; optimization and linking of the activities performed by the drivers with the time windows for the execution of orders at the fuel stations visited). The system must also function in a situation where some customers place orders for goods, while the rest are served according to a VMI (Vendor Managed Inventory) strategy. Petrol stations operated in accordance with the VMI concept dominate the Polish fuel market. Here, the supplier's task is to set the dates and quantities of deliveries so that no supply problems arise, but at the same time stock levels and, therefore, storage costs are kept as low as possible. The proposed solutions must also consider aspects such as the requirements for fuel transport resulting from SENT (pl: System Elektronicznego Nadzoru Transportu) records and CO_2 emissions.

The authors' experience has shown that the model that determines IRP supply routes and takes CO_2 emissions into account in its basic version is non-linear, which makes it difficult to search for an optimal solution.

The correctly implemented VMI allows for increased efficiency both on the customer's side (reducing the costs of inventory control, reducing the number of errors in the ordering process thanks to automation, improving the level of service) and on the supplier's side (reducing distribution costs thanks to reducing restrictions during its planning, optimizing delivery processes, unified communication, faster access to information, and the ability to dynamically react to changes). The research included in this paper is a response to the needs resulting from practice. Over the last few years, the authors have observed the results of cooperation between a specialized transport company providing services in hazardous cargo with some leading suppliers of cryogenic gases, LPG, and liquid fuels in Poland and Europe. One of the major findings of the observations performed was that if the company could use the available resources (mainly LPG truck trailers) more effectively, that is, by optimizing their routes, the hauler could significantly reduce distribution costs. An important step to do so is to correctly define the stations to be delivered on the next day/days. Such an approach leads to an important stage of the planning and scheduling process, the so-called preprocessing. The purpose of the preprocessing is to establish a set of urgent stations and a set of complementary stations using the MILP model. For this purpose, parameters related to the level of forecast demand on given days at all recipients in the LPG stations network are used.

The remainder of the paper is organized as follows. In the next section, related works from literature are studied. In Sect. 3, the problem formulations are presented. Then, the preliminary assumptions are described in Sect. 4. The next section presents the proposed MILP model for designating a gas station to replenish stocks. The results obtained from computational experiments are presented and discussed in Sect. 6. Finally, conclusions and proposals for future work are drawn up in the last section.

2 Related Studies

In the literature, the problem of integrated inventory planning with vehicle routing was pioneered 40 years ago with the work of Bell et al. This problem originates from the concept of Vendor Managed Inventory (VMI) integrating simultaneously the issues of inventory management, vehicle routing, and delivery planning. In the literature, it includes a class of problems called Inventory Routing Problems (IRP). The first work on IRP was an extension of models developed for the Vehicle Routing Problem (VRP – Vehicle Routing Problem). The IRP approach involves integrated planning of inventory levels, vehicle routes and deliveries. The inventory routing problem has been presented in detail in the literature and in many scientific papers [1, 6, 16]. An analysis of the literature makes it possible to indicate that the methods developed within the IRP models work well in theory (and are often presented by researchers as results of simulation model studies). However, they do not meet the expectations of solving the problem of sales variability that occurs in practice in, for example, the industry of domestic distribution of LPG to gas station networks.

Archetti and Ljubić [1] investigated various formulations for the inventory routing problem (IRP). They focused on aggregate formulations, i.e. those in which the variables do not have an index associated with the vehicles, and we analyse the relationship between compact formulations and their counterparts containing exponentially many constraints. In contrast, Avila-Torres et al. [2] presented the problem of inventory routing with prioritised and fixed heterogeneous fleets. An integer programming model was proposed to solve this problem, which minimises the total distribution cost given inventory levels, priority constraints and a fixed heterogeneous fleet.

Some of the work analysed addresses the issue of inventory routing for fuel supply, including LPG gas [3–5, 8, 9, 15, 17]. The analysis shows that there are many items in the literature dealing with IRP problems under constant demand. Models considering stochastically variable demand are less frequently presented. In the paper of Cho et al., the authors investigated the problem of controlling the production and stocks of liquefied natural gas (LNG) and the routing of ships under severe weather conditions. For this purpose, they developed two mathematical optimisation models to deal with potential disruptions. The work of Juan et al. [10] and Karkula and Kaleta [11], on the other hand, use a combination of optimisation methods and heuristic algorithms with simulation techniques to solve a stochastic variant of the IRP problem. Roldán et al. [14] presented a review of the literature on stock coordination and distribution problems, with particular emphasis on those problems involving stochastic demand and stochastic lead times.

The last decade has also seen several works in which environmental criteria seeking to minimise harmful emissions are also included in the model of the IRP problem in the objective function. In the thesis of Muthu [13], a study is presented in which the author considered the carbon emissions associated with product distribution in traditional routing problems and inventory routing problems. On the other hand, Mirzapour Al-e-Hashem and Rekik [12] developed a MILP model for a multi-product, multi-period inventory routing problem (IRP) in which green logistics problems were considered, taking into account the interrelationship between transport costs and greenhouse gas emissions.

To summarise this analysis of current work on the IRP problem, it can be concluded that, despite the large number of publications related to the IRP problem, there is a lack of studies that present detailed models and algorithms for the selection of gas stations for planning based on demand forecasts and replenishment rules to multiple VMI customers within a given planning horizon and in a highly variable demand situation. The present work is intended to fill the research gap in this area.

3 Problem Statement

In this section the problem formulations are presented. In practice, fuel supply planning in a distributed network of service stations and LPG depots is a complex decision-making problem that requires the integration of many key components. Figure 1 illustrates the concept of a multi-layered architecture for a supply planning support system. The key components of the architecture are:

- master data module with a database containing information on stations, historical planning, vehicles, demand data, among others,
- optimization module, which will solve optimization tasks and transfer the planning results to the operator/planner module,
- operator/planner graphic module, which will select data for planning, display and visualize planning results, as well as parameters and states of vehicles and monitored LPG tanks,
- map module, which will feed the planning data with up-to-date distance data considering the requirements for trucks and hazardous cargo rules,
- delivery monitor module, which will be installed directly in the tanks,
- delivery monitor module - mobile application (for Android systems), through which communication with the driver takes place and updates of delivery plans are sent, and in the other direction the driver sends delivery confirmations and some related parameters (temperature, density, etc.) or other events,
- module for communication with the measurement system of individual gas stations,
- integration layer, enabling communication between distributed architecture elements based on web services and REST APIs.

In this work, we consider a research problem on optimization of LPG fuel supply routes in a network of gas stations. The basic goal of optimization is to minimize the distance needed to transport a unit of goods (LPG fuel). We solve the issue using a modified Vehicle Routing Problem with Time Windows (VRPTW) linear programming model. The original VRPTW model solution ensures the shortest routes for vehicles to reach the recipients of the goods. However, the application of the model in practice requires considering the specificity of the Polish fuel market, the resources of the transport company, the driver's working hours (the necessary breaks at work or other restrictions) and the legal requirements resulting from the transport of goods monitored by the SENT system. We can also include other goals in the model, e.g., minimizing CO_2 emissions during transport.

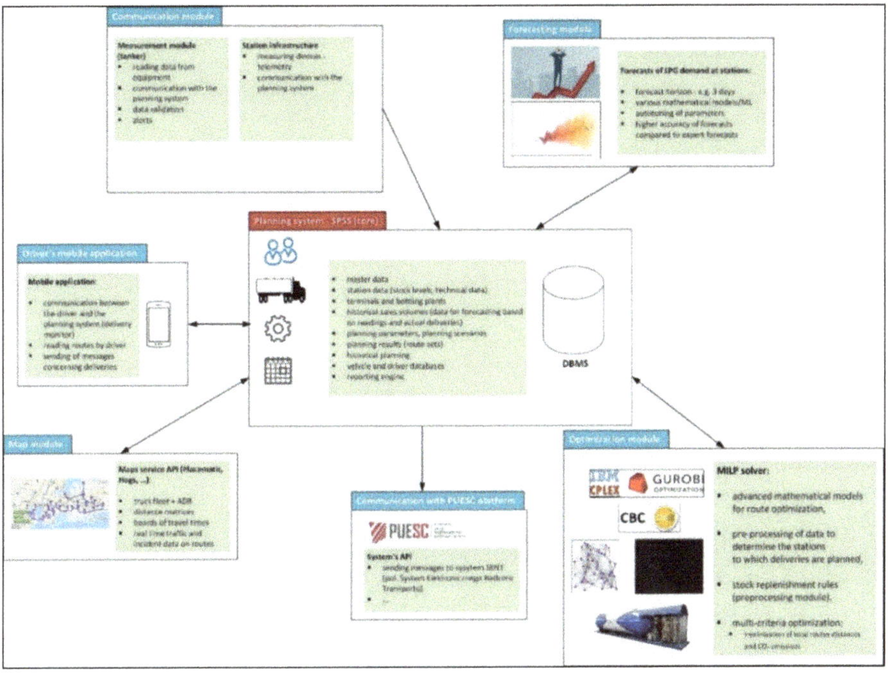

Fig. 1. Complexity of the LPG distribution planning process. *Source* Own elaboration

4 Data Preprocessing for Operational Planning and Tank Truck Routing

4.1 Objectives Achieved During the LPG Supply Planning Process

In the research, we assume that customers are served in accordance with the Vendor Managed Inventory (VMI) strategy. Gas stations operated according to the VMI concept dominate the Polish fuel market. Here, the supplier's task is to set the dates and quantities of products delivered so that there are no supply problems, but at the same time the stock levels, and therefore the storage costs, are as low as possible. Serving clients according to the VMI concept allows for greater flexibility in setting delivery dates and volumes, which allows for more effective planning of delivery routes. However, the supplier of goods must monitor the recipients' inventory levels and make demand forecasts. In the article, we ignore the issue of demand forecast but focus on the remaining activities necessary to perform at the data preprocessing stage. Our goal is to properly prepare the data for the VRPTW model.

A company that delivers LPG to gas stations in accordance with the VMI concept establishes a delivery plan in such a way as to avoid stock-out situations among recipients and to minimize transport costs (minimizing transport costs measured in the long term is equivalent to minimizing the distance needed to transport a unit of goods). A simplified scheme of the fuel distribution planning process is presented in Fig. 2 and the key stages of the data pre-processing algorithm for the designation of urgent and additional stations to visit are shown in Fig. 3.

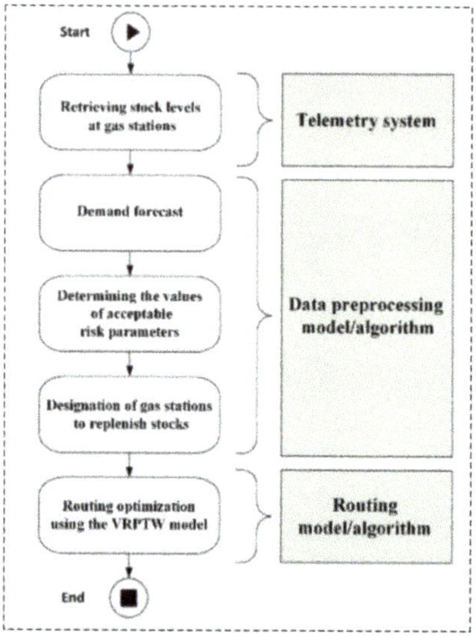

Fig. 2. Simplified scheme of the LPG fuel distribution planning process. *Source* Own elaboration.

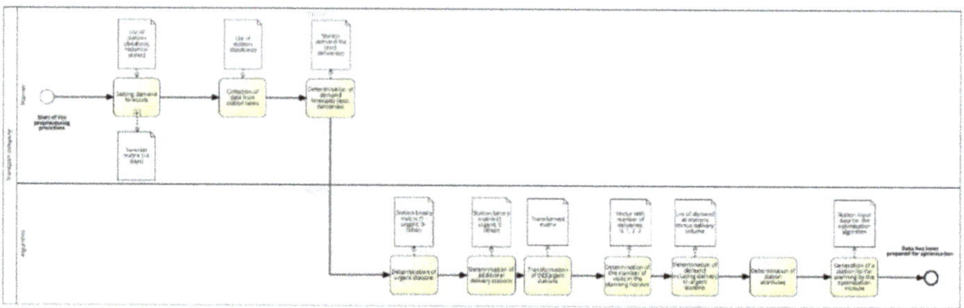

Fig. 3. Key stages of the data preprocessing algorithm for the designation of urgent and additional stations. *Source* Own elaboration.

The purpose of data preprocessing is to determine a list of urgent stations (the tanker must go to the gas station) and a list of additional stations (the tanker can go to the gas station) for a planning horizon of several days. An important element of preliminary data analysis is also determining the number of tank trucks needed to service a given area on a given day and to ensure, within the list of urgent and additional stations, that the tank trucks will be able to set off on their route with maximal possible used load capacity used.

4.2 Stages of the LPG Supply Planning Process

The selection of urgent stations is determined by telemetry readings, the quality of forecasts, and the accepted level of acceptable risk of a stock-out situation. Additional stations are selected based on various criteria: inventory volume, maximum delivery volume, likelihood of needing to replenish stocks within the next few days, distance from urgent stations, etc.

Our proposal concerns making decisions based on parameters indicating the level of acceptable risk.

1. Safety level α – a gas station gets a "very urgent" label for a day *i* if its predicted fuel supplies for day $i + 1$ are less than $\alpha\%$ of the maximal supply at this station.
2. Safety level β – a gas station gets an "additional" label for a day *i* if its predicted fuel supplies for day $i + 1$ are less than $\beta\%$ of the maximal supply at this station.
3. Safety level γ – a gas station gets a label 'very urgent with faster replenishment possibility' for a day *i* if it was earlier labeled as "very urgent" and its predicted fuel supplies for day i-1 are lower than $\gamma\%$ of the maximum supply on this station (the station is automatically removed from a "very urgent" list).

In our investigation, we assumed that $\gamma = \beta$ (every gas station marked as 'very urgent' on day i changes the status to 'very urgent with faster replenishment possibility' on this day if it meets condition 2 on day i–1).

4.3 Assignment of Categories to Urgent Fuel Stations

The algorithm for data preprocessing for the delivery route optimization process assigns gas stations to the following categories (sets):

K1: very urgent stations on day 1 (stock level must be replenished on the first day for which the delivery plan is created) – if the tank forecast for the stock level of tanks at the station on day 2 does not exceed the minimum level plus the safety level α,

K2: very urgent stations on day 2 without the possibility of acceleration (the stock level must be replenished on the second day for which the delivery plan is created) – if the forecast for the stock level of tanks at the station on day 3 does not exceed the minimum level plus the safety level α,

K3: very urgent stations on day 2, where the stock of tanks can be replenished the day before (the stock level must be replenished no later than on the second day for which the delivery plan is created, but it can be done one day in advance) – if the criteria are met:

- the gas station has been assigned to the collection of urgent stations on day 2;
- the station was not assigned to the set of urgent stations on day 1;
- the forecast for the tank stock level at the station on day 1 does not exceed the minimum level plus the safety level γ. The station assigned to the K3 category is removed from K2,

K4: very urgent stations on day 3 without the possibility of acceleration (the stock level must be replenished on the third day for which the delivery plan is created) – if the forecast for the stock level of tanks at the station on day 4 does not exceed the minimum level plus the safety level α,

K5: very urgent stations on day 3, where the stock of tanks can be replenished the day before (the stock level must be replenished no later than on the third day for which the delivery plan is created, but it can be done one day in advance) if the criteria are met:

- the gas station has been assigned to the collection of urgent stations on day 3;
- the station was not assigned to the set of urgent stations on day 2;
- forecast for the tank stock level at the station on day 2 does not exceed the minimum level plus the safety level γ. The station assigned to the K5 category is removed from K4.

In the research, we assume that the stock levels at the gas station can be replenished no more than once a day. For stations where the stock levels must be replenished more than once a day, one needs to create separate rules.

4.4 Selection of Additional Stations

The set of additional stations for a given day includes stations for which the threshold of inventory forecast for a given day does not exceed the safety level β. The number of additional stations designated for a specific day can be arbitrarily large, but we need the demand generated at them greater than the rest of the LPG resources being transported on tank trucks (after considering all urgent stations).

Later, at the route calculation stage, the VRPTW model will select from the list of additional stations those stations that, when added to the urgent stations, will create routes with the shortest total length possible (within a three-day horizon). Therefore, not all stations assigned to the set of additional stations have to be included in the generated routes.

5 MILP Model Performing the Task of Data Preprocessing

Before formulating the Mixed Integer Linear Programming (MILP) model, we defined the goals to be achieved. Primary objectives are as follows.

- transport cost minimization;
- the risk of stock-out event minimization.

Additional (technical) objectives are as follows:

- maximization of tank truck loading space utilization;
- minimization of the number of gas stations visited during one route;
- minimizing CO_2 emissions during the transport of a unit of goods (within a specified area of gas stations);
- minimization of the coefficient expressing the distance needed to transport 10,000 [l] LPG gas (within a specified area of gas stations).

Additionally, the preprocessing model should return the values of parameters needed by the VRPTW model:

1. The lists: a) of necessary (very urgent), b) necessary possible for faster replenishment, and c) additional gas stations served for three days planning horizon;
2. The number of tank trucks for each day.

The description of the preprocessing optimization model is given below and the nomenclature for the MILP model is presented in Table 1.

The objective function is defined to minimize the loading space of unused tank trucks for a three-day planning horizon (expressed in liters):

$$z = \left[3D - \sum_{k=1}^{3} (v_k - (y_k - 1)D) \right] \Omega \rightarrow min \tag{1}$$

subject to constraints:

$$\bigwedge_{i=1,\ldots,I} \bigwedge_{k=1,2,3} x_{ik} \geq p_{ik}^{*} \tag{2}$$

$$\bigwedge_{i=1,\ldots,I} \bigwedge_{k=1,2,3} x_{ik} \leq p_{ik}^{+} \tag{3}$$

$$\bigwedge_{i=1,2,\ldots,I} x_{i1} + x_{i2} + x_{i3} \leq p_i^{mx} \tag{4}$$

$$\sum_{i=1}^{I} x_{i1} c_{i1} \geq v_1 \tag{5}$$

$$\sum_{i=1}^{I} x_{i2} c_{i2}^{mod} \geq v_2 \tag{6}$$

$$\sum_{i=1}^{I} x_{i3} c_{i3}^{mod} \geq v_3 \tag{7}$$

$$\bigwedge_{i=1,\ldots,I} c_{i2}^{mod} = c_{i2} - c_{i1} \left(p_{i1}^{*} + u_{12} \right) \tag{8}$$

$$\bigwedge_{i=1,\ldots,I} c_{i3}^{mod} = c_{i3} - c_{i2}^{mod} \left(p_{i2}^{*} + u_{23} \right) \tag{9}$$

$$\bigwedge_{i=1,\ldots,I} x_{i1} + x_{i2} \geq p_i^{*12} \tag{10}$$

$$\bigwedge_{i=1,\ldots,I} x_{i2} + x_{i3} \geq p_i^{*23} \tag{11}$$

$$Dy_1 \geq \sum_{i=1}^{I} c_{i1} \left(p_{i1}^{*} + p_i^{*12} u_{12} \right) \tag{12}$$

Table 1. Nomenclature for the proposed MILP model

Indices	
i	gas station number, $i = 0, 1, 2, \ldots, I$
k	day number in planning horizon, $k = 1, 2, 3$
Parameters	
Ω	planning accuracy (e.g. $\Omega = 100[l]$)
I	number of gas stations in an area
P^+	binary matrix with information on "very urgent" or "additional" gas stations for each day of the planning horizon, $P^+ = \left[p_{ik}^+\right], i = 1, 2, \ldots, I; k = 1, 2, 3,$ $p_{ik}^+ = \begin{cases} 1 - \text{a gas station } i \text{ is identified as "very urgent" or additional for day } k, \\ 0 - \text{otherwise} \end{cases}$
P^*	binary matrix with information on "very urgent" (without faster replenishment possibility) gas stations for each day of the planning horizon,$P^* = [p_{ik}^*], i = 1, 2, \ldots, I; k = 1, 2, 3$ $p_{ik}^* = \begin{cases} 1 - \text{a gas station } i \text{ is identified as "very urgent" for day } k, \\ 0 - \text{otherwise} \end{cases}$
P^{*12}	binary vector with information on gas stations identified as "very urgent with faster replenishment possibility" for the second day of the planning horizon,$P^{*12} = \left[p_i^{*12}\right], i = 1, 2, \ldots, I$
P^{*23}	binary vector with information on gas stations identified as "very urgent with faster replenishment possibility" for the third day of the planning horizon, $P^{*23} = \left[p_i^{*23}\right], i = 1, 2, \ldots, I$
P^{mx}	binary vector with information on the number of possible replenishments for each gas station during the three-day planning horizon, $P^{mx} = [p_i^{mx}], i = 1, 2, \ldots, I$
C	predicted demand for an LPG gas at the station (in units of planning accuracy Ω),$C = \left[c_{ik}\right], i = 1, 2, \ldots, I; k = 1, 2, 3,$
D	maximum capacity of the tank truck (in units of planning accuracy Ω), e.g. for a tank truck with a capacity of 39000 [l] and $\Omega = 100[l]$:$D = 390\Omega$
Decision variables	
x_{ik}	fuel supplies at the gas station i can be replenished on day k, $x_{ik} = \begin{cases} 1 - \text{fuel supplies at the gas station } i \text{ can be replenished on day } k, \\ 0 - \text{otherwise} \end{cases}$

(*continued*)

Table 1. (*continued*)

y_k	number of tank trucks needed for a day k
v_k	total amount of LPG gas delivered on day k (expressed in planning accuracy units Ω)
u_{12}	acceleration of deliveries to gas stations identified as "very urgent with faster replenishment possibility" for the second day of the planning horizon $u_{12} =$ $\begin{cases} 1 - \text{accelerate delivery for all possible "very urgent" identified for the second day,} \\ 0 - \text{otherwise} \end{cases}$
u_{23}	acceleration of deliveries to gas stations identified as "very urgent with faster replenishment possibility" for the third day of the planning horizon $u_{23} =$ $\begin{cases} 1 - \text{accelerate delivery for all possible"very urgent" identified for the second day,} \\ 0 - \text{otherwise} \end{cases}$

$$Dy_2 \geq \sum_{i=1}^{I} c_{i2}^{mod}\left(p_{i2}^{*} + p_i^{*12}(1 - u_{12}) + p_i^{*23}u_{23}\right) \tag{13}$$

$$Dy_3 \geq \sum_{i=1}^{I} c_{i3}^{mod}\left(p_{i3}^{*} + p_i^{*23}(1 - u_{23})\right) \tag{14}$$

$$\bigwedge_{k=1,2,3} v_k \leq Dy_k \tag{15}$$

$$\bigwedge_{i=1,\ldots,I}\ \bigwedge_{k=1,2,3} x_{ik} \in \{0, 1\} \tag{16}$$

$$\bigwedge_{k=1,2,3} y_k \in \mathbb{N} \tag{17}$$

$$\bigwedge_{k=1,2,3} v_k \in \mathbb{N} \tag{18}$$

$$u_{12} \in \{0, 1\} \tag{19}$$

$$u_{23} \in \{0, 1\} \tag{20}$$

In the model, we use small simplification and assume that: (I) each day only one tank truck cannot be fully loaded; (II) the demand on gas stations can be accomplished even by two deliveries. Such assumptions simplify the calculations at the pre-processing stage, but are not valid during the launching of the proper VRPTW model.

The objective function (1) expresses the simplified value of the loading space of unused tank trucks for a three-day planning horizon. Constraint (2) ensures that gas stations from a set of "very urgent" stations are replenished on a specific day. Similarly, constraint (3) provides that tank trucks can visit only gas stations from urgent or additional lists. Constraint (4) guarantees that the maximum number of replenishments for each gas station is not exceeded. Constraints (5)–(7) bind the decision variables that express the total delivery for each day with the demand for the gas stations. Equations (8) and (9) ensure that in the case of delivery, the predicted next day demand at the gas station decreases by the amount of the delivery (at this stage, we assume that each day only one delivery to a gas station is possible, so the delivery cannot be fastened if another delivery is planned for this day: $p^*_{i1} \neq u_{12}; p^*_{i2} \neq u_{23}$; in practice two deliveries during one day are possible but the demand has to be calculated differently). Constraint (10) ensures that deliveries to gas stations identified as urgent on day 2 with the possibility of earlier delivery must be made on day 1 or 2. Similarly, constraint (11) guarantees that, if possible, deliveries to gas stations from the set of urgents for day 3 can be executed during day 2 or 3. Constraints (12)–(14) calculate the number of tank trucks for each day of the planning horizon. Constraint (15) limits the summarized quantity of LPG delivered each day to the gas stations by the capacity of tank trucks planned for routing. The model is completed by the boundary constraints (16)–(20).

6 Computational Experiments

In order to verify the impact of parameters that express the acceptable level of risk on the quality of solutions, we performed experiments. Both the data pre-processing model and the VRPTW model were used in the research. Finally, we evaluated the delivery routes obtained after applying the VRPTW model. The calculations were performed using Python-MIP library (https://www.python-mip.com/), and the Gurobi solver with exact solutions for the data preprocessing model, and a high final quality solution setting for routing (modified VRPTW with a gap parameter set on 5%).

We made the following assumptions when carrying out the calculation experiments:

- in our research, the VRPTW model used to determine tank truck routes has a single optimization criterion: it minimizes the traveled distance. In fact, the prepared system provides the possibility of considering an additional optimization criterion: minimizing CO_2 emissions or creating a meta-criterion taking into account both goals. In the research presented in the article, the CO_2 emission value is determined only for the objective function minimizing the traveled distance,
- the ambient temperature does not change and does not affect the supply of LPG transported by tanker,
- all LPG resources transported by tank truck must be delivered to fuel stations,
- the prepared forecasts of LPG demand at gas stations are always accurate (the problem of forecasting stock levels in tanks at gas stations is extensive and requires a separate study).

The basic parameter allowing the evaluation of the effectiveness of the obtained solutions is the coefficient expressing the number of kilometers traveled by tank trucks needed to transport 1 ton of LPG gas (in the experiments conducted, for simplicity, instead of the mass of the transported goods, we analyze the capacity: 10,000 dm3 of LPG gas[1]).

The coefficient value is largely influenced by, among others, the distance of the gas station from the bottling plant and the distance of the gas stations from each other. It is therefore measured independently in each of the serviced areas. Table 2 shows the distances between gas stations in the areas analyzed in the paper.

Table 2. Distances between visited locations in the areas analyzed in the research

Area	No. gas stations	Distance from a base to gas stations [km]			Distance between gas stations [km]		
		Average	Min	Max	Average	Min	Max
R01	44	17.98	2.68	41.17	16.28	0.11	53.13
R02	22	69.81	4.61	149.18	70.31	2.37	196.07
R03	18	165.45	78.12	327.84	60.77	2.58	168.65

The frequency of gas deliveries to gas stations depends on the volume of demand and the tank parameters. The tanks have a capacity of 4850 dm^3 to 20,000 dm^3, but the amount of fuel for sale (presented in Table 3) also depends on the pump capacity and the maximum filling level, which is usually about 80%.

Table 3. Maximal supply of LPG on gas station

Area	Maximal supply of LPG gas on gas station [l]		
	Average	Min	Max
R01	10,825	6,578	17,740
R02	7,939	3,130	13,600
R03	8,570	6,256	13,600

The aim of the investigation is to check the impact of the parameters α, β, and γ on the distance necessary to transport a unit of LPG and the amount of CO_2 emissions. The research was carried out for the following parameter values: $\alpha \epsilon \{-0.1, -0.05, 0, 0.05, 0.1\}$ and $\beta \epsilon \{0.4, 0.6, 0.8\}$, with additional condition: $\gamma = \beta$.

Figure 4 shows the distance needed to transport 10,000 [l] LPG gas within areas R1, R2, and R3. Research shows that an increase in the level of acceptable risk of a stock-out

[1] The volume of LPG changes depending on the ambient temperature.

situation (the α parameter decreases) allows for much more flexible delivery planning. This is especially visible in the R1 area, where gas stations have the smallest tanks that need to be refilled frequently. In the R3 region, where there are most gas stations and no very small tanks, the solution is more influenced by the β parameter. Increasing the value of the β parameter significantly increases the number of possible route combinations by increasing additional and urgent stations with the possibility of acceleration.

Figure 5 shows CO_2 emissions during the transport of 1000 dm^3 LPG within analyzed areas.

In the R1 region, the value of the α parameter has the greatest impact on the results, while in the remaining regions, the β parameter is more important. Especially in the R3 region, greater flexibility in the selection of additional stations translates into much lower CO_2 emission values. In the R2 region, the emission rate is significantly influenced by both α and β.

7 Conclusions and Further Works

The issue of planning LPG fuel deliveries to gas stations is complicated and requires a lot of additional work. The paper presents a model that performs the task of data processing and shows, using real data, the difficulties associated with determining the values of parameters affecting the quality of the solution. However, one may be tempted to draw more general conclusions. Negative values of the safety level parameter α imply a higher level of risk acceptance. Furthermore, a greater number of "very urgent" gas stations are classified as "very urgent with faster replenishment possibilities." When combined with moderate values of the safety level parameters β and γ (ranging from 0.4 to 0.6), this results in higher deliveries and less frequent restocking. In cases where there is a limited number of gas stations with moderate supply capabilities in the area (R02, R03), it becomes necessary to flexibly determine the set of urgent stations ($\alpha < 0$). This means that replenishing the stock can occur later. A station will be labeled as urgent a day later, but its inclusion in the schedule will depend on the cumulative demand from other stations in the area. It is possible that, in such cases, stocks at these stations may be replenished a day earlier, following the originally planned schedule, with a lower acceptable level of risk.

The issues presented in the article do not cover the topic of planning deliveries to gas stations. The problem of forecasting demand at gas stations is extremely important and is omitted in this article. This issue is complicated due to large seasonal fluctuations and should be the subject of further analysis. Further work should take into account other factors affecting the quality of planning in the longer term: the degree of replenishment of stocks at customers (may be important in the event of temporarily increased demand that exceeds the supplier's capabilities), tank truck loading level, and greater flexibility in determining LPG deliveries to additional stations (with limited summarized demand at gas stations in the region), demand variability for two-day delivery routes.

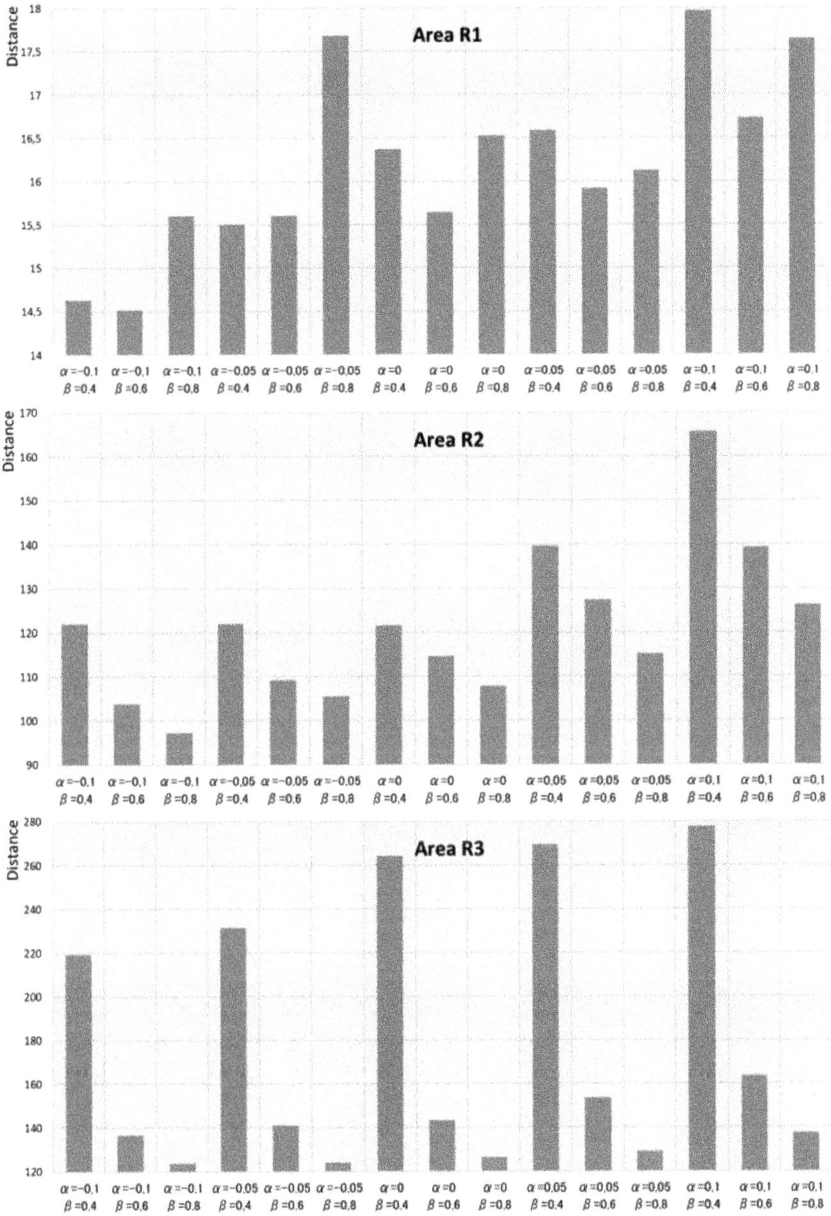

Fig. 4. Distance needed to transport 10,000 l. LPG gas within the areas analyzed

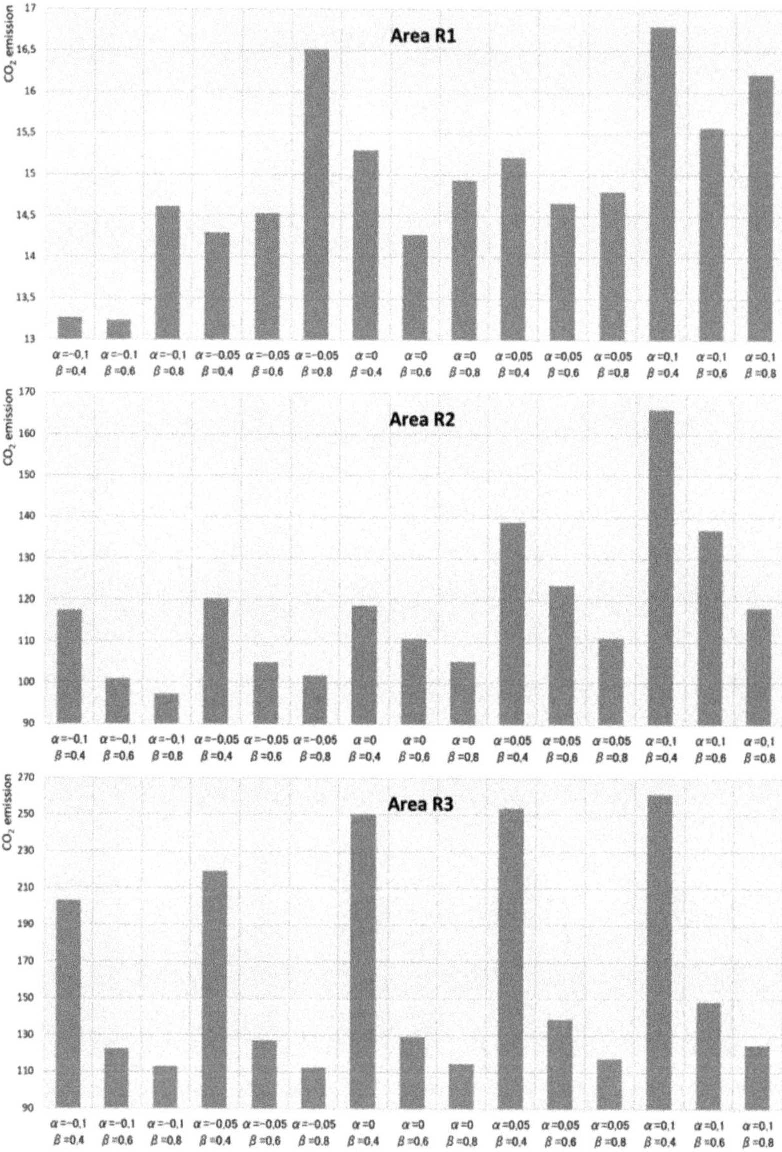

Fig. 5. CO_2 emissions during the transport of 1000dm^3 LPG within analyzed areas

Acknowledgements. The study was partially funded by the National Centre for Research and Development (NCBR), Poland under the POIR.01.01.01–00-0265/21 programme – R&D project titled "Inventory routing optimization for petrol stations".

References

1. Archetti, C., Ljubić, I.: Comparison of formulations for the inventory routing problem. Eur. J. Oper. Res. **303**(3), 997–1008 (2022). https://doi.org/10.1016/j.ejor.2021.12.051
2. Avila-Torres, P.A., Arratia-Martinez, N.M., Ruiz-y-Ruiz, E.: The inventory routing problem with priorities and fixed heterogeneous fleet. Appl. Sci. **10**(10) (2020). https://doi.org/10.3390/app10103502
3. Bani, A., El Hallaoui, I., Corréa, A.I., Tahir, A.: Solving a real-world multi-depot multi-period petrol replenishment problem with complex loading constraints. Eur. J. Oper. Res. **311**(1), 154–172 (2023). https://doi.org/10.1016/j.ejor.2023.04.016
4. Bell, W., et al.: Improving the distribution of industrial gases with an on-line computerized routing and scheduling optimizer. Interfaces **13**(6), 4–23 (1983). https://doi.org/10.1287/inte.13.6.4
5. Benantar, A., Ouafi, R., Boukachour, J.: A petrol station replenishment problem: new variant and formulation. Logist. Res. **9**(6), 1–18 (2016). https://doi.org/10.1007/s12159-016-0133-z
6. Bertazzi, L., Speranza, M.G.: Inventory routing problems with multiple customers. EURO J. Transp. Logist. **2**(3), 255–275 (2013). https://doi.org/10.1007/s13676-013-0027-z
7. Campbell, A., Clarke, L., Kleywegt, A., Savelsbergh, M.: The inventory routing problem. In: Crainic, T.G., Laporte, G. (eds.) Fleet Management and Logistics, pp. 95–113. Springer, Boston (1998). https://doi.org/10.1007/978-1-4615-5755-5_4
8. Cho, J., Lim, G.J., Kim, S.J., Biobaku, T.: Liquefied natural gas inventory routing problem under uncertain weather conditions. Int. J. Prod. Econ. **204**, 18–29 (2018). https://doi.org/10.1016/j.ijpe.2018.07.014
9. Dror, M.: Routing propane deliveries. In: Langevin, A., Riopel, D. (eds.) Logistics Systems: Design and Optimization, pp. 299–322. Springer, Boston (2005). https://doi.org/10.1007/0-387-24977-X_10
10. Juan, A.A., Grasman, S.E., Jose Caceres-Cruz, J., Bektaş, T.: A simheuristic algorithm for the single-period stochastic inventory-routing problem with stock-outs. Simul. Model. Pract. Theory **46**, 40–52 (2014). https://doi.org/10.1016/j.simpat.2013.11.008
11. Karkula, M., Kaleta, J.: Simulation methods application for LPG deliveries planning and scheduling to the network of stations under demand uncertainty. Logist. Transp. **34**, 63–78 (2017)
12. Mirzapour Al-e-Hashem, S.M.J., Rekik, Y.: Multi-product multi-period inventory routing problem with a transshipment option: a green approach. Int. J. Prod. Econ. 157, 80–88 (2014). https://doi.org/10.1016/j.ijpe.2013.09.005
13. Muthu, K.P.: An investigation of inventory routing problem with carbon emission and sustainability consideration. Master Thesis, State University of New York at Buffalo (2021)
14. Roldán, R., Basagoiti, R., Coelho, L.C.: A survey on the inventory-routing problem with stochastic lead times and demands. J. Appl. Log. **24**, 15–24 (2017). https://doi.org/10.1016/j.jal.2016.11.010
15. Singh, T., Arbogast, J.E., Neagu, N.: An incremental approach using local-search heuristic for inventory routing problem in industrial gases. Comput. Chem. Eng. **80**, 199–210 (2015). https://doi.org/10.1016/j.compchemeng.2015.05.023
16. Tang, Z., Jiao, Y., Ravi, R.: Combinatorial Heuristics for inventory routing problems. INFORMS J. Comput. **34**(1), 370–384 (2022). https://doi.org/10.1287/ijoc.2021.1064
17. Vidović, M., Popović, D., Ratković, B.: Mixed integer and heuristics model for the inventory routing problem in fuel delivery. Int. J. Prod. Econ. **147**(Part C), 593–604 (2014). https://doi.org/10.1016/j.ijpe.2013.04.034

Socioeconomic Aspects of Hyperloop Development Based on The Central Communication Port (CPK)

Rafał Rumin(✉), Olena Stryhunivska, Kinga Bieniek, Dawid Pękała, Szymon Gach, and Mateusz Milczanowski

AGH University of Krakow, Krakow, Poland
rumin@agh.edu.pl

Abstract. This paper examines the socioeconomic implications of implementing Hyperloop technology in Poland, focusing on its integration with the Central Communication Port (CPK). The development of Hyperloop, a high-speed vacuum train system, offers transformative potential for long-distance transportation by significantly reducing travel times, enhancing logistics, and promoting energy efficiency. This paper explores the design, and simulation of the Hyperloop stations tailored to Polish transport infrastructure. Simulations conducted in FlexSim software demonstrate the potential of Hyperloop stations to streamline passenger flow, logistics, and safety protocols.

Keywords: Hyperloop · Central Communication Port (CPK) · transport integration · FlexSim simulation · socioeconomic development

1 Introduction

Contemporary transportation systems are undergoing dynamic transformation, driven by the demands for increased speed, efficiency, integration, and reduced environmental impact. As technology develops and socioeconomic conditions evolve, there is a growing demand for innovative mobility solutions capable of addressing the challenges of sustainable development and intensifying urbanization. In this context, the emergence of Hyperloop technology, a vacuum tube transportation system enabling capsules to travel at speeds exceeding 1000 km/h, represents a potentially breakthrough stage in the evolution of both passenger and freight transportation [1, 7].

Hyperloop technology combines characteristics of high-speed rail, air transport, and automated logistics, establishing a new paradigm for fast and energy-efficient mobility. By reducing aerodynamic drag within a low-pressure environment and utilizing linear propulsion and magnetic levitation, this system allows for significant energy savings compared to traditional modes of transport. Numerous studies indicate that implementation of Hyperloop technology could deliver tangible economic and environmental benefits, especially in countries with growing demand for fast, low-emission interregional connections.

J. Feliks et al. (Eds.): CLC2024 2024, 2026.
https://doi.org/10.1007/978-3-032-28931-5_30

Poland, as one of the key transit countries in Central and Eastern Europe, faces a unique opportunity to leverage this technology in the development of an integrated public transport system. The Central Communication Port (CPK) project is of relevance, this multimodal hub is designed to connect air, rail, and road transport and become a central element of the national mobility network. Integrating the Hyperloop system with CPK infrastructure could significantly shorten travel times between major urban centres, increase spatial accessibility of peripheral regions, and contribute to enhancing national economic competitiveness.

This article presents an analysis of the potential socioeconomic aspects of Hyperloop system implementation in Poland, focusing on its integration with the CPK project. The paper provides an overview of the Hyperloop concept and the current state of its global development, followed by a description of the design and functional assumptions of CPK as a key transportation node [2, 5, 6].

Subsequent sections introduce a conceptual model for integrating Hyperloop into the CPK network, along with results from passenger and cargo flow simulations conducted in the FlexSim environment, allowing for assessment of operational performance and safety of the proposed solution. The analysis also includes evaluation of potential socioeconomic effects such as increased transport accessibility, job creation, energy efficiency, and environmental and social acceptance of the new technology.

2 Overview of Hyperloop Technology

Hyperloop technology represents one of the most innovative transportation concepts of the 21st century, integrating elements of high-speed rail, aerodynamics, electromagnetism, and industrial automation. The idea was first comprehensively introduced in the "Hyperloop Alpha" report published by Elon Musk in 2013, which proposed a system of capsules traveling at very high speeds through tunnels with reduced air pressure. Since then, the concept has become the subject of intensive scientific research and technological experimentation conducted by research institutes and enterprises worldwide.

The fundamental operating principle of the Hyperloop system is the minimization of aerodynamic resistance, achieved by maintaining the air pressure inside the transport tube at a level close to vacuum (approximately 100 Pa). Under such conditions, the capsules (pods) travel using linear or electromagnetic propulsion, reaching speeds in the range of 800–1200 km/h. Additionally, the use of magnetic levitation (maglev) enables virtually contactless movement, dramatically reducing friction. Combined with intelligent traffic control systems and predictive safety algorithms, Hyperloop can deliver very high throughput as well as exceptional operational stability [1, 7, 15, 17].

From a technical perspective, the development of Hyperloop may be divided into several major research and development areas:

- Propulsion and power systems, which encompass linear drives, induction and synchronous motors, as well as integration with renewable energy sources such as photovoltaics or hydrogen energy storage [1, 4, 19–21].
- Tunnel and tube infrastructure technology, focusing on high-strength materials (e.g., stainless steel, carbon composites) and solutions to ensure vacuum integrity and operational safety [4, 18, 22, 23].

- Safety and traffic control, involving system redundancies, pressure, temperature, and acceleration sensors, along with automated evacuation procedures [12, 18, 22, 28].
- Environmental and economic aspects, concerning energy efficiency, carbon emissions, construction and operation costs, and integration with existing transport infrastructure [3, 8–11, 29–31].

Numerous projects worldwide aim to implement this technology under real-world conditions. The most advanced initiatives include Hardt Hyperloop (Netherlands), Zeleros (Spain), TransPod (Canada), and Nevomo (Poland), formerly Hyper Poland, which is developing the MagRail system as an intermediate step between conventional magnetic rail and a full vacuum system. Testing conducted by these entities enables verification of design assumptions and assessment of the technology readiness level (TRL).

From an engineering standpoint, one of the critical issues remains the system's power supply and its impact on the electrical grid. According to the report by the U.S. Department of Energy (2021), a hypothetical 1000-km Hyperloop system may generate a power demand of approximately 100–600 MW for passenger transport and as much as 2000 MW for freight transport. This underscores the need to design highly flexible energy infrastructure, utilizing smart grids and advanced energy storage.

Scientific literature emphasizes that Hyperloop's potential is not limited to speed and travel time but also encompasses a systemic shift in the mobility paradigm, including integration with multimodal transport, Logistics 4.0, and Smart Cities. Deployment of such solutions may help reduce congestion on traditional rail and air networks, as well as diminish greenhouse gas emissions in the transport sector.

Despite significant advances, Hyperloop technology remains in the pre-commercial phase (TRL 4–7), and its widespread use requires further research regarding safety, operational economics, environmental impact, and regulatory frameworks. Nevertheless, over the longer term, the development of this technology may become a catalyst for infrastructural and socio-economic transformation, particularly in countries striving to align transport integration with the objectives of the European Green Deal and Sustainable Mobility Policy.

3 Central Communication Port (CPK) as a Strategic Transport Node

The Central Communication Port (CPK) project is one of the most significant infrastructure undertakings in modern Polish history. Its objective is to establish an integrated, multimodal transportation hub of national and international importance, connecting air, rail, and road transport into a cohesive system for passenger and logistics services. Located in the central region of the country, between Warsaw and Łódź, CPK serves both as a new intercontinental airport and as the core of a modern rail system encompassing most of Poland's major metropolitan areas.

Strategically, CPK aims to strengthen Poland's position as a key transport node in Central and Eastern Europe, supporting efficient connectivity within the Trans-European Transport Network (TEN-T). The project aligns with the European Union's long term objectives for sustainable mobility, greenhouse gas emission reduction, and the development of low-emission infrastructure. In accordance with the priorities of the European Green Deal, the focus is on promoting integrated and efficient transport systems that shift dependence away from road and air travel towards faster, greener modes such as rail and next-generation tube transport.

The aviation component of CPK envisions the construction of a new passenger terminal with an initial capacity for approximately 32 million passengers per year, expanding to 40 million in subsequent phases. The airport is intended to become one of the largest transfer hubs in Europe, linking domestic, European, and intercontinental flights. The development strategy also includes the establishment of a cargo hub with throughput of around 1 million tons annually crucial for the logistics sector and Poland's standing in global supply chains [5, 13, 26].

An integral element of the project is the railway component, comprising approximately 2,000 km of new railway lines, including high-speed connections under the CPK Railway Port program. The rail network will connect Warsaw to Poland's largest cities (e.g., Łódź, Wrocław, Poznań, Katowice, Kraków, Gdańsk, Szczecin), reducing travel times between them to approximately two hours. The system will also facilitate links to major transport hubs in Central Europe, including Prague, Bratislava, Vienna, and Berlin, reinforcing Poland's integration with the European rail and economic system.

CPK is also intended to function as a next-generation logistics centre, integrating land, air, and potentially in the future, tube transport (Hyperloop). The application of advanced technologies for automation and digitalization of logistics processes (including the Internet of Things, goods flow management systems, and predictive planning algorithms) positions the project as a model example of modern smart transport hub infrastructure.

From a regional development perspective, CPK has the potential to stimulate economic growth in central Poland through job creation, attraction of investment, growth in the construction and technology sectors, and improved transport accessibility for previously peripheral areas. Studies conducted by the World Bank and OECD indicate that transport accessibility is among the key determinants of regional growth dynamics, contributing to the diffusion of innovation, human capital mobility, and increased regional competitiveness.

The CPK project also has significant social and environmental dimensions. On one hand, it is anticipated to improve citizens' quality of life by reducing travel times, increasing access to public services, and enhancing interregional mobility. On the other hand, the project requires a sustainable approach to spatial planning and environmental protection, including ecological compensation, noise and emission mitigation, and the safeguarding of agricultural and natural areas. In this context, applying principles of sustainable infrastructure design and life cycle assessment (LCA) is critical.

Finally, in terms of long-term transport development strategy, CPK has the potential to become a platform for the integration of future technologies such as autonomous transport, logistics drones, artificial intelligence systems, and, indeed, Hyperloop. Incorporating these technologies into the functional structures of the transport hub would enable the creation of a mobility system that combines high efficiency with low emissions and a high level of safety.

In summary, the Central Communication Port should be viewed not only as an infrastructure investment, but as a strategic component of Poland's transport system transformation, aligned with European and global sustainable mobility trends. Its potential integration with Hyperloop technology constitutes a natural step towards establishing next-generation transport links, combining speed, efficiency, and innovation with the objectives of climate and social policy.

4 Integration of the Hyperloop System with the Central Communication Port (CPK) Network

The integration of Hyperloop technology with the infrastructure of the Central Communication Port (CPK) constitutes a key element in the concept of building a next-generation, integrated transportation system in Poland. The objective of such integration is to create a synergistic arrangement in which Hyperloop serves as an ultra-high-speed component of the national and European transport network, complementing high-speed rail (HSR) and air connections. Incorporating vacuum transport into CPK's functional structures would not only result in substantially reduced travel times between major metropolitan centres but would also foster the development of a new passenger-logistics model based on energy efficiency, safety, and sustainability [14, 16]

Within the proposed concept, the Hyperloop terminal would be located adjacent to the main airport complex and CPK rail hub, enabling full integration of air, railway, and tube transport at a single interchange point. The terminal should serve both passenger and cargo functions, acting as a modern multimodal node for domestic and international traffic.

The integration model also assumes the construction of a tunnel connecting the CPK terminal with the main Hyperloop corridors linking Warsaw, Łódź, Wrocław, and Poznań. In the long term, it would be possible to expand the network with international sections towards Berlin, Prague, and Vienna, thus positioning Poland within a pan-European next-generation transport system.

The integrated Hyperloop-CPK terminal would require advanced traffic control and process automation technologies. SCADA (Supervisory Control and Data Acquisition) and ITS (Intelligent Transport Systems) could play a pivotal role in monitoring vacuum environment parameters, managing pods, overseeing safety, and synchronizing with air and rail traffic. Furthermore, the inclusion of predictive analytics tools (e.g., artificial intelligence algorithms) would enable dynamic route planning, minimization of downtime, and optimization of energy resource usage.

4.1 Design Assumptions and Technical Parameters

For Polish conditions, it is assumed that the Hyperloop system would form a transport network of 300 to 500 km in length, connecting CPK with major metropolitan areas. Passenger pods are projected to operate at cruising speeds of 800–1000 km/h, with a minimum headway of approximately 2 min. Consequently, travel time between Warsaw and Wrocław could be reduced to around 25 min, and between Warsaw and Poznań to about 30 min.

Regarding technical infrastructure, the application of redundant systems would be crucial to ensure continuous operation of propulsion, power, and pressure maintenance systems, as well as modular compression and evacuation stations compliant with European tube transport safety standards.

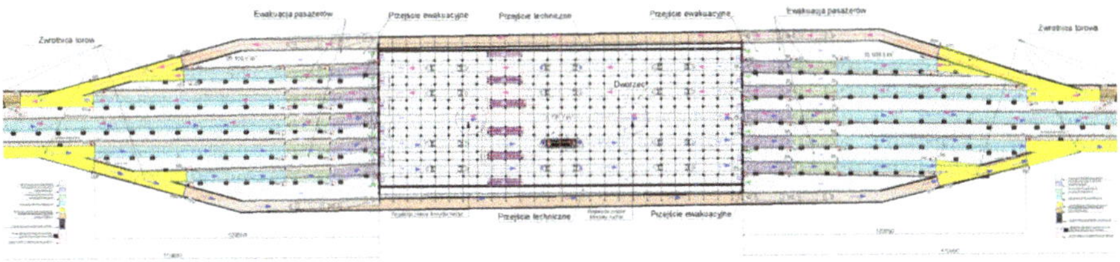

Fig. 1. Concept of the Hyperloop underground station.

To analyse the functionality and operational efficiency of the designed system, computer simulations were performed in the FlexSim environment, software dedicated to modelling passenger, logistics, and industrial flows. Key model parameters included peak-hour passenger numbers, average service times, transition times between terminal zones, capsule travel speeds, and evacuation procedures (Fig. 1).

4.2 The Significance of Hyperloop-CPK Integration for Transport System Development

Integrating Hyperloop technology with CPK is a step towards transforming the Polish transport system into a sustainable mobility model. Incorporating vacuum transport into the port's functional structure enables the reduction of functional distances between regions, supports economic development in peripheral areas, and improves national competitiveness in the European context.

From a long-term strategy perspective, the Hyperloop-CPK system could become part of the Trans-European Transport Network (TEN-T), connecting Poland with north-south and east-west corridors. Moreover, deploying this technology aligns with the European Union's climate objectives, particularly in terms of greenhouse gas reduction in the transport sector and increasing the share of renewable energy in critical infrastructure supply.

Integrating Hyperloop with the CPK network is not merely a technological innovation, but a systemic approach to future transport planning, combining engineering, logistics, economics, and public policy. In the medium and long term, this solution may serve as the foundation for an intelligent, safe, and low-emission mobility system in Poland and Central Europe.

5 Simulation and Analysis of Passenger and Freight Flows

The conceptual model for the Hyperloop–CPK connection encompasses comprehensive architectural, logistical, and simulation modelling of the vacuum station in ArchiCAD, SolidWorks, and FlexSim environments. This model serves as a study of integrating vacuum transport with the multimodal infrastructure of the Central Communication Port (CPK) in Baranów, accounting for passenger, cargo, energy, and operational safety flows.

The Hyperloop terminal is designed as a four-level underground station fully integrated with the CPK air and rail terminal.

5.1 Operational and Design Parameters

The operational parameters adopted for the Hyperloop–CPK system consider realistic assumptions for the Warsaw–CPK–Łódź–Wrocław corridor (approx. 500 km):

- Maximum capsule speed: 1,000 km/h
- Acceleration: 2.5 m/s^2
- Capsule capacity: 28 passengers (or 20 t of cargo)
- Departure frequency: every 120 s
- Passenger boarding/alighting time: 20–150 s
- Pressure lock transition time: 60 s.

A dual-module variant was also designed (connection of two passenger or freight capsules), allowing for doubled capacity and flexible traffic management (Fig. 2).

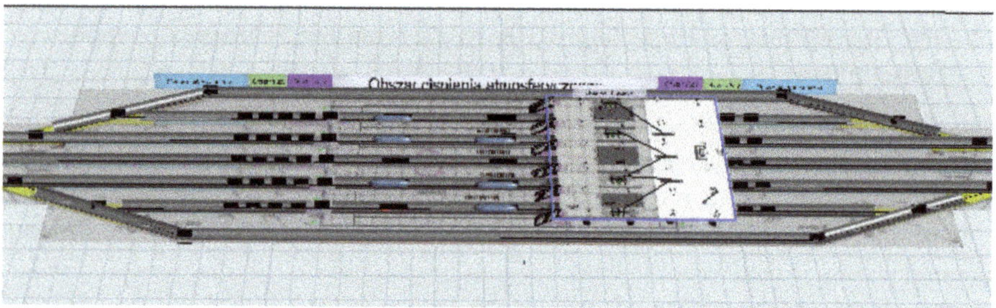

Fig. 2. FlexSim model of the Hyperloop underground station.

The model was constructed as a dynamic system comprising three layers: passenger, logistics, and infrastructural-technical.

The passenger process was mapped as: entry → check-in → security control → boarding → capsule ride → exit.

The logistics process included: cargo receipt → sorting → loading → transport → unloading → storage.

5.2 Simulation Results

The obtained results confirmed high operational efficiency of the system:

- Average passenger service time: 134 s
- Number of people in the system: 524
- Cargo flow: up to 1.4 million t/year

- Station throughput: 43,893 people/day, ca. 16 million/year
- Operational efficiency: 88–96%
- Traffic overloads during: 7:00–9:00 and 15:00–21:00
- Full saturation (100%) in periods: 09:00–15:00 and 21:00–05:00
- Warsaw–CPK travel time: approx. 120 s.

The conducted analysis demonstrated that capsule departure frequency is the most significant parameter influencing operation fluidity.

The integrated simulation of passenger and logistics flows within the Hyperloop–CPK system constituted a critical verification stage of the conceptual model, including the underground Hyperloop station located adjacent to the CPK terminal. The use of discrete-event simulation proved to be an effective tool for assessing the system's efficiency, reliability, and safety. The results obtained provide a solid foundation for subsequent project stages, such as life cycle cost analysis (LCCA), pilot projects, and implementation of technology in TEN-T Plus corridors.

6 Socioeconomic Impacts of Hyperloop Implementation

The integration of Hyperloop connections within the structure of the Central Communication Port (CPK) should be considered as part of a systemic transformation of mobility in Poland, wherein air, rail, and tube transport function complementarily within a single interchange hub in Baranów. CPK's design parameters, a major rail hub, expansive airport, and service-logistics facilities in an "airport city" formula, naturally anchor demand for frequent, short-distance "last mile" connections between the airport and long-distance networks.

From an economic perspective, Hyperloop–CPK integration amplifies the multiplier effect of infrastructure investments. The scale and intensity of traffic generated by CPK, augmented by a high-speed vacuum link, translates into more effective use of terminal and fleet assets, shorter operational cycles, and higher infrastructure productivity. The scope of the endeavour creates lasting demand for advanced engineering competencies (automation, control systems, airlock safety, deep station maintenance) and project-consulting services, strengthening the domestic R&D ecosystem around future mobility [2, 8].

Labor market implications encompass both investment and operational phases. Besides demand for construction and installation work, persistent needs concern operation of systems under varying pressure regimes (atmospheric, pressure reduction, vacuum zones), maintenance of critical devices (airlocks, power supply, evacuation), and operational data analysis and cybersecurity. This necessitates updates to educational programs (automation, robotics, intelligent transport systems) and further development of dual education in partnership with industry [24, 29, 31].

Environmental and energy dimensions manifest both directly and indirectly. Shorter transfer sequences and high service frequency may reduce car traffic in the airport zone, lowering external costs. On the other hand, the adopted station concepts (zone separation, redundant airlocks, efficient evacuation) enhance process safety with minimal impact on the surface environment. Coupled with anticipated renewable energy supply and storage, this supports efficiency and climate targets without deteriorating service quality [25, 27, 29].

A fundamental element is social cohesion and accessibility. The multi-level station layout and service parameters enhance travel conditions for diverse user groups, shortening real "door-to-door" times across mixed transport relationships. This increases daily mobility, decreases logistical peripherality of adjoining municipalities, and expands the effective range of labour markets [30, 31].

7 Conclusions

The conducted conceptual and simulation studies confirm the technical and operational feasibility of integrating the Hyperloop system with the Central Communication Port (CPK) hub. The designed multi-level underground station, featuring pressure zone separation and redundant airlocks, provides a rational foundation for phased implementation of the solution.

Flow simulations demonstrated system stability under high-frequency operations, shortened travel and transfer times, as well as increased capacity and reliability of the entire hub. The model maintained fluid flow with average demand of around 40,000 passengers per day, equivalent to approximately 16 million trips annually. The minimum departure interval was about 2.5 min, and the number of modules served in both directions reached 16 per hour.

Incorporating the Hyperloop link into the CPK program enhances multimodal coherence, improves infrastructure accessibility and productivity, and supports environmental objectives through potential integration with renewable energy sources. The prerequisite for transition to implementation is the completion of regulatory frameworks for the low-pressure environment, clarification of admission and safety procedures, and operational verification in a pilot with progressive scaling and rigorous life cycle cost control.

In summary, Hyperloop holds the potential to become a fast and high-frequency connector within the multimodal CPK ecosystem. The solution increases hub productivity and shortens travel times while maintaining safety and service quality standards. Advancing to the implementation phase requires comprehensive regulatory and technical finalization for the low-pressure environment, refinement of financial structuring, and initiation of a pilot that empirically confirms the parameters derived from the simulation model. Only such an approach will allow for safe and efficient deployment of the system within critical infrastructure settings.

Acknowledgements. Research project supported/partly supported by program "Excellence initiative—research university" for the AGH University of Krakow.

References

1. Musk, E.: Hyperloop Alpha. SpaceX, Hawthorne, CA (2013)
2. Kowal, B., et al.: Demand for passenger capsules for hyperloop high-speed transportation system-case study from Poland. IEEE Trans. Transp. Electrification. **8**(1), 565–589 (2021)

3. Majchrzak, A., et al.: Catalyst organizations as a new organization design for innovation: the case of hyperloop transportation technologies. Acad. Manag. Discov. **4**(4), 472–496 (2018)
4. Malecha, Z., et al.: Analiza gotowości technologicznej systemu transportu wykorzystującego pojazdy poruszające się z dużą prędkością w przestrzeni zamkniętej z obniżonym ciśnieniem. Narodowe Centrum Badań i Rozwoju, Warszawa (2018)
5. Stryhunivska, O., Gdowska, K., Rumin, R.: A concept of integration of a vactrain underground station with the solidarity transport hub Poland. Energies. **13**(21), 5737 (2020)
6. Rumin, R., Duda, J., Blaut, J., Pękała, D., Merolla, T.: Utilizing Artificial Intelligence for Energy-Efficient Route Planning in Hyperloop Low-Pressure Capsule Transit: a Study in Alignment with Sustainable Development Goals. Zeszyty Naukowe. Organizacja i Zarządzanie/Politechnika Śląska, z. 204 Współczesne zarządzanie = Contemporary Management, 541–552 (2024)
7. Premsagar, S., Kenworthy, J.: A critical review of hyperloop (ultra-high speed rail) technology: urban and transport planning, technical, environmental, economic, and human considerations. Front. Sustain. Cities **4**, p. 842245 (2022)
8. Vesjolijs, A., Stukalina, Y. And Zervina, O. A framework for quantifying Hyperloop's socioeconomic impact in smart cities using GDP modeling. Economies, 13(8), p. 228 (2025).
9. Gkoumas, K. And Christou, M. A triple-helix approach for the assessment of hyperloop potential in Europe. Sustainability, 12(19), p. 7868 (2020).
10. Lach, Ł., Suder, M., Kusa, R. And Duda, J. Forecasting the economic impact of a vacuum tube high-speed transport system in Poland: an input-output approach. Managerial Economics, 22(2) (2021).
11. Jacyna, M., Wasiak, M., Lewczuk, K. And Kłodawski, M. Simulation model of transport system of Poland as a tool for developing sustainable transport. Arch. Transp., 31 (2014).
12. Blaut, J., Duda, J., Rumin, R., Pękała, D. And Merolla, T. Enhancing efficiency through hydrogen-powered Hyperloop capsules: AI-assisted route planning in accordance with circular economy principles Manag. Syst.Prod. Eng., (4 (32), pp. 464–473 (2024).
13. Węglinski, B.: Central transport hub in Poland's airport system. J. Mod. Sci. **62**(2) (2025)
14. Rudowski, M.: Intermodalny transport kapsuł Hyperloop–koncepcja, wymagania, korzyści. Problemy Kolejnictwa (2018)
15. Belova, O.V. And Vulf, M.D. Pneumatic capsule transport. Procedia Eng., 152, pp. 276–280 (2016).
16. Rydzek, M.: Kolej niskociśnieniowa–nowa generacja transportu. Ocena możliwości wdrożenia koncepcji podróżowania z wykorzystaniem tuneli próżniowych. TTS Technika Transportu Szynowego, 26 (2019).
17. Błażejczyk, S., Różycka, Z.: Hyperloop-analiza szans i zagrożeń związanych z rozwojem nowoczesnego środka transportu. J. TransLogistics. **4**(1), 213–222 (2018)
18. Kisilowski, J. And Kowalik, R. Displacements of the levitation systems in the vehicle hyperloop. Energies, 13(24), p. 6595 (2020).
19. Yang, Y., Wang, H., Benedict, M., Coleman, D.: Aerodynamic simulation of high-speed capsule in the Hyperloop system. In: 35th AIAA Applied Aerodynamics Conference, p. 3741 (2017)
20. Tudor, D. And Paolone, M. Optimal design of the propulsion system of a hyperloop capsule. IEEE Trans. Transp. Electrification, 5(4), pp. 1406–1418 (2019).
21. Pradhan, R., Katyayan, A.: Vehicle dynamics of permanent-magnet levitation based hyperloop capsules. In: Dynamic Systems and Control Conference. American Society of Mechanical Engineers 51906, p. V002T22A004 (2018)
22. Lluesma-Rodríguez, F., González, T., Hoyas, S.: CFD simulation of a hyperloop capsule inside a closed environment. Res. Engineering. **9**, 1–3 (2021)
23. Bose, A. And Viswanathan, V.K. Mitigating the piston effect in high-speed hyperloop transportation: a study on the use of aerofoils. Energies, 14(2), p. 464 (2021).

24. Polak, K.: Technologia Hyperloop i perspektywy jej zastosowania. Prace Instytutu Kolejnictwa. **156**, 28–32 (2017)
25. Kupriyanovsky, V., Klimov, A., Alenkov, V., Pokusaev, O. Dobrynin, A.: Hyperloop-current status and future challenges. Int. J. Open Inf. Technol. **8**(7), pp. 129–144 (2020)
26. Duda, J., Kusa, R., Rumin, R., Suder, M. And Feliks, J. Identifying the determinants of vacuum tube high-speed train development with technology roadmapping–a study from Poland. Eur. Plan. Stud., 30(2), pp. 405–424 (2022).
27. Department Of Energy (Doe): Effect of Hyperloop Technologies on the Electric Grid and Transportation Energy. Washington, DC: U.S. Department of Energy [online]. Available from:https://www.energy.gov/sites/prod/files/2021/02/f82/Effect%20of%20Hyperloop%20Technologies%20on%20the%20Electric%20Grid%20and%20Transportation%20Energy%20-%20Jan%202021.pdf [viewed: 2025-10-18]
28. Kang, S.E.: Risks and benefits of future travel with Hyperloop: a multi-analytical approach. J. Travel Res., 00472875241313366 (2025)
29. Momenitabar, M., Ebrahimi, Z.D. Arani, M.A: Systematic and Analytical Review of the Socioeconomic and Environmental Impact of the Deployed High-Speed Rail (HSR) Systems on the World. arXiv preprint https://arxiv.org/abs/2003.04452 (2020)
30. Planing, P., Hilser, J. Aljovic, A.: Acceptance of hyperloop: developing a model for hyperloop acceptance based on an empirical study in the Netherlands. Travel Behav. Soc. **38**, p. 100887 (2025)
31. Vesjolijs, A., Stukalina, Y. And Zervina, O. Assessment of socio-economic impacts of Hyperloop technology on European trade routes. Economies, 13(3), p. 65 (2025).

Author Index

MISC
Špirková, Simona 199
Świerczek, Justyna 47

B
Baraniecka, Anna 78
Bieniek, Kinga 270
Blahová, Jarmila 105
Burchart, Dorota 63

C
Cheba, Katarzyna 86
Czapkowski, Szymon 237

D
Dolejšová, Venuše 38
Dudek, Magdalena 211

E
Elçi, Tuğçe 169

F
Feliks, Jerzy 93
Fibiš, Dalibor 38
Fidlerova, Helena 105
Friedrich, Christian 176

G
Gach, Szymon 270
Gdowska, Katarzyna 11, 70, 211
Giel, Robert 53, 115
Gnacy, Natalia 53
Güney, Hasan 169

H
Hake, Cornelius 176
Holman, David 38
Horák, Josef 63
Horizral, Lukas 147

J
Jagoda, Agnieszka 78, 86

K
Kačmáry, Peter 138, 189, 199
Kaleta, Jacek 253
Kantar, Deniz 169
Kapłan, Radosław 211
Karkula, Marek 155, 253
Kleinová, Kristína 189, 199
Kliment, Marek 28
Kołakowski, Tomasz 86
Kovalcik, Jakub 124
Kovalčík, Jakub 138
Kronova, Jana 28
Krupa, Marian 155
Książek, Roger 11, 211

L
Lenort, Radim 3, 63
Lyszkiewicz–Potaczek, Joanna 78

M
Makyšová, Helena 105
Malčic, Tomáš 219
Mesároš, Maroš 199
Milczanowski, Mateusz 270
Muchová, Patrícia 228
Mucowska, Magdalena 86

O
Ondov, Marek 147, 228
Osielczak, Karolina 11, 70

P
Paliienko, Olena 93
Pękała, Dawid 270
Pekarcikova, Miriam 28
Poturaj, Honorata 132
Przytuła, Iga 63

R
Raźniak, Andrzej 211
Reichert, Frederik 176
Rosova, Andrea 147

J. Feliks et al. (Eds.): CLC2024 2024, 2026.
https://doi.org/10.1007/978-3-032-28931-5

Rumin, Rafał 93, 270
Rydlewski, Mateusz 237

S
Skupień, Emilia T. 244
Sofranko, Marian 147
Spirkova, Simona 124
Staka, Martin 138
Staniek, Marcin 63
Straka, Martin 124, 189
Stryhunivska, Olena 93, 270

T
Tarczyński, Grzegorz 253
Trebuna, Peter 28
Tubis, Agnieszka 47
Türker, Ahmet Yesevi 169

U
Ustaoğlu, Ahmet 169

V
Végsöová, Oľga Glova 19

W
Weigele, Jonas 176
Werbinska-Wojciechowska, Sylwia 53, 115
Wicher, Pavel 3
Winiarska, Klaudia 115

Z
Záruba, Jiří 219

GPSR Compliance
The European Union's (EU) General Product Safety Regulation (GPSR) is a set of rules that requires consumer products to be safe and our obligations to ensure this.

If you have any concerns about our products, you can contact us on

ProductSafety@springernature.com

In case Publisher is established outside the EU, the EU authorized representative is:

Springer Nature Customer Service Center GmbH
Europaplatz 3
69115 Heidelberg, Germany

www.ingramcontent.com/pod-product-compliance
Ingram Content Group UK Ltd.
Pitfield, Milton Keynes, MK11 3LW, UK
UKHW020436270926
475829UK00013B/40
9783032289308